THEORIES OF
GEOGRAPHIC

Ontological Approaches to Semantic Integration

CONCEPTS

THEORIES OF GEOGRAPHIC

Ontological Approaches to Semantic Integration

CONCEPTS

Marinos Kavouras **Margarita Kokla**

National Technical Univesity of Athens
Greece

CRC Press
Taylor & Francis Group
Boca Raton London New York

CRC Press is an imprint of the
Taylor & Francis Group, an **informa** business

CRC Press
Taylor & Francis Group
6000 Broken Sound Parkway NW, Suite 300
Boca Raton, FL 33487-2742

First issued in paperback 2019

© 2008 by Taylor & Francis Group, LLC
CRC Press is an imprint of Taylor & Francis Group, an Informa business

No claim to original U.S. Government works

ISBN-13: 978-0-8493-3089-6 (hbk)
ISBN-13: 978-0-367-38791-4 (pbk)

Library of Congress Cataloging-in-Publication Data

Kavouras, Marinos.
    Theories of geographic concepts : ontological approaches to semantic integration / authors, Marinos Kavouras and Margarita Kokla.
        p. cm.
    ISBN 978-0-8493-3089-6 (alk. paper)
    1. Geographic information systems 2. Geographical perception. 3. Ontology. 4. Semantics (Philosophy) I. Kokla, Margarita. II. Title.

G70.212.K38 2008
910.01--dc22   2007024097

Visit the Taylor & Francis Web site at
http://www.taylorandfrancis.com

and the CRC Press Web site at
http://www.crcpress.com

To my wife Nancy,
and to my sons, Demitris and Vangelis

—Marinos Kavouras

To my parents, Nikos and Agathi
and to my sister Ioanna

—Margarita Kokla

Contents

PART 1 *The Context*

PART 2   *Theoretical Foundations*

PART 3 Formal Approaches

PART 4   Ontology Integration

PART 5   Post-Review

Preface

All men by nature desire to know.

Aristotle's *Metaphysics*

The basic incentive for writing this book was a genuine—no matter how futile—human desire, existing in most of us (not only in philosophers), to understand our world. It is primarily this innate desire that led us to write this book and you to read it.

As a child, at about the age of nine, I, M. Kavouras, got one of those crazy notions children often get.* I started wondering if everything around me was an illusion or if it really existed. I wondered, for example, if the things I would see around the corner existed before I made the turn, or if somebody invisible constructed them on-the-fly, depending on spur of the moment decisions, resulting in a realistic experience. Soon, I aborted these anti-realistic thoughts, which, as I discovered much later, are vaguely similar to the negative tenets of nominalism or solipsism. This was because, first, they appeared pointless, not to mention egocentric, second, because something of this kind would have been very difficult to implement (early engineering concerns about real-time systems!); and third, because I would simply sound nuts (something that kids are told to avoid!)

Six years of high school with heavy "classic majors" of ancient Greek and Latin went unappreciated (if not a complete waste of time) at the time, but proved useful much later. For many years, we were taught to divide science (επιστήμη) into sciences, separating them into theoretical and exact sciences. This proved to be one of the most stultifying and narrow views.

We, both writers, acquired formal surveying engineering education and postgraduate specialization in geoinformatics, although in different spatiotemporal contexts. This has helped us immensely to experience, comprehend, measure, and present some of the geographic realities, and realize how differently people view, understand, and communicate geographic knowledge. The subsequent long involvement in what is currently called "geographic information science—GIScience" revealed the importance of "representing" appropriately our knowledge about the world. Work and involvement in theoretical cartography have contributed to this interest.

In the 1990s, during the beginning of interoperability efforts, we started getting seriously involved in GIS standardization. This, as most standardization activities, appeared necessary and useful on one hand, but dull enough research-wise; until one had to deal with the real differences in representing geographic knowledge, by focusing on the meaning of concepts, that is, their semantics. Specific and practical problems of integrating different collections of real geodata created a number of

---

* According to extensive literature on "Philosophy for Children," "Are Children Capable of Philosophical Thinking?", etc., such thoughts are quite common (*Stanford Encyclopedia of Philosophy*, 2007).

problems and questions. Most available approaches to semantic integration provided ad hoc, nonsystematic, subjective manual mappings (in the best case) often leading to Procrustean* amalgamations (in the worst) to fit the target standard.

As most of the geographic information research community, we were not happy with any of the available methods, and even less with the approaches available to geographic knowledge representation, particularly knowledge related to concepts and semantics. Addressing and dealing with these issues became central and essential to the progress of GIScience. I, Margarita Kokla, through my doctoral research, was involved in the theories and approaches to a systematic semantic integration of geospatial data. Inevitably, this motivated us to follow the daedal but fascinating scientific paths of philosophy, semantics, linguistics, and cognition and realize the depth and the weight that notions such as "concept," "category," and "ontology" have carried along throughout the centuries of human thought and inquiry. Thus, terms that until then were used rather naturally and superficially acquired a different essence, followed by the conception that knowledge is not isolated but connected through a visible or sometimes invisible nexus. Therefore, to resolve issues relative to meaning and semantics, it is necessary to explore this nexus and grasp the knowledge it contains.

After conducting research on the subject for some years, we felt that the time was right to systematize and integrate this knowledge. These facts in conjunction with the ontological research conducted at the National Technical University of Athens (NTUA) and the teaching needs of three courses in the NTUA Post-Graduate Programme in Geoinformatics led us both to the decision to write this book—a book that would address theoretical, formal, and pragmatic issues of geographic knowledge representation and integration, based on an ontological approach. This was the second and more practical reason for writing this book.

Thus, many years later, we have developed an increasing appreciation of metaphysics, among other perspectives (cognitive, linguistic, semantic), as the basis of questioning what is or what we can know about reality, whether we refer to reality or our thoughts about it, and eventually how we should represent our knowledge of it. In all these attempts, one thing remained constant—the pursuit of universal approaches and theories for the representation of meaning. Under such a holistic approach, there is only one science, not many. Special "sciences" are harder and harder to separate. Geographic information science is one domain in great need of good theories for the representation of concepts such as geographic "beings" and "kinds." The fact that a metaphysical framework is essential for the success of science, the general science of being, and in consequence (specifically to the success) of GI science is, in our opinion, beyond debate.

ABOUT THE TITLE

The book is about theories of geographic concepts—partitions of the geographic world regarding the process of forming knowledge. Concepts are understood via

---

* Procrustean: Producing or designed to produce strict conformity by ruthless or arbitrary means (*The American Heritage Dictionary of the English Language*, 2000).

their semantics usually encapsulated, elucidated, and formalized by ontologies. We integrate ontologies (which are comprised of concepts and their in-between relations) based on the concepts' semantics. Therefore, we focus on a semantics-based ontology integration. In this context, the term "ontology-based semantic integration," which is often met in the literature, is not that accurate. It better fits approaches that are based on a shared ontology (e.g., a top-level or general ontology) in order to integrate local schemata or classifications. Concerning the subtitle, the term "ontological approaches" is not (technically) used to denote different methodologies about semantic integration. It is rather used to denote a general focus of the book on issues related to *ontological research*, such as concepts (their nature, definition, formalization, etc.), hierarchical structure, semantic properties and relations, etc.

Although ontology integration may facilitate data integration, the latter isn't within the scope of the book. Also, the book does not address, at least not directly, the topic of building new ontologies, known as ontology engineering. It rather puts emphasis on the real issue of integrating existing geo-ontologies. Of course, the process of comparing ontologies and resolving conflicts reveals various problems and imperfections of existing ontologies—an essential basis to avoid the repetition of such mistakes when designing a new ontology.

FOCUS AND READERSHIP

The book focuses on theoretical aspects that are valuable to researchers. Formal approaches, which make it possible to materialize the theoretical foundations, are included so there is also a practical usability of the underlying theories. Technology-dependent issues are not the focus of this book. The primary readership for this book would be postgraduate students and researchers in the field of geographic information science. Because the book reaches a level of pragmatic issues and implementation, a secondary readership would be database integration specialists in spatial data infrastructures.

ORGANIZATION

The book consists of seventeen chapters organized into five parts. The introductory Part 1 sets the context by emphasizing the importance of philosophical, cognitive, and formal theories in preserving the semantics of geographic concepts in knowledge representation, ontology development, and integration. Part 2 unfolds important theoretical issues related to the subject and purpose of the book. The realization of ontological approaches necessitates a number of formal tools and conceptual structures, which are introduced in Part 3. Part 4 covers ontology integration from the implementation point of view, presenting approaches, guidelines, cases, and examples. Finally, Part 5 attempts a post-review of the contemporary situation of ontological research in GI science, with emphasis on issues giving way to further research. In order to make reading easier, a rather flat structure with minimal nesting was selected. Cross-referencing between chapters was also avoided. This was made possible by introducing some necessary but minimal explanations locally, resulting in more self-contained chapters.

ACKNOWLEDGMENTS

Several people have assisted, directly or indirectly, in the completion of this book. First, we thank all members of the OntoGEO Research Group at NTUA, including doctoral students Achilleas Koutsioumbas and George Panopoulos, for their kind assistance. Especially we thank Sofia Kontaxaki for her constructive input, Katerina Koliou for correcting the manuscript, and Stella Boroutzi for designing the book cover.

We are grateful to Jean-Paul Cheylan, Nick Chrisman, Sabine Timph, and the anonymous fourth reviewer for their valuable suggestions, which we seriously tried to take into consideration. In addition, many prominent GI scientists have influenced our work with discussions throughout the years. Appreciating the risk of being unjust to various colleagues and friends, we want to specifically mention: Werner Kuhn, Andrew Frank, Stephan Winter, Max Egenhofer, David Mark, Helen Couclelis, Tapani Sarjakoski, Erik Stubjaer, Timos Sellis, the staff of the NTUA Cartography Laboratory, the members of the ISPRS WG II/6, and many others. E. J. Lowe and J. F. Sowa were very kind to grant permission to use one of their figures in Part 2, and we very much appreciate it. We would like to thank our publisher, Taylor & Francis Group, for its assistance and patience in producing a book that is editorially very close to what we envisioned.

Finally, no words could possibly describe the tremendous patience and support of our families without which this book would not have been what it is. Acknowledging our gratitude and thankfulness to them is the least we can do.

Marinos Kavouras and Margarita Kokla
Athens, Greece

The Authors

Prof. Marinos Kavouras, P.Eng., obtained a diploma in rural and surveying engineering (1979) from the National Technical University of Athens (NTUA) in Greece. He received his M.Sc.E. (1982) and his Ph.D. (1987) from the University of New Brunswick in Canada. He is currently the dean of the School of Rural & Surveying Engineering of NTUA and the director of the NTUA Geoinformatics Graduate Programme, in parallel to his position as a professor of GIScience and Cartography at the NTUA. He is also the vice president of the Hellenic Cadastral Agency and the chair of Working Group II/6–System Integration and Interoperability (2004-08) of the International Society for Photogrammetry and Remote Sensing. He has served on numerous organizing and scientific committees, and as a reviewer in scientific journals. Furthermore, he has undertaken various research programmes in Geoinformatics and has formed OntoGEO, a group conducting ontological research in GIScience. His research interests include, among others, theoretical issues in GIScience, knowledge representation, geospatial ontologies, semantic interoperability, spatiotemporal modeling, concept mapping, and theoretical cartography. He has coauthored over 100 publications in scientific journals and conference proceedings.

Dr. Margarita Kokla, P.Eng., holds a Diploma in Rural and Surveying Engineering (1998) from the National Technical University of Athens (NTUA). She also obtained an M.Sc.E. (2000) in Geoinformatics and a Ph.D. (2005) in Semantic Interoperability in Geographic Information Science from the same school. In 2005, she received the Dimitris N. Chorafas Foundation 2005 Prize in recognition of her work in knowledge representation and engineering in geospatial domain. She is also the secretary of Working Group II/6–System Integration and Interoperability (2004-08) of the International Society for Photogrammetry and Remote Sensing. Furthermore, she has participated in various research programs in geoinformatics. Her research interests include geospatial semantics, ontologies, semantic integration and interoperability, and knowledge representation. She has coauthored several publications in scientific journals and conference proceedings and has also served as a reviewer in scientific journals.

Part 1

The Context

1 Introduction

1.1 GEOGRAPHIC REALITY, CONCEPTS, AND KNOWLEDGE REPRESENTATION

Geographic Information Science (GIScience) constitutes a new scientific discipline concerned with the systematic study of information and knowledge about geographic reality. Probably the most fundamental question that can arise about GIScience is how we conceptualize geographic reality and, consequently, how the formed concepts accurately represent the world or our knowledge about the world. Conceptualizations form geographic concepts and eventually knowledge. Their associated representations are commonly organized into collections, nomenclatures, taxonomies, or ontologies. For various reasons explained later, these conceptualizations differ. Thus, they create *partial*, *imprecise*, or *conflicting* portrayals (representations) of geographic reality. In short, *partial* refers to a nonexhaustive coverage of reality. *Imprecise* refers to how truly or closely the portrayal represents the part of reality. Lastly, *conflicting* refers to inconsistent portrayals of the same part of reality. Thus, commonly encountered geographic concepts, such as "forest," "river," "administrative unit," "cadastral parcel," "transportation network," "road," "bank," "utility," "mixed-urban fabric," or "semi-natural area," may differ considerably from one taxonomy to another. This creates serious problems when geographic knowledge is to be communicated, reused, or combined. Sometimes this becomes immediately apparent, other times it is discovered later or when it is already too late. Four exemplary cases/scenarios are introduced below to illustrate the above points:

Case 1: Three organizations—a planning agency, an environmental agency, and a cadastral organization—are to exchange geographic information about land use for a particular area. During the process they discover that they have to deal with a multitude of interrelated problems, such as:

1. Different understanding of homonymous concepts (polysemy). For instance, what "forest" means to forestry engineering, biodiversity analysis, landscape protection, or agro-pastoralism; or what "parcel" means to cadastre, land use, land cover, agriculture, etc.
2. Different names used for similar concepts (synonymy).
3. Overlapping partitionings, concepts, and coverage of the domain (land use).
4. Differences in the way common concepts are related or hierarchically structured.

5. Unbalanced hierarchies due to finer distinctions of certain branches of the tree structure.
6. Common instances across databases (e.g., specific land areas) assigned to different concepts in different taxonomies.
7. Common instances allocated to a more general class in one hierarchy than in another.
8. Different spatial, thematic, and temporal reference systems used to describe concepts and the associated instances.
9. Equivalent concepts described differently.
10. Equivalent concepts formalized differently.
11. Equivalent concepts explicated differently.

Case 2: A statistical agency needs to conduct the 2011 agricultural survey on land cover but with a technologically more accurate methodology. A new and more detailed classification is to be established, which, however, needs to satisfy the following constraints:

1. Ability to compare to the coarser classifications of the previous surveys of 2001 and 1991 to rationalize about diachronic land cover evolution.
2. Ability to compare/link to land cover classification systems enforced by standardization bodies.
3. Ability to incorporate information from other preceding or parallel surveys based on similar yet different nomenclatures.

Case 3: A European directive (e.g., INSPIRE) demands the integration of national or local area Spatial Data Infrastructures (SDIs) into a pan-European geospatial database. Besides the problems of Case 1, these SDIs exhibit different spatial and thematic granularity. Moreover, the issue of cultural or language differences applies in this case as well.

Case 4: An agent is interested in developing an application that incorporates all pertinent geospatial information that is available on the web. Neither the information nor its sources or their associated context are known a priori. The first goal is to capture the intended meaning of the agent's request and match it to those pieces of information that meaningfully fit it. The next step is to provide the agent with a semantically unified view of the requested domain, without worrying about explication details, as if all information was from a single consistent data source.

Case 1 touches upon a number of problems at the semantic (1, 2, 3, 5, 6, 7, 9), structural (4), and explication (11) levels. The use of different reference systems (8) affects all levels, but especially explication. Case 2 touches upon two issues: (1) that of associating information of different epochs, and (2) that of conforming to central standards. Case 3 introduces the issue of different granularity among information sources, as well as the issue of cultural or language differences in conceptualizing and expressing geospatial concepts. Case 4 refers to upon-request retrieval and integration of semantically context-related geospatial information from heterogeneous sources on the web.

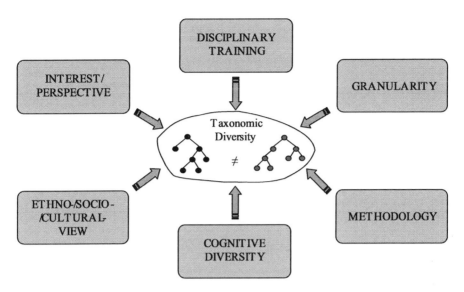

FIGURE 1.1 Causes of taxonomic diversity.

Conceptualization differences and the resulting taxonomic diversion are attributable to the following reasons (Figure 1.1):

1. *Perspective/interest.* Often what determines the concepts and taxonomies to be designed or adopted is the application needs. As a result, different application needs create different taxonomies.
2. *Disciplinary training.* Disciplines tend to develop a common understanding of their domain knowledge. Usually this is specialized and not just general or common sense knowledge. As a result, such knowledge is usually valid across countries/cultures.
3. *Methodology.* In the scientific context, the methods we employ often determine what we see and how we partition reality to a great extent. A common example is that of determining land cover nomenclatures according to the data sources and interpretation method used (e.g., in remote sensing). This also relates to the subject of epistemology.
4. *Granularity.* The scale of analysis not only determines the taxonomical detail but may also create completely different taxonomies. For example, a 1:100,000 land cover nomenclature differs considerably from that of 1:5,000.
5. *Ethno-/cultural-/socio-based view.* Many geographic concepts of a domain are the result of constructive social agreement and partial consensus. Due to the social construction of sense and concepts, there may be significant differences in the way different countries, societies, or cultures partition a domain, creating thus, different taxonomies.
6. *Human cognitive diversity.* When people work autonomously, they perceive and conceptualize geospace differently, thus creating their own cognitive taxonomies.

As the above reasons are mostly orthogonal, their combination results into more complicated taxonomic conflicts. Examining why, when, and how conceptualizations and representations of geographic reality occur is half of the problem. The other half is how to identify possible contradictions and effectively deal with them. Both issues are addressed in this book. The terminology used above denotes concepts as used or understood in a loose commonsensical way. These shall be defined more rigorously and according to different perspectives later on.

The above examples highlight the variety of differences in conceptualizing and representing geographic reality, and the problems this may cause in understanding the intended meaning of information so that it is properly reused. Therefore, our main concern and focus is put on the *semantics* of geographic concepts and the associated representations. It becomes, thus, essential that we discuss the identity and characteristics of geographic concepts.

1.2 GEOGRAPHIC CONCEPTS

In a rather extensive—still not vast—literature, one can refer to some noteworthy attempts in defining geographic concepts. Smith and Varzi (2000) distinguish between fiat and bona fide boundaries of geographic objects. Varzi (2001a) examines whether geographic entities can survive without a physical territory and boundaries. Couclelis (1996) introduces a typology of an intuitive ontology of the boundaries of geographic objects. Frank (2001) proposes a 5-tier ontology, consisting of tier 0: physical reality, tier 1: observable reality, tier 2: object world, tier 3: social reality, and tier 4: cognitive agents. Casati, Smith, and Varzi (1998) introduce a toolkit for geographic ontologies based on topology, location, and mereology. Kuhn (2001) proposes a methodology for building geographic ontologies, which would take into account activities of spatiotemporal reference. Boundaries of geographic objects are subject to vagueness, which complicates their representation. Kulik (2001) proposes a geometric theory for modeling such boundaries based on supervaluation, Bennett (2001) explores different senses of "vague" geographic concepts, while Varzi (2001b) captures the theoretical issues behind the notion of vagueness as applied in the geographic domain. Egenhofer (1993) and Fonseca and Egenhofer (1999) also address the special character of geographic concepts.

There is still, however, an unspoken or unconscious confusion as to what defines a geographic concept. We shall briefly present our postulation for it is necessary in setting the context. First of all, there are concepts that possess a spatial extent or spatial characteristics among other attributes; they are called *geographic* or *geospatial* (mountains, rivers, roads, land property, boundaries, etc.). The same concepts in different contexts can be treated neglecting any of the above geospatial characteristics, appearing therefore, as aspatial. This has created the position that geographic concepts are no different from other concepts and there is nothing special about geospatial. On the opposite, there is of course the generalist's view that most of the reality has a geographic/geospatial reference; at least the physical one (vehicles, persons, buildings, municipalities), but also the nonphysical, which relates to a physical one (e.g., traffic load, public views, average income, population density), being in this world. Therefore, almost everything is geographic.

Both views have some right as they are only partial accounts. First, it is true that most concepts, being in this world, can be assigned a spatial reference. Second, it is clear that there are some concepts whose existence, at least in a certain context, depends vitally on their geospatial character. So, is there a geographic concept as such? Is everything with a spatial reference geographic? Does everything depend only on context? To these rightful questions, we shall attempt to give some definite answers, used thereinafter.

First and most important, there is some difference between concepts as beings (which lie in this world), and concepts as representations about beings (by our thought system), which are often (intentionally or accidentally) partial and contextual. In other words, although ontological concepts may have a geospatial nature, their representation counterparts (serving a specific task) may not need to. This may at times seem irrelevant or practically not very useful, given that we do not know reality itself, or may consider our thoughts to be "accurate" representations of it. Nevertheless, the approach taken here is similar to that of a realist (Section 1.5.1), believing that there are many things, including geographic ones, that do exist and we can approach their knowledge. This has an important merit, for many different views about reality, no matter how distorted at times, are likely to be more similar as they are grounded* on the same reference.

Second, we do not consider as geographic concepts all that may be assigned a geospatial characteristic, but only those which, in a global or local context, possess a geospatial property essential to their existence. Concepts whose "geospatiality" is essential in any context (i.e., context independent) are more prominent (e.g., geographic boundary, landmark, waypoint, marked land property, etc.). These are mostly related to physical reality, and they present similarities to the *bona fide concepts/ entities* of Smith and Varzi (2000). These concepts are usually perceptually and cognitively rich (easily identified or agreed upon). Less frequently, there are also other context-dependent geographic concepts (vehicles in a traffic control system, ships in electronic charts, etc.). On the other hand, there are geographic concepts that are not readily evident but are constructed by a social agreement and partial consensus. Such typical concepts are administrative units, census tracks, agglomerations, and other spatial socioeconomic units (SEUs) (Frank, Raper, and Cheylan, 2001), and are akin to the *fiat concepts/entities* of Smith and Varzi (2000).

As explained in Chapter 5, the meaning of concepts is commonly described by (a) their intension, that is, a definition referring to the concept's properties, or (b) their extension, that is, the collection of the individuals belonging to this concept. Geographic concepts usually refer to an unlimited number of individuals; therefore, they are commonly described intensionally. Exceptions to this, applicable to narrow closed-set domains (e.g., in concepts: U.S. State, E.U. Member, etc.), are also possible but uncommon. Intensional definitions are also important for they approximate the intended meaning of the concept.

In the literature, the term *category* is sometimes used synonymously to that of *concept*. There are varying uses of these terms as well as of others (e.g., classes,

---

* *Grounding*, or *symbol grounding* (Harnad, 1990), is simply stated as the connection of meaningless symbols (which represent concepts about the world) to reality or data about reality, in order to provide the necessary meaning.

sorts, etc.) so that an attempt to rigorously define their meaning is difficult (Smith, 2004) and even claimed to be impossible (Frumkina and Mikhejev, 1996). Theoretical aspects of the concepts are discussed in Chapter 5. In short, and according to the terminology employed here, *categories* and *concepts* are not equivalent notions. A category is a collection of instances, which are treated as if they were the same, whereas a concept refers to all the knowledge that one has about a category. By this distinction, a category refers to the concept's reference, while a concept retains its intensional meaning.

To sum it up, the general theme of the book addresses geographic concepts, ontologies, and semantic integration. The emphasis is put on the semantics of concepts, and those relations and properties, which are essential to integration. In this context, characteristics, spatial or aspatial, are taken into consideration only if considered semantically important.

1.3 DIMENSIONS OF GEOGRAPHIC CONCEPTS

When referring to geographic reality, we include both prominent classical geographic concepts from the physical world, as well as constructed ones. Semantics and characteristics of concepts in general theoretical terms are discussed in Part 2. It is important, however, to introduce here a framework of the principal dimensions used in dealing with geographic concepts. These can be organized into four groups, (a) reference-container, (b) semantics, (c) semiotics/pragmatics, and (d) quality, as follows (Figures 1.2 and 1.3).

Reference-container
- *Spatial frame.* This is a datum functioning as container of space, used to represent (explicitly or implicitly) the location of geospatial entities, as well as other spatial properties and relations.
- *Temporal frame.* This is the temporal reference system. All temporal properties and relations are projected in this frame.
- *Thematic frame.* This reference system is the projection space of all thematic information of geospatial semantics.

Semantics
- *Context.* This expresses a restricted conceptual milieu, also called perspective, framework, or situation, in which concepts are formed, and information is interpreted, obtains meaning, and becomes relevant.
- *Term.* This refers to the name of a concept or entity, which by and large is considered as an essential expression of identity.
- *Internal properties.* Geospatial concepts possess internal properties independent of the existence of or relation to other concepts. These can be spatial or aspatial. Spatial properties include, for example, shape, form, structure, complexity, absolute location, area and volume, velocity, spatial change characteristics, etc. Aspatial properties include embodiment (material appearance), affordance, purpose, behavior, temporality (temporal location, life, existence, periodicity), etc.

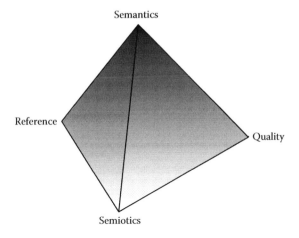

FIGURE 1.2   The tetrahedron of a geographic concept.

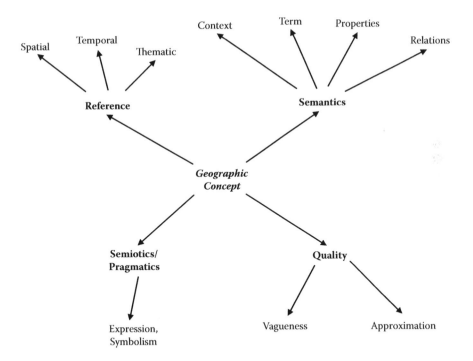

FIGURE 1.3   Dimensions of geographic concepts.

- *External relations.* These refer to relations among two or more concepts. Spatial relations include, for example, relative position, distance,
  orientation, adjacency, proximity, containment, etc. Aspatial relations
  include subsumption-inheritance, dependency-association, partonomy-
  meronymy, role, concurrency, etc.

Semiotics/pragmatics
- *Expression/symbolism.* Geographic concepts are associated with signs (images, words, symbols, etc.), used to capture and show their intended meaning.

Quality
- *Vagueness.* This refers to the degree of inexactness, fuzziness, or indeterminable character of geographic concepts, of their properties and relations (including spatial and aspatial), as well as of reasoning based on vague concepts. Uncertainty, randomness, and ignorance (Sowa, 2000, 348) are important factors contributing to this parameter.
- *Approximation.* This refers to the detail or granularity of (a) the conceptualization, (b) its representation, and eventually (c) its visualization. The first refers to the way the geographic world can be perceived and mentally abstracted from different "distances." The second refers to representations of spatial and thematic content at different levels of detail. Finally, the third refers to the way detail is symbolized and portrayed. The last one has long concerned the research direction of cartographic generalization, too.

1.4   SEMANTICS AND ONTOLOGIES

Concepts mentally represent all possible things (real or abstract) that exist or may exist. *Representations* are not the things themselves. The relation between the *representation* (representamen, abstraction, symbol) and the *represented* thing has been metaphorically expressed by Alfred Korzybski (1994) in one of his most legendary premises, the *map-territory relation*, in the sense that the map is not the territory it stands for. How this relation works is the central problem of *semantics*. The *semantics* of concepts of a given domain is usually encapsulated, elucidated, and specified by an *ontology*. As a consequence, any semantic-based comparison, association, or integration of different domain knowledge necessitates the ability to integrate the respective ontologies.

As with *semantics*, *ontologies* have also gained a great deal of attention by almost every scientific domain during the last decade, with GIScience being no exception (Agarwal, 2005). Philosophers have always pursued (or dreamed of) universal ways for sorting everything that existed and developed ontologies. These ontological theories, while being extremely interesting, had very little practical impact on daily life or engineering activities; not, of course, that they intended to have any. Recently, with the convergence of many separate disciplines of the past and the development of new information and knowledge-based technologies, the need for universal and theoretically sound methods for organizing knowledge led to a rediscovery of work existing for centuries in philosophy and metaphysics. On the other hand, this interest was not directed to "what exists," "what might exist," "an objective account of reality," and related genuine metaphysical concerns, but rather to specified taxonomies about a given domain.

Despite, however, the vast amount of related literature, there is not a universal view or consensus about what ontologies are, what they can be used for, whether they indeed bring something new and useful or they constitute just another catchphrase or hype, where they differ from the existing conceptual design approaches, and so on. Contributing to this is the fact that approaches and views about ontologies are often presented in the literature in a vague, incomplete, and shoddy way, addressing what seem to be fragmented and unrelated knowledge issues with very limited real implementation. It is very difficult to make them commensurable, assess their suitability to a particular task, or combine them with another approach (in a modular fashion) dealing with an adjacent knowledge problem. In a way, the central problem of integrating heterogeneous information moves up to the more difficult problem of associating diverse integration approaches!

To be fair to most of these attempts, however, one has to admit that in various scientific fields, GIScience included, most effort has been devoted to explication issues, while there is a lack of deep knowledge about the meaning of representations, that is, about the concepts and semantics involved. Without a consensus on these issues, it is inevitable that ontological research experiences a limited applicability. Therefore, ontological work, especially due to its limited application, has helped to reveal preexisting insufficiencies in fundamental knowledge. In the absence of a standard ontological consensus for a particular domain, another related impediment is that theories/methods developed or proposed are difficult to compare/evaluate. The latter is a more universal issue and associates to the famous but also controversial issue of *incommensurability* (semantic/taxonomic), as initially presented (but also refined later) by Thomas Kuhn (Hoyningen-Huene and Sankey, 2001).

In the relatively new and also interdisciplinary field of GIScience, the previous issue becomes quite prominent, that is, a bounded and accepted body of knowledge is far from being achieved or agreed upon. Also, at the practical level, there have been concerns and often confusion about the practical role or usability of ontologies (Hunter, 2002). Given the wide acceptance of this notion, Winter (2001) speculates about the use of ontology as a paradigm shift or just another buzzword.

Nevertheless and despite the above concerns, there has been notable progress in the use of ontologies in knowledge representation and integration. Ontologies, depending on how they are understood and employed, provide an indispensable conceptual vehicle for a number of uses, such as formal representations of data, alignment and integration of ontologies, knowledge discovery and elicitation, data mining, intelligent reasoning, semantic-based query processing, similarity comparisons, web semantics, etc. The central focus of this book, however, limits itself to: (a) the understanding of geographic concepts, (b) the analysis of their semantic dimensions, and (c) the ability to create bodies of geographic knowledge represented by ontologies. The long objective is the ability to meaningfully exchange, integrate, and reuse geographic information in an interoperable manner without having to know in advance the meaning of the information. Pursuing such an ambitious objective necessitates that geo-ontologies elicit deeper knowledge of geographic concepts, where the focus is not on the specified representation per se but always on what is being represented (what exists or may exist).

As a consequence, the ontology view adopted in this book may not (and could not) claim to be at the pure philosophical (metaphysical) level of a single Ontology defined as "what constitutes reality" or as "the science of being" (Smith, 1995, 1998), but it is certainly above Gruber's (1993) specification-centered view of ontology ("a specification of a conceptualization") adopted by artificial intelligence (AI) or information science. Taking into real concern more fundamental theoretical issues on geographic concepts and creating the widest possible consensus on the body of geographic knowledge is the only way to solve the semantic interoperability problem. Any deviation from such an objective, prioritizing specification issues instead, just pushes the actual interoperability problem to be dealt with at a higher level and later.

Also, because there is a huge amount of inconsistent accounts of semantics and ontologies, constantly growing, any attempt to digest and integrate every minor proposed view/approach would be hopeless, meaningless, and useless. It would be more practical instead to limit our focus on the most important issues presenting the issue at hand in a self-contained and comprehensive way, with guidelines and examples, so that it first has some practical value, and second, it is directed towards open research issues where understanding and problem solving is not mature enough.

It has been made clear so far that the reason for employing ontologies and ontological approaches is to encapsulate, elucidate, and specify the semantics of concepts of a given domain. Although this would be considered only natural, it appears that it is not always the case. It has been stated (Agarwal, 2005, 508) "that an ontology can also be designed without including semantics." Such a case can be called "ontology" only euphemistically, while it resembles more "taxonomy." At least in the present context, the notion *ontology*, no matter what it is constituted of, is supposed to bear meaning, and thus, it always associates to some semantics, implicitly or explicitly. There is a notable difference between "deep" or "rich" semantics, referring to the semantic properties and relations, and "shallow" or "specification" semantics, associated with all sorts of database attributes usually addressed at the explication/specification level. This, in its turn, characterizes the associated ontologies. High-level ontologies (such as metaphysical ones) bear concepts with rich semantics, whereas low-level ontologies bear concepts with shallow semantics. Uschold and Gruninger (2002) also distinguish between implicit and explicit semantics, associated with the degree of formality along a "semantic continuum" as well (see Chapter 6 on Semantics).

The characteristics of geographic ontologies, in the way they are encountered in practice (e.g., taxonomies, standards, nomenclatures, etc.), are introduced in Chapter 2, and general theoretical aspects about ontologies are discussed in Chapter 4.

1.5   PERSPECTIVES

Despite the advances in ontological research and information science, there is great confusion on a number of issues. First of all there is a different meaning of ontological concepts (the ontology of ontologies) in different scientific niches. Notions such as "concepts," "categories," "semantics," "relations," "properties," "attributes," "integration," and many others are used very differently. Most of the time this difference is due to two distinct perspectives (Figure 1.4):

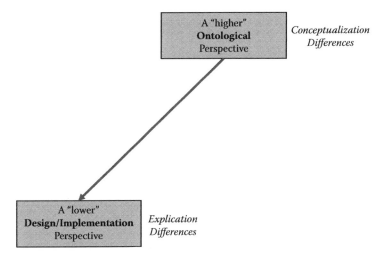

FIGURE 1.4 Conceptualization and explication perspectives: ontological and semantic notions are used differently according to two perspectives.

- A "higher" ontological perspective with an interest in representing "appropriately" reality (which is geographic in our case), or more precisely our knowledge about reality. Issues such as concepts, meaning, semantics, and representation are very important in this perspective, as they relate to domain and conceptualization differences.
- A "lower" design/specification/implementation perspective with an interest in formalizing, processing, and associating existing information or data. Very important here are database issues, attributes, structural, schematic, and syntactic issues, as they relate to explication differences.

Although the two perspectives can and should be complementary, there is a systematic distortion of the notions from the first perspective to the second. As a result, in various approaches and projects, the "ontological concepts" and "semantics" are often database elements with data-type definitions. This is usually treated with some sort of formalization coupled with elementary to more sophisticated conversion utilities. Consequently, the real needs of the first perspective (semantic integration) are not fulfilled.

From all perspectives that can contribute to geographic knowledge and the understanding of geospatial concepts, one cannot overlook philosophy and metaphysics, linguistics, cognitive science, and AI. Each one of them constitutes an entire scientific discipline with many subfields and a multitude of important aspects and views, even in the specific subject of geographic reality. Even a moderate outline of what has been advocated in the literature would do great injustice to the approaches. For the better understanding of these issues, the reader is directed to more appropriate resources. Furthermore, it has not been possible yet to strongly argue or support the supremacy of a specific philosophical theory over another in explaining this world. Notice that in our geographic case mere understanding is indispensable, yet insufficient. We further need to be able to apply these conceptual views to our representations in order to create and reason about geographic knowledge. It is important to bear in mind that this book only touches upon a number of intricate and unsettled issues as long as they provide a better ground for justifying some engineering choices.

1.5.1   THE PHILOSOPHICAL PERSPECTIVE

Different sciences, domains, applications, or users partition reality differently, aiming at creating accurate yet partial accounts of it. Accepting the realists' position that "reality is one and truth indivisible" (Lowe, 2001), the accuracy (or truth) of various accounts of reality is tested on the ground of their fitting together to represent reality as a whole. Therefore, we could talk about partial but globally *accurate representations.* If they are to be true with respect to the single reality, they are bound to be internally consistent as well. The principle is that new representations to be included will also fit the whole view and will not affect the previous. Despite referring to partial representations of reality, the interest of the realists is not in an integrated representation of reality but in reality itself.

Partial or different accounts of the world also exist in the mind of the nonrealists. These, however, do not have to fit a single reality or truth. Since it is accepted that we cannot know anything about reality, we accept that representations are solely based on our thoughts about it, and they are expected to differ likewise. Communication or the need for cooperation and common understanding is the driving force behind attempting to match partial representations of the world. Here, we could talk about *precise representations.* They do not reveal how different the represented parts of reality are but how our thoughts differ. As a result, it is only natural that they exhibit differences that need reconciliation. If representations fit together, this only indicates an internal agreement for the union of the representations considered, and nothing ensures that this will also hold when new representations are included.

The above distinction is conceptually interesting and intellectually stimulating. To philosophers, it is essential—to engineers, however, subtle or immaterial. This is because in practice, some of its importance diminishes for a few reasons. First of all, in the first case, we cannot fully conceive reality and the indivisible truth. Reality becomes apparent to us only through our thoughts and the representations we construct of it. There is no global external way to test how accurately or truly our representations fit reality, but only how well they fit together. Second, referring to the nonrealist position, our thoughts about reality and reality itself are generally not independent but rather interweaved and inseparable. Furthermore, for many reasons—one being that according to Aristotle (in his *Politics*) we are by nature social animals—our thoughts about reality are likely to relate or coincide. This is also very similar to the belief that concepts link up with reality (Swoyer, 2006) and it is this linkage that grounds "intersubjectivity," which is essential to concept combination. Therefore, whether one assumes the first or the second position, we can only deal with representations. What differs is the role attributed to the representations by the two different positions.

1.5.2   THE LINGUISTIC PERSPECTIVE

Geospatial knowledge, concepts, and semantics have always used linguistic expressions as their vehicle. Despite the fact that there is amazing variation in the way language structures space, it is undeniable that language: (a) is extremely powerful in semantically representing geographic concepts, (b) is the only available way of properly defining things, (c) has been the most natural way of communication for humans for centuries, (d) is interweaved with human perception, cognition, and thought, and (e) contains a wealth of existing knowledge waiting to be extracted. The role of language in constructing reality became very prominent in the twentieth century, leading to the so-called "linguistic turn." Although times have changed (Swoyer, 2006) regarding the belief of the supremacy of language, its importance at least remains in all of the above points.

In GIScience, the importance of language and cognition came into stage in the late eighties to the early nineties, having as a landmark the NATO ASI Conference at Las Navas, Spain, on "Cognitive and Linguistic Aspects of Geographic Space" (Mark and Frank, 1991). Many things have changed since. Primarily, language has been investigated with respect to space (Bloom et al., 1999). Nevertheless, as noted above in Section 1.2, spatial relations may contribute only part of the meaning of geographic concepts, as they do not usually concern universals but particulars. That is, spatial relations (e.g., topology, direction) expressed in language may not be essential when referring to the definition/identity of a geographic concept (e.g., "forest" in general), but rather when referring to its instances (e.g., specific "forests"). Language, on the other hand, also provides other valuable semantic information about geographic concepts that are not necessarily spatial. Common textual sources of important geosemantics are lexicons and natural language definitions mostly as structured text, and less often as free text.

Dealing with natural language in a man-machine communication has obvious advantages but also presents difficulties. Such an achievement would require breakthroughs in a number of important problem areas. Two of them seem to be pivotal and also relate to the objective of this book. These are: (a) the ability to transform imprecise natural language into another form of knowledge representation that may be understood by the machine; and (b) the ability to extract the semantics expressed by the text in order to maintain and convey the intended meaning. Notice that these two objectives may appear somewhat dependent. On one hand, a systematic and effective extraction of semantics requires some—preliminary at least—structure and formality; on the other hand, knowing the semantic gist of a piece of text may be used to create a better, formal, but also semantically rich structure. Anyway, it is necessary to develop methods and tools for a systematic analysis of natural language descriptions in order to extract semantic information. Once the semantics has been extracted from text, it can be elucidated via an ontology. This subsequently enables a systematic semantic comparison of different concepts and classifications. Semantic information extraction is more effective and useful in the case of a structured text, such as one of lexical definitions. Such approaches belong to the field of natural language processing (NLP), which is presented in Chapter 12. Working with free text is also a very interesting case in a broader and slightly different application domain. Free text can be formalized and represented via a conceptual structure (e.g., a conceptual graph).

In order, however, to take advantage of communicating in a natural language, yet maintaining the necessary level of formality and control, a solution would be the creation of controlled "natural-like" languages (Fuchs, Schwertel, and Schwitter, 1999; Sowa, 2004) that constitute subsets of the natural ones with vocabulary and syntax constraints. Controlled languages can be directly transformed into any kind of structured logical expression (e.g., First-Order-Logic, programming languages, etc.). There have also been recent attempts to develop such a common logic controlled language for geospatial information (Kontaxaki and Kavouras, 2005).

1.5.3 THE COGNITIVE PERSPECTIVE

A very important perspective in representing reality is that involving human understanding. This has led to the study of mind and intelligence, that is, the highly interdisciplinary field of *cognitive science*, which has been flourishing since the middle of the previous century. This has also influenced research in GIScience on all topics dealing with human involvement. With no particular sequence or order, one could mention human perception, concept formation, categorization, knowledge representation, communication of geospatial information, theoretical cartography, reasoning, human-machine interaction (interface design), spatial analysis, navigation-wayfinding, societal issues, etc.

Cognitive science has influenced or has been influenced by other perspectives, resulting in various sub-branches. *Cognitive linguistics* focuses on lexical studies as they relate to human cognitive capacity. A special branch of cognitive linguistics is *cognitive* (or *conceptualist*) *semantics* (Lakoff, 1987; Gärdenfors, 2000; Talmy 2000). Cognitive semantics is concerned with meaning-construction and knowledge representation, based on the tenet that lexical meaning is conceptual and semantics is subjective.

Cognitive semantics is against *realist views* of semantics, according to which concepts and meaning are not a psychological construct but items existing out there in the real world independently of the existence of humans (and their minds). Therefore, according to the realists' position in metaphysics, conceptions (i.e., cognitive-based) are mere approximations of metaphysical concepts, which are Platonic Essences (Rey, 1985). In other words, cognitive-based conceptions (cognitive concepts) are the point of departure in cognitive semantics (or epistemology), and the point of arrival in metaphysics. Figure 1.5 illustrates the association between metaphysical and human space.

There are also other theories in contrast to realism. *Nominalism* rejects universals (Section 4.3), supporting that ideas represented by words are only in our mind. *Solipsism*—an extreme form of *skepticism* originating from the Greek sophist Georgias—is renowned for its three negative tenets or "trilemma":* (a) nothing exists, (b) even if something exists, it cannot be known, and (c) even if something could be known, it cannot be communicated. It is quite obvious that Georgias would have no interest whatsoever in reading this book or any related literature on knowledge representation, interoperability, and integration!

In GIScience, cognitive issues have been addressed by many (Mark, Smith, and Tversky, 1999; Mennis, Peuquet, and Qian, 2000). The theoretical work also done in

---

* *Internet Encyclopedia of Philosophy,* http://www.iep.utm.edu/g/gorgias.htm

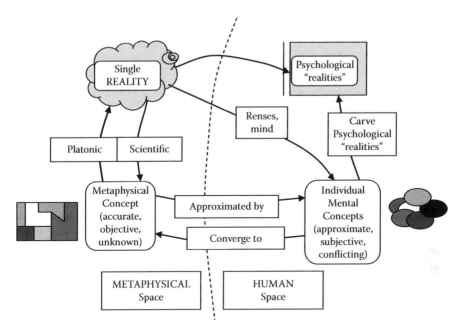

FIGURE 1.5 Association between metaphysical and human space.

cartography poses very interesting cognitive issues, but the focus is on how symbols are perceived and used; thus, the emphasis is put on *semiotics* (Peirce, 1894; Bertin, 1983; MacEachren, 1995). Finally, Kuhn (2003) has written an interesting position paper on the possible contribution of cognitive semantics to information systems and vice versa.

1.5.4 THE ARTIFICIAL INTELLIGENCE/INFORMATION SCIENCE PERSPECTIVE

In the related fields of information science (IS), knowledge representation (KR), and artificial intelligence (AI), there is a shift in focus from the reality itself (and the way it is conceptualized, or represented), to the specification of its representations. According to this perspective, whose main interest is in the explication of the signifier and not in the signified, the meaning of concepts is not given considerable thought. It is rather considered as something given or somewhat previously defined. That is, no matter how concepts have been semantically dimensioned in a semantic framework, as the one above (Section 1.3), the AI/IS perspective only sees the respective specifications. Two concepts are alike when their specifications match, and different when they do not. In case there is a question of degree of matching/resemblance, usually the deep problem is covered up with the involvement of an "expert" who is going to carry the weight of making an "expert" decision. This has led to another view of what semantics is and what role it plays in comparison and integration. Of course, systematizing, formalizing, and eventually implementing interoperability processes is something we cannot live without. Nevertheless, while essential at this level, it leaves the deep semantic comparison to be decided upon at a higher level.

Such an objective has also led to the adoption of new interesting notions, such as *information objects* (Priss, 2004), which are neither purely physical (i.e., tangible with a spatiotemporal reference) nor purely abstract (such as ideas or numbers). They may have some secondary (probably unimportant) physical characteristics (e.g., physical means of storage, book, word, etc.) but, more importantly, they have abstract characteristics such as their information content (Priss, 2004). Considering these information objects compatible to the signs defined by Peirce has led Priss to the development of a semiotic-conceptual framework with the objective of explaining mathematically differences between programming and ontology languages.

1.6   DATA, INFORMATION, KNOWLEDGE (AND WISDOM)

Some other notions subtly connoted above, but without a rigorous definition, is that of *data*, *information*, and *knowledge*. These, together with *wisdom* have formed the DIKW chain. We shall continue avoiding it to a certain extent. First, there is a whole literature on the subject: some precise, some informal, and definitely not universally accepted (Zeleny, 1987; Wiig, 1993; Ackoff, 1997; Davenport, 1997; Quigley and Debons, 1999; Pomerol and Brézillon, 2001; Stenmark, 2002; Bellinger, Castro, and Mills, 2004). Second, in real life, namely, in everyday communication (even scientific ones), these notions are used very loosely. Third, it does not really make much difference (i.e., it does not have serious practical implications) to our objective set above. The last one is somewhat a good excuse for the second reason. It could be argued, and it has indeed (Ackoff, 1997; Bellinger, Castro, and Mills, 2004), that a simplified, yet universally accepted, sequence of notions from the semantically poorer to the semantically richer or superior, is:

data → information → knowledge → wisdom

Some (Zeleny, 1987) support that data, information, knowledge, and wisdom answer the *know-nothing, know-what, know-how,* and *know-why* questions, respectively. This chain, however, has been confronted by others. "*Not only are the definitions of the three entities* (Notice: wisdom is not included) *vague and imprecise: the relationships between them, although non-trivial, are not sufficiently dealt with. It is unwise trying to define these entities in terms of each other since such definitions seem to further confuse the picture*" (Stenmark, 2002, 3; italics added). In this statement, Stenmark does not refer to "wisdom." Additional problems are introduced by the connoted linearity of the sequence and the distance between the notions (which deceivingly suggests that the stair from data to information is equivalent to that from information to knowledge). It is also implied that the only way is up, that is, from data to knowledge, and not the other way around. Against the latter are those (e.g.,Tuomi, 1999) who support that there are no raw data *in vitro*; data always carry the influence of the knowledge that produced them in a certain context, therefore the sequence is the other way around. This view has also been criticized (Stenmark, 2002) for making the reverse mistake.

What is accepted here, in short, is that these notions are interwoven, and that information does not exist per se (by itself) outside a particular context involving also a particular cognitive process. Therefore, one's information is the next person's data until the necessary cognitive process takes place, which will eventually convert it to information. This process is facilitated by knowledge. The important issue here is that if two contexts/cognitive processes on the interpretation of the same data differ, information also differs accordingly. A simple example follows that may be of some help and hopefully does not confuse further the notions used.

data = facts as symbols without additional meaning
   P1(39,000, 4,700)   P2(40,000, 4,700)

information = data + meaning (semantics in certain context)
   Horizontal Cartesian coordinates (X, Y), of the position of two points in a reference system R, expressed in meters.

knowledge = a deterministic process based on information, insight, experience, and reasoning.
   The two points are only 1,000 meters apart on the same parallel, and P2 is east of P1. There is no elevation information about the points, or an estimate of the accuracy of this position.

wisdom = a complex, extrapolated, not-deterministic process, based on accumulated knowledge, addressing future actions.
   I need the corresponding coordinates in another reference system and some redundant information in order to be able to convert positional information between the systems as well as with a certainty estimate.

A metaphorical view of Data, Information, Knowledge, and Wisdom has been presented by Hey (2004) and Sharma (2005). Also, the reader being interested in a rigorous, formal, yet compact presentation of the semantic conceptions of information is directed to the "Semantic Conceptions of Information" by Floridi (2005) in the *Stanford Encyclopedia of Philosophy*.

1.7   THE CORE OF THE MATTER

Similar to other domains, research on concepts, semantics, and ontologies has also found its place in geographic information science. Despite, however, the increasing amount of publications, no one can claim that research (with respect to practical results) is thriving. The current state of affairs shows that we still have a long way to go.

In this book, without sacrificing generality, we share the position that there is a single reality. We may not know it in its entirety, but nothing prevents us verging on it in order to know something about it. As a consequence, we create various representations, which, depending on the partitioning of the world and the perspective/interest, are going to be partial and subjective. Because they refer to a single reality, however, they are likely to be somewhat consistent with each other. Any inconsistency (conceptual always) is likely due to our limited knowledge of reality, our limited representational ability, or our social predisposition. Their integration to build a consistent ontological view is more likely to succeed if we consider them as referring to or emanating from the reality represented, rather than focusing on the representations per se, namely disregarding what they refer to.

In an interesting position, Smith (2004, 3; italics added) admits that the term *concept* has not been sufficiently defined, one reason being *"the fact that these terms deal with matters so fundamental to our cognitive architecture (comparable in this respect to terms like 'identity' or 'object') that attempts to define them are characteristically marked by the feature of circularity."* In fact, he argues against the use of concepts, for they, being constructions of human cognition, carry partiality and refer to (or even create) a user's ontology/reality. Such a use of concepts has dominated knowledge representation approaches. He proposes the use of objective scientific knowledge about reality based on *universals* and *particulars* (see Section 4.3), instead of dealing with subjective forms of "human convention and agreement." His approach, however, applies only to the objectively known physical geographic reality—in other words to the representation of bona fide crisp "things" mentioned already above. The rest of geographic "things" constitute fiat-type things (e.g., administrative units, which unavoidably are human constructs, that is, a result of human cognition or consensus).

In the context of the present book, it is recognized that both perspectives, that is,

1. The cognitive, user-centered, principally subjective, knowledge-representation notion of *concept*, which in turn creates different subjective and often conflicting "realities" (and associated "ontologies")
2. The scientific-centered, realist's perspective, objective (to the current state of scientific knowledge) notion of a single Ontology (with a capital "O"), based on universals and particulars

have a merit as they address (or should address) different types of geographic realities. In fact, in the pragmatic context of this book, the notion of *concept*—a constituent of an ontology—is used in both cases and in the technical sense as a representation tool (e.g., similar but not quite the same as *class, category*) expressing how the world is partitioned. Partitioning reality is also known as carving nature at its joints (Swoyer, 2006). Usage of *concept* as a tool is also observed and accepted by Smith (2004), although he indicates the danger of a "shift in focus" towards concepts. In the first case, *concepts* are defined/created cognitively, whereas in the second, they are defined scientifically. Of course, there is a difference between how ontologies and concepts should be theoretically, and how they have been in fact created. Integration has often to deal with existing imperfect situations making the best of it. However, Smith's position on a single true reality and the use of objective scientific knowledge may find limited application in GIScience, because (a) there is a lot of existing knowledge, which is a result of a cognitive process or arbitrary consensus, and (b) scientific knowledge on geographic categories is not as straightforward as in other scientific domains.

Concepts are, thus, employed as convenient instruments, being the constituents (building blocks) of an ontological representation. They are the result of partitioning reality, signifying the resulting parts. Whether this partitioning is a product of a strict scientific endeavor (referring to an objective reality, as in natural sciences), a constructive social agreement and partial consensus, or an arbitrary cognitive process (where humans create concepts and "realities")—because any of these may be the case—is essential to know and consider. Nevertheless, this does not change the fact that existing geo-ontologies, while suffering from all these imperfections or partialities, contain important knowledge that needs to be integrated in order to be useful. On the other hand, it is due to this same heterogeneity and difference in perspective that we gain a better account and understanding of an intrinsic and multidimensional geographic reality. Understanding the theoretical dimensions of geographic concepts is an essential step towards the development of appropriate methods for semantic integration, which in turn relates to the book title. As explained above, geographic concepts are mental constructs—linked to geographic reality—formed from the physical (objective-scientific), human (cognitive), and social (consensual) perspectives.

In our endeavor, and accepting that reality at least exists, there are three different relative metaphors bearing importance. First, as Aristotelian humans (with a desire to know), we cannot escape by the guiding metaphor of "discovery" (see Swoyer, 2006 on realism) in exploring the natural joints of geographic reality. Second, as conceptualists, we are guided by an "alien" metaphor, in carving the world from the exterior, according to concepts based on our senses and cognitive system. Third, as pragmatists, we are guided by an "engineering" (problem solving) metaphor, considering that (geographic) knowledge is only what is important/useful to some end; the rest is irrelevant or worthless. It is obvious that the first metaphysical perspective differs from the other two.

1.8   UPSHOT

To an often-asked question, as to whether ontological research is another hype, the answer is definitely a no; at least with regard to ontological research dealing with deep semantics. Ontologies provided the chance to re-address fundamental issues that have been avoided in the past and which are at the heart of the interoperability issue. The failure of standards and the requirement for advanced (semantic) interoperability opened the back door to ontological research.

On the other hand, no one can deny that there has been a misuse in ontology-based approaches and a careless use, not to say abuse, of the term itself. This is not strange but rather a rule in every aspect of research, not to say life in general. This artificial interest diminishes once other hypes make their appearance. There have already been proposals to go beyond ontologies and there will be more in the years to come. Nevertheless, one thing cannot change: the need to understand (a) what geographic reality is, (b) how we can systematically represent it, (c) how the representations carry meaning about what they represent, (d) how from a single reality we end up with inconsistent representations, and (e) how existing representations can be reused to reason not about the representations but about what is being represented. Due to these very critical, rightful, and far from being settled issues, research is actually just starting. Even if another catchy term is introduced, the questions stated above will still demand answers—answers that are not only theoretically fascinating but also practically possible to implement.

In this milieu, it is of the essence to identify, address, and possibly seek practical solutions to a number of core issues/questions. We shall outline them here, for they pervade the rest of the book.

1. What are the characteristics of existing geographic ontologies?
2. What are the needs for geographic information reuse and integration?
3. How is (or should) geographic reality (be) partitioned into concepts?
4. How can we elucidate the meaning of geographic concepts, that is, semantics?
5. Are semantic notions used similarly across different scientific contexts?
6. What are good sources of semantics for geographic concepts? How can we extract and formalize semantic information from these sources? In particular, how can we elicit valuable semantic information from natural language descriptions/definitions?
7. How can ontologies express the semantics of concepts?
8. Can there be a "closed" set of fundamental semantic properties expressing the meaning of geoconcepts?
9. Which semantic properties are spatial? Can there be a universally accepted ontology of spatial relations?
10. What happens to concepts when their properties change?
11. Which are appropriate conceptual structures for geographic knowledge?
12. How can concepts/ontologies be compared/integrated?
13. What is similarity between concepts? How can it be measured? Where can it be utilized? What is the role of context in concept comparison?
14. Are there appropriate approaches/tools for achieving semantic interoperability?
15. Are there general (top-level) ontologies that can be used as reference to integrate geographic ontologies?
16. How can user involvement be minimized, objectifying thus, the integration method?
17. How can a user be properly guided in making the right choices in an integration endeavour?
18. What are open research topics in ontology-based semantic integration?

All these and other issues are discussed thereinafter.

REFERENCES

Ackoff, R. L. 1997. Transformational consulting. *Management Consulting Times* 28(6).

Agarwal, P. 2005. Ontological considerations in GIScience. *International Journal of Geographical Information Science* 19(5): 501–536.

Bellinger, G., D. Castro, and A. Mills. 2004. Data, information, knowledge, and wisdom. http://www.systems-thinking.org/dikw/dikw.htm (accessed 1 May 2007).

Bennett, B. 1983. *Semiology of graphics: Diagrams, networks, maps.* Trans. W. J. Berg (from original work published 1967). Madison: University of Wisconsin Press.

Bertin, J. 1983. *Semiology of graphics: Diagrams, networks, maps.* Trans. W. J. Berg (from original work published 1967). Madison: University of Wisconsin Press.

Bloom, P., M. A. Peterson, L. Nadel, and M. F. Garrett, eds. 1999. *Language and Space,* Cambridge, MA: MIT Press.

Casati, R., B. Smith, and A. C. Varzi. 1998. Ontological tools for geographic representation. In *Formal ontology in information systems,* ed. N. Guarino, 77–85. Amsterdam, The Netherlands: IOS Press.

Couclelis, H. 1996. A typology of geographic entities with ill-defined boundaries. In *Geographic objects with indeterminate boundaries,* ed. P. A. Burrough and A. U. Frank, 45-56. New York: Taylor & Francis.

Davenport, T. H. 1997. *Information ecology.* New York: Oxford University Press.

Egenhofer, M. 1993. What's special about spatial?: Database requirements for vehicle navigation in geographic space. In *Proc. ACM SIGMOD international conference on management of data,* ed. P. Buneman and S. Jajodia, 398-402. Washington, DC: ACM Press.

Floridi, L. 2005. Semantic conceptions of information. In *The Stanford Encyclopedia of Philosophy,* Winter 2005 Edition, ed. N. Zalta. http://plato.stanford.edu/archives/win2005/entries/information-semantic/ (accessed 1 May 2007).

Fonseca, F. and M. Egenhofer. 1999. Ontology-driven geographic information systems. In *Proc. 7th ACM Symposium on Advances in Geographic Information Systems,* ed. C. B. Medeiros, 14-19. Kansas City, MO: ACM Press.

Frank, A., J. Raper, and J. P., Cheylan, eds. 2001. *Life and motion of socio-economic units: GISDATA (Gisdata, 8).* London: Taylor & Francis.

Frank, A. U. 2001. Tiers of ontology and consistency constraints in geographic information systems. *International Journal of Geographical Information Science* 15(7): 667–678.

Frumkina, R. M. and A. V. Mikhejev. 1996. *Meaning and categorization.* New York: Nova Science Publishers.

Fuchs, N. E., U. Schwertel, and R. Schwitter. 1999. Attempto Controlled English: Not just another logic specification language. In *Logic-Based Program Synthesis and Transformation, Eighth International Workshop LOPSTR'98,* Manchester, UK, ed. P. Fiener, Lecture Notes in Computer Science 1559: 1–20. Berlin, Heidelberg and New York: Springer-Verlag.

Gärdenfors, P. 2000. *Conceptual spaces.* Cambridge, MA: MIT Press.

Gruber, T. R. 1993. A translation approach to portable ontologies. *Knowledge Acquisition* 5(2): 199–220.

Harnad, S. 1990. The symbol grounding problem. *Physica D: Nonlinear Phenomena* 42: 335–346.

Hey, J. 2004. The data, information, knowledge, wisdom chain: The metaphorical link. Intergovernmental Oceanographic Commission (UNESCO). http://ioc.unesco.org/Oceanteacher/OceanTeacher2/02_InfTchSciCmm/DIKWchain.pdf (accessed 1 May 2007).

Hoyningen-Huene, P. and H. Sankey, eds. 2001. *Incommensurability and related matters.* Boston Studies in the Philosophy of Science. Dordrecht: Kluwer Academic Publishers.

Hunter, G. 2002. Understanding semantics and ontologies: They're quite simple really—if you know what I mean! *Transactions in GIS* 6(2): 83–87.

Kontaxaki, S. and M. Kavouras. 2005. Geo-Q: Spatial knowledge extraction based on a Controlled English Query Language and conceptual graphs. In *Proc. GIS Planet 2005, International Conference and Exhibition on Geographic Information,* May 29–June 2, Estoril, Portugal..

Korzybski, A. 1994. *Science and sanity: An introduction to non-Aristotelian systems and general semantics,* with a Preface by Robert P. Pula. 5th ed. Fort Worth, TX: Institute of General Semantics.

Kuhn, W. 2001. Ontologies in support for activities in geographic space. *International Journal of Geographical Information Science* 15(7): 613–631.

Kuhn, W. 2003. Why information science needs cognitive semantics—and what it has to offer in return. Draft Paper, MUSIL Lab, Institute for Geoinformatics, University of Münster. Germany, http://musil.uni-muenster.de/documents/WhyCogLingv1.pdf (accessed 1 May 2007).

Kulik, L. 2001. A geometric theory of vague boundaries based on supervaluation. In *Spatial Information Theory, International Conference COSIT '01*, ed. D. R. Montello, Lecture Notes in Computer Science 2205: 44–59. Berlin Heidelberg and New York: Springer-Verlag.

Lakoff, G. 1987. *Women, fire and dangerous things. What categories reveal about the mind.* Chicago: University of Chicago Press.

Lowe, E. J. 2001. Recent advances in metaphysics. In Proc. *2nd International Conference on Formal Ontology in Information Systems, FOIS 2001*, Ogunquit, Maine, October 17–19. New York: ACM Press. http://www.cs.vassar.edu/~weltyc/fois/fois-2001/keynote/ (accessed May 1, 2007).

MacEachren, A. M. 1995. *How maps work: Representation, visualization and design.* New York: Guilford Press.

Mark, D. M. and A. U. Frank, eds. 1991. *Cognitive and linguistic aspects of geographic space.* NATO ASI Series. Dordrecht, The Netherlands: Kluwer Academic Press.

Mark, D. M., B. Smith, and B. Tversky. 1999. Ontology and geographic objects: An empirical study of cognitive categorization. In *Spatial Information Theory, Cognitive and Computational Foundations of Geographic Information Science,* ed. C. Freksa and D. M. Mark, Lecture Notes in Computer Science 1661: 283–298. Berlin: Springer.

Mennis, J. L., D. J. Peuquet, and L. Qian. 2000. A conceptual framework for incorporating cognitive principles into geographical database representation. *International Journal of Geographical Information Science* 14(6): 501–520.

Peirce, C. 1894. What is a sign? In *The Essential Peirce. Selected Philosophical Writings.* Vol. 2. (18931913), ed. N. Houser and C. Kloesel. Bloomington: Indiana University Press .

Pomerol, J.-C. and P. Brézillon. 2001. About some relationships between knowledge and context. In *Proc. 5th International and Interdisciplinary Conference on Modeling and Using Context*, ed. P. Bouquet, L. Serafini, P. Brézillon, M. Benerecetti, and F. Castellani, 461–464. New York: Springer-Verlag.

Priss, U. 2004. A semiotic-conceptual framework for knowledge representation. In *Proc. 8th International ISKO Conference, Knowledge Organization and the Global Information Society,* ed. I. C. McIlwaine, 91–96. Würzburg, Germany: Ergon Verlag.

Quigley, E. J. and A. Debons. 1999. Interrogative theory of information and knowledge. In *Proc. SIGCPR '99*, ed. R. Agarwal and J. Prasad, 4–10. New Orleans, LA: ACM Press.

Rey, G. 1985. Concepts and Conceptions: A reply to Smith, Medin and Rips. *Cognition* 19: 297–303.

Sharma, N. 2005. http://www-personal.si.umich.edu/~nsharma/dikw_origin.htm (accessed 1 May 2007).

Smith, B. 1995. Formal ontology, common sense and cognitive science. *International Journal of Human-Computer Studies* 43: 641–667.

Smith, B. 1998. An introduction to ontology. In *The ontology of fields*, ed. D. Peuquet, B. Smith, and B. Brogaard, 10–14. Santa Barbara, CA: National Center for Geographic Information and Analysis.

Smith, B. 2004. Beyond concepts: Ontology as reality representation. *In Proc. Third International Conference on Formal Ontology In Information Systems (FOIS 2004)*, ed. A. Verzi and L. Vieu, 73-84. Amsterdam: IOS Press.

Smith, B. and A. C. Varzi. 2000. Fiat and bona fide boundaries. *Philosophy and Phenomenological Research* 60: 401–420.

Sowa, J. F. 2000. *Knowledge representation: Logical, philosophical and computational foundations*. Pacific Grove, CA: Brooks Cole Publishing Co.

Sowa, J. 2004. Common Logic Controlled English Specifications. http://www.jfsowa.com/clce/specs.htm (accessed 1 May 2007).

Stenmark, D. 2002. Information vs. knowledge: The role of intranets in knowledge management. In *Proc. HICSS-35*, Hawaii: IEEE Press.

Swoyer, C. 2006. Conceptualism. In *Universals, concepts and qualities*, ed. P. F. Strawson and A. Chakrabarti, 127–154. Burlington, VT: Ashgate Publishing.

Talmy, L. 2000. *Toward a cognitive semantics*. Cambridge, MA: MIT Press.

Tuomi, I. 1999. Data is more than knowledge: Implications of the reversed knowledge hierarchy for knowledge management and organizational memory. *Journal of Management Information Systems* 16(3): 107–121.

Uschold, M. and M. Gruninger. 2002. Creating semantically integrated communities on the World Wide Web. Invited talk in Workshop on the Semantic Web, Co-located with WWW 2002, Honolulu, HI. http://semanticweb2002.aifb.uni-karlsruhe.de/USCHOLD-Hawaii-InvitedTalk2002.pdf (accessed 1 May 2007).

Varzi, A. C. 2001a. Philosophical issues in geography: An introduction. *Topoi* 20: 119–130.

Varzi, A. C. 2001b. Vagueness in geography. *Philosophy & Geography*, 4: 49–65.

Wiig, K. M. 1993. *Knowledge management foundations: Thinking about thinking—how people and organizations create, represent, and use knowledge*. Arlington, TX: Schema Press.

Winter, S. 2001. Ontology: Buzzword or paradigm shift in GI science? [Editorial] *International Journal for Geographical Information Science* Special Issue on Ontology in the Geographic Domain 15(7): 587–590.

Zeleny, M. 1987. Management support systems: Towards integrated knowledge management. *Human Systems Management* 7(1): 59–70.

2 Geographic Ontologies

2.1 NOTIONS AND PERSPECTIVES

As already introduced in Section 1.4, geographic ontologies, also called geospatial or geo-ontologies, elucidate the explicit knowledge of the geographic domain they describe. They achieve that by capturing the semantics of the concepts involved. The objective of the present chapter is not to justify or question the effectiveness of an ontological approach to semantic-based integration, nor to refer to ontologies in general. These two issues are addressed in Chapters 1 and 5, respectively. What is more important here is to realize the variety of existing geo-ontologies and understand their principal characteristics and limitations, so that users know what they are confronted with. On the other hand, understanding the deficiencies and the consequential limitations of existing geo-ontologies facilitates their possible enhancement or the building of new (and hopefully better) ones.

It must be realized from the start that in the geographic context, the notion of "ontology" (geo-ontology in our case) has been used rather freely. It refers to the systematization of all kinds of knowledge with varying degree of formality. Without any particular order, one can thus talk about ontologies of:

space, spatial relations, spatiotemporal entities, location, etc.
place
geographic boundaries
semantic relations and properties
affordances
physical entities
a particular geo-thematic domain (e.g., cadastre, transportation, land use, planning, census, etc.)
spatial operators
thematic operators
etc.

And these are only some general, common, and proper uses of the notion, referring to the meaning of the concepts involved. Very often (but also very erroneously) people use "ontology" to denote the explicit documentation of the database of a specific application. Thus, without any deep concern about what is being represented, these "ontologies" describe narrow and application-specific views of "reality" of limited pertinence. Even with such an improper use of "ontology" to denote something that does not associate to semantics and meaning, no one can deny the usefulness of properly documenting geo-databases at the specification level, in view of possible reuse. It remains, however, outside the scope of this book.

One of course has to admit that the notion of ontology in information science is relatively recent and definitely after "ontologies" (under different names) had already been established. In other words, many of the existing geo-ontologies were not known as such when they were originally created. Thus, it is to some degree understandable why such "ontologies" differ.

Some additional clarifications are also necessary at this point. In the incomplete list of partial ontologies above, it seems possible and certainly useful to attempt to clarify, systematize, organize, and present all collections of GI-related knowledge. It is also true that every time we attempt to define a concept, by properly avoiding circular definitions, we introduce additional fuzziness because the defining terms have to be defined themselves, and so on. From the practical point of view, however, the starting point to any geo-information repository is that of concepts-categories being selected/designed so that geo-instances can be properly organized—conceptually and physically. As such, the most pivotal issue of ontology content is that of handling geographic concepts-categories. Wider possible uses of the notion "concept"—under which everything can be a concept, even the concept itself—are not addressed here. They constitute an important but secondary issue, on which sufficient work has not been reported so far. In our case, this means that in addition to defining geographic geoconcepts/categories, we also need to sufficiently define the semantic relations/properties of these concepts.

2.2 GEOGRAPHIC ONTOLOGY TYPES AND RELATED ISSUES

2.2.1 Ontology Types

It is now time to examine the typology of the so-called geographic ontologies. These include:

> General spatial data standards
> Project-related nomenclatures
> Geographic taxonomies
> Feature catalogues—data dictionaries
> Classification schemata
> Top-level ontologies (geospatial portions)
> Ontologies as maps

People often use these notions (ontology, taxonomy, nomenclature, classification, etc.) interchangeably. Usually such a loose use is harmless, yet uninformative by itself. The user does not really know what to expect behind such a notion. Among very many, van Rees (2003) discusses and compares the usage of the terms ontology, taxonomy, classification, and thesauri, identifying similarities and differences. This is not particular to the geographic domain, but of a wider concern (dealt with in Chapter 4).

2.2.2   ONTOLOGY COMPONENTS

Ontologies in general (see Chapter 4) may consist of four components: *concepts, relations, axioms, and instances.* The minimum that is expected from a geo-ontology is a set of *concepts* expressed as a vocabulary of the *terms* used, a specification of the term meaning (commonly expressed by a definition), their *properties*, and the *relations* among concepts or concepts' properties. This is considered a light and informal ontology. There are also cases where concepts are expressed only by terms (definitions are missing), and the specification of the intended meaning can only be implicitly inferred, using common sense, domain/expert knowledge, or available structural knowledge of the associated concepts (e.g., a hierarchy). Relations, alike concepts, may also have their own properties. Geo-ontologies are rarely axiomatized. *Axioms* specify constraints or rules about the values of properties, relations, properties of relations, and instances. Since an ontology elucidates a domain of knowledge, *instances*, that is, "things" represented by a concept, are catachrestically considered as an ontology component, and indeed they are not common in geo-ontologies. One should not confuse instances with subconcepts. Nevertheless, geodefinitions may contain (see *ostensive* definitions below) few instances as examples. Finally, instances, accompanied by an ontology, constitute what is known as a *knowledge base.*

2.2.3   FORMALITY

Geographic ontologies come in many suits. Sometimes the application demands a very precise, explicit, and formal specification of a limited number of well-known entities having a universally accepted meaning. A good example would be the feature catalog or ISO standard of hydrographic or aeronautical charts. For reasons of supporting an international service that facilitates navigation and provides safety in a universal manner, flexibility and expressivity are limited so that the resulting ontology is very specific and formal, with concepts, relations, and constraints. Other times, the geo-ontology just consists of a list of flat terms, without any definition, structure, relations, constraints, axioms, instances, etc. Here, a lot is implied and a possible excuse is that either the concepts involved are very straightforward, or they are directed to a very narrow domain of knowledgeable people who know what is behind the concept terms.

It is not difficult to refute both assumptions as wrong. It is not at all easy to determine concepts that are straightforward to understand. This may sound too strong; nevertheless, the most commonly accepted geographic concepts are probably the hardest to agree upon a single definition. From simple, seemingly commonsensical concepts, for example, forest, road, city, canal, hill, to more complex ones, such as cadastral parcel, territory, continental shelf, people realize how differently they interpret the same term. Even domain experts disagree on specialized geoconcepts; therefore, mere terms are not sufficient in specifying domain knowledge.

2.2.4   STRUCTURE

Along this spectrum, from the heavyweight to the very light ontologies, there are of course geo-ontologies with various characteristics and formality. In addition to terms, some also provide *structure*—usually hierarchical with super/subconcepts. Structure is very useful, as it helps us understand better a concept thanks to its neighbors. Structure can be used to compare concepts of different ontologies by comparing their associated concepts or the depth of the tree. Other *associative relations* among concepts are also possible; however, they are not systematically employed in GI databases.

2.2.5 IS-A HIERARCHIES

Hierarchical structure is mostly based on IS-A relations. IS-A hierarchies are normally based on *properties* shared by similar concepts. General (super) concepts have fewer properties, and special concepts inherit the properties of their superconcepts. For example, if the superconcept is "agricultural area" and two of its subconcepts are "arable land" and "permanent crops," both subconcepts inherit all properties of the superconcept, that is, they are what that is, plus something more that differentiates them. This, trivial though it may sound, is not always the case in the multitude of classifications met in the GI domain. They present problems such as: (a) there are nominally different concepts but with no distinguishing characteristics; (b) subconcepts sometimes do not inherit all superconcept's properties. Notice that the issue of relations-properties is elaborated further below, and more theoretically in Chapter 5. Ontology structure is used to facilitate comparison of individual concepts in two hierarchies on the basis of associated concepts in the corresponding structures. Also, the distance of a concept node from the root of the tree is also used as a basis of comparison and similarity measuring. It is also used in structural comparison/matching in different ways. For instance, one can identify corresponding nodes in the two structures, establish semantic bridges and use them to reach structure-based deductions about the rest of the concepts, especially about the in-between ones. Examples of structure correspondence/matching can be found in schema matching techniques (Rahm and Bernstein, 2001), similarity models (Rodríguez and Egenhofer, 2003; Yang et al., 2005), AnchorPROMPT (Noy and Musen, 2003), conceptual graph matching (Myaeng and Lopez-Lopez, 1992), and many others.

2.2.6 PARTS

Another set of relations, sometimes associated to geoconcepts, is PART-OF relations. For example, the concept PORT may be described to consist of (a) BUILDINGS, (b) STRUCTURES, and (c) ASSOCIATED LAND. Knowing a concept's *parts* may be very helpful in understanding also its *whole*. *Mereology*—a notion from logic, mathematics, and metaphysics, known as the theory of part-whole relations—may prove useful here. *Mereology* (from the Greek word *meros = part*) relates to *meronymy* or *partonomy*, a notion used by the linguistics community. Mixing IS-A relations and PART-OF relations in a conceptual structure—hierarchies included—is theoretically possible but should be used with great caution, and it should better be avoided because it creates confusion. An example of such an improper use is shown in Figure 2.1.

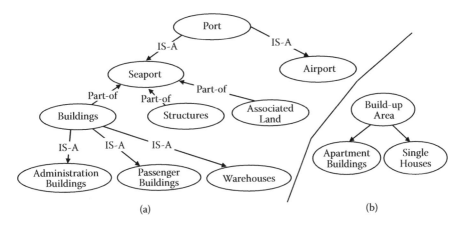

FIGURE 2.1   Mixing IS-A and PART-OF relations.

2.2.7   MULTIPLE INHERITANCE

The hierarchical IS-A structure followed by most original geo-ontologies (specifying a single view of reality) is a tree structure; that is, each child node (subconcept) has only one parent node (superconcept). There are cases, however, where this is not sufficient. One such case is when there are hybrid views with subconcepts belonging to more than one superconcept. Such *multiple inheritance* hierarchies are sometimes inevitable when one is merging two different trees and needs to maintain both views of the world. For example, the concept CANAL may be a subconcept of two superconcepts: TRANSPORTATION LINE and IRRIGATION LINE. Another case where multiple inheritance applies is in part-of relations. Here, the same subconcept may easily be part of more than one superconcept. For example, the concept PARKING may be part-of many other superconcepts. A third case requiring multiple inheritance is when instances are related to concepts, in which case an instance may easily belong to various superconcepts. A specific person, for example, David Smith, may be an instance of the concepts LAND OWNER and POWER ATTORNEY in the same cadastral database.

2.2.8   COVER, USE, AND USABILITY

Another very common problem in geoclassifications is that of mixing certain kinds of geoproperties, such as *land cover* and *land use*, which are admittedly difficult to distinguish at times. From the general point of view, the first refers to the "material" a concept is made of, including its "outside" (vegetation structures, etc.); while *use* clearly refers to the way land is (economically) used (by humans). It is not rare, however, that these two notions are somewhat interdependent or exchangeable (Anderson et al., 1976; NCGIA, 1997; CARA, 2006). A VINEYARD for instance, can be claimed to refer to both. It is "a piece of land where grapevines are grown," telling us that it is full of grapevines (cover), which are used to make wine (use).

Instead of attempting to at least clarify the issue, certain nomenclatures used by Statistical Services elude the problem by naming their classification "Land cover/use Classification," leaving it to the user to decide accordingly. Also, the well-known Anderson's classification (Anderson et al., 1976) attempts to combine land use and land cover.

Notice that the notion of *use* refers to the existing or designated use of a thing (land in the present case) and not to the possible use or functionality of things. The latter is expressed as *usability* and leads to the complex notion of *affordances* (Gibson 1979; Norman, 1988; Gaver 1991; McGrenere and Ho, 2000), also adopted in GIScience (Jordan et al., 1998; Kuhn, 2002). Although there are different definitions of *affordance*, it can be quickly viewed as a property of a thing that indicates how one can interface with that thing. Flat open land, for instance, affords the possibility of parking your car in it, or playing soccer. A wide sidewalk with no obstacles on it affords movement with a double stroller.

The issue of land cover/use also relates to the resolution/scale level, as well as to methodology. It is common that small scales refer more to land cover than to land use. The latter requires not only more spatial detail and interpretation (NCGIA 1997) but also additional external (and often not visible) information. Remotely sensed satellite data, for example, has proven to be more suitable for land cover information, at least in the times when spatial resolution was limited to scales 1:50,000 to 1:100,000. Given all the possible differences between land cover/use classifications, it becomes obvious how complex their comparison/integration can be.

2.2.9 DEFINITIONS

An essential source of semantic information in geo-ontologies is that of available *definitions*. As explained rigorously in Chapter 6, a definition accompanies the concept term and delimits the meaning of the concept. Unless provided by another source (such as a database), distinguishing (essential) properties are described in the concept's definition. Parts—when applicable—can also be described in definitions. Definitions are, thus, fundamental in understanding, formalizing, and comparing geoconcepts. However, there are cases where definitions are not paid proper attention and are compiled rather poorly. This is why definitions, properties, attributes, etc. have been criticized by many as to whether they do really describe semantically the concepts defined. This, however, is not an inherent problem of the definitions but of those making inappropriate use of them. On the other hand, proper analysis and comparison of definitions may reveal conceptual conflicts. A definition discovered to present problems is an extremely useful outcome that helps rectification. It is also supported (Gahegan, Takatsuka, and Dai, 2001) that although categories are defined (constructed) externally and a priori in GIS, a posteriori knowledge acquired in situ, especially in field studies, results in some refinement of the definition. In the remaining section, three issues are addressed always with respect to geoconcepts: (a) essential differences in definitions, (b) different kinds of definitions, and (c) how to make better definitions.

When comparing definitions of geoconcepts, several subtle but also essential problems may arise. A very crucial one is the possible influence of locality (local views) on the perceived geographic reality. With respect to physical reality, it is well known that there are concepts that cannot be universally defined. Even seemingly simple (or commonsensical) concepts such as LAKE, CANAL, FOREST are likely to differ from place to place. For instance, LAKE, FOREST, OLIVE GROVE, GLACIER, or TUNDRA are understood and detailed differently (some may not be understood at all) in Canada, in Spain, or in Kenya. With respect to socioeconomic reality, differences in concepts may be even greater. A third issue, related to the previous, is the local geospatial language used in the definition of concepts.

Geospatial definitions are derived in different ways. Some, few unfortunately, are the result of strict scientific endeavor, describing reality in an "objective" manner. Some rely on human cognition to define concepts, being unavoidably partial. Some are the result of consensus, delimiting concepts arbitrarily. Under this case, we can also include definitions, which delimit concepts on the basis of the methodology/technology used to acquire this information. Such typical examples are the definitions of the CORINE Land Cover nomenclature (CLC2000, 2000) about FOREST: "Tree heights under normal climatic conditions are higher than 5 m," or about URBAN FABRIC: "The discrimination between CONTINUOUS and DISCONTINUOUS URBAN FABRIC is set from the presence of vegetation visible in the satellite image." Another case is when the consensus is wider and not ad hoc, becoming a convention or a standard of a particular domain.

There are also different kinds of definitions for geoconcepts. Most are *intensional* in the sense that they provide all necessary and sufficient properties of the members of a concept. *Extensional or enumerative* definitions are not convenient for they need to specify the concept's extension, namely, every object (instance) that falls under the definition. Providing an exhaustive list of all objects under a concept is suitable only for small and finite data sets, and this is not the case with geographic ontologies. There are also *ostensive* definitions which do not use extensions but point to some examples in order to express the meaning of a concept and what is included. It is common to describe cases that are excluded, too. One should be certain, however, whether inclusion or exclusion examples used in definitions refer to subconcepts or to instances. That is, in the definition of the CLC concept BROAD-LEAVED FOREST, it is stated that it includes "plantations of, e.g., eucalyptus, poplars," "walnut trees and chestnut trees used for wood production included in forest area context," "evergreen broad-leaved woodlands composed of sclerophyllous trees," etc. It would be different if it included specific instances as typical examples (of inclusion or exclusion), such as "the Valia Calda forest."

2.2.10 RELATIONS AND PROPERTIES

A researchable question of importance here is how the semantic information contained in definitions can be extracted in a systematic and objective way. The first question is whether there is a unique, accepted, universal ontology about spatial relations and properties. Such an ontology could find immediate application in a number of spatial reasoning tasks. There have been numerous attempts to derive such ontologies in the traditional spatial information sciences and standardization bodies (e.g., ISO, FGDC, OGC, etc.), or in other scientific areas also dealing with space (psychology, cognitive science, linguistics, etc.); but such a well-accepted ontology has not been established yet. More critical to our context, however, is the question of whether there are spatial relations-properties that are essential to the definition of concepts. Such spatial relations usually relate two or more concepts, and as such it is unlikely that they refer to absolute position (location) as instances usually do. Topological relations are more common. Here, there is important work on *mereotopology* (see also Chapter 4), a formal theory combining *mereology* and *topology*, on the relationships among wholes, parts, and boundaries (Smith, 1996; Varzi, 1996).

Looking more carefully into the relations/properties used in intensional definitions of geoconcepts, some interesting realizations can be made. A thorough analysis of existing descriptions of geoconcepts is potentially very useful, for it can contribute to the establishment of a reference system for semantically delineating all geoconcepts. Such an analysis has shown (see Chapter 12) that semantic properties met in definitions include: *purpose, agent, cover, location, time, duration, frequency, size, shape.* Semantic relations include: *is-a, part-of, has-part, connectivity, adjacency, overlap, intersection, relative position, containment, surroundness, direction, proximity.* A further question of interest is on the interaction and possible weighted combination of these properties in providing identity to geoconcepts, as well as whether something like that can be possible to establish.

2.2.11 Top-Level Ontologies

The objective of top-level (upper, foundation) ontologies is to create a single ontology describing all general concepts across domains. Such an ontology would potentially help associate domain ontologies via the general concepts. As with the standards, there are several such attempts (Cyc, DOLCE, WordNet, BFO, SUMO, etc.) (Chapter 4), with no one becoming a de facto standard yet. Furthermore, there are positions for and against the feasibility of such efforts, while there is also a strong commercial interest. What is of interest here is whether top-level ontologies contain sufficient general geospatial concepts. Analysis has shown that there is definitely a need to accommodate richer geospatial concepts. Here one can mention the work of the Institute for Formal Ontology and Medical Information Science (IFOMIS), with the objective, amongst others, to "build a top-level ontology capable of supporting domain geo-ontologies." Their SNAP-SPAN approach (Grenon and Smith, 2003) aims at addressing the spatiotemporal aspects of geographic concepts. In order to enrich WordNet categories, the CGKAT top-level concept ontology has been proposed (Martin, 1995) with two hundred relation types, among which there are spatial and temporal ones.

2.2.12 Maps as Ontologies

Geographic concepts are the central content of maps. Therefore, a map, among its various other roles, can also be seen as an ontology. In fact, almost all notions expressed so far (concepts, instances, semantics, context, vagueness, properties, relations, etc.) also apply to maps. Notice that here we are not considering the *ontology of maps*, which specifies explicitly map elements associated to map design and compilation, but rather *maps as ontologies* of geographic reality. In the first case, ontological concepts refer to the map construing elements. In the second (also our) case, concepts refer to real geographic things.

A map can be seen as another representation, in fact a very powerful one. Compared to other forms of machine representation of geographic reality, such as various geospatial databases, it presents space constraints and employs abstraction and symbolism to elicit, express, and convey meaning; this limitation, however, becomes cartography's advantage as a representation means. In other words, without the luxury of storing an unlimited number of geospatial entities, properties, and relations, maps inevitably rely on the very essential elements of information, leaving out the nonsemantic ones. Examining a map as a repository of geospatial information, the map concepts (and possibly their hierarchy) are briefly described in the map legend. This description is usually just the terms and the corresponding map symbols. Terms are supposed to be self-explanatory. The map content provides the instances of the concepts.

There are radically different map types (a hydrographic chart, a land cover map, a cadastral map, a population density thematic map, etc.), denoting analogous differences when seen as ontologies. They specifically present differences at least at two levels: (a) their semantic expressivity and (b) their formalization. The first denotes the certainty of what is being represented by the map representation. The second denotes not the semantics of the representation, but its formality. A hydrographic chart, for instance, is characterized by high formalization of semantically very specific concepts, based on a consensus and without leaving room for ambiguity. A standardized land cover map may exhibit high formalization; nevertheless, the level of semantic certainty—as to what a concept (e.g., *forest* or *semi-natural area*) actually denotes in reality—may be limited for various, quite understandable reasons. Generally speaking, formalization precedes semantic completeness and is easier to accomplish, for example, by complying with a standard.

Geoconcepts represented in maps are also supposed to convey their meaning contextually with respect to space. If, for instance, the concept of PORT is semantically clear and portrayed as an instance at various locations on the map, it is expected that spatially related features will be associated to PORT concepts, such as terminals, parking space, structures, offices, etc. Therefore, they do not need to be explicitly described, for they obtain their meaning contextually. This, however, is not a case related to maps only, but to any database. Yet, undoubtedly the visual association of spatial features and contextual reasoning performed by humans is greatly facilitated by maps.

2.2.13 METHODOLOGY-DRIVEN ONTOLOGIES

As stated in Section 1.1, one important reason that results in taxonomic diversions, even when considering ontologies of the same (geographic) domain, is the methodology used. Geoinformation is collected, interpreted, and classified using complex methodologies, different sensors, apparatus, and technologies. Although some consider the level of observation/measurement as being neutral, objective, and prior to any interpretation judgment, it is not exactly so. What one measures largely depends on the anticipated subsequent use of the observation.

Generally speaking, one can distinguish at least three approaches towards knowledge acquisition. The first approach, and the most proper one, specifies the ontology of the desired knowledge repository and then, based on its needs, selects and adapts data acquisition methods and technologies. For example, first define properly and rigorously what a FOREST is in a specific context, and then find ways that will safely lead to its delineation. The second approach again specifies the ontology of the desired knowledge repository and then attempts to make the best use out of existing data repositories to fulfill its requirements and derive the initially specified concepts. In the same example, after the definition of FOREST, see how one can best approximate its delineation using available information. The third approach begins with what data exists or what can be acquired using available methods and technologies and then creates an "implementable" ontology that is certainly attainable. For example, see what or how much one can interpret from LANDSAT™ imagery via remote sensing techniques, and then define FOREST accordingly. Such an approach, according to which we define/accept what we can see/derive/approach/attain, is not strange to scientific or engineering practice and should not necessarily be considered as "setting the expectations low." It sometimes facilitates better quality control of geographic information acquisition projects, since both data sources and methodologies are straightforward. Such "pragmatic" approaches, however, create methodology-driven ontologies, which make the use of metadata with explicit semantic descriptions absolutely essential in order to make proper use of the resulting data repositories.

2.3   ONTOLOGICAL VAGUENESS

An important issue associated with geographic concepts is vagueness. Vagueness itself is a very broad topic (Russell , 1923; Williamson, 1994; Keefe, 2000) seen from various perspectives, such as the linguistic, the cognitive, and the philosophical one. A common description of vagueness of a concept involves the existence of borderline cases (Sanford, 1975; Sorensen, 1997; Nobis, 2000); that is, the existence of some elements that cannot be safely assigned to the concept. Vagueness is mostly expressed by vague predicates (Bueno and Colyvan, 2007). If one excludes a narrow set of crisp (not vague) concepts from exact domains (e.g., mathematics and exact sciences), the rest of the concepts in everyday use are diluted by vagueness. Most geographic concepts are no exception to this rule (Couclelis, 1996).

Vagueness has troubled philosophers since ancient times.* Hampton (2007; 355) states: "How, it is asked, can we claim to have certain knowledge of the world when the very words that we use to express that knowledge are prone to such vagueness?" Indeed, scientists find it difficult to effectively deal with vague knowledge. Gottlob Frege, the renowned German mathematician, logician, and philosopher, saw vagueness as a serious and pervasive weakness of natural language that had to be avoided by all means; for example, by using artificial languages with completely

---

* Among the oldest vagueness puzzles is *sorites* (heap paradox) attributed to the Greek logician Eubulides of Miletus (Keefe and Smith, 1997). There is not a clear-cut point or border along a continuum of gradual change at which the characterization takes place, for example, by adding one grain of wheat at a time, it is not possible to find the point when you make a heap. A sorites argument starts with seemingly true premises but ends up with a seemingly false conclusion (Bueno and Colyvan, 2007).

precise predicates (Pelletier and Berkeley, 1995). He believed that vague concepts are unacceptable; concepts should not have borderline cases but sharp boundaries instead (Varzi, 2001). Frege's position on abolishing every vague term out of logic, however, would make natural language completely inexpressive. The difficulty in distinguishing vague from not vague concepts (Varzi, 2001), even before one decides what to do with them, worsens the problem. Although Russel (1923) (Pelletier and Berkeley, 1995) shares similar concerns with Frege, he accepts the inevitable effect of vagueness on language which can be of advantage in ordinary use. Russell (1923; 89) also refers to vague representations: "a representation is *vague* when the relation of the representing system to the represented system is not one-one, but one-many." Wittgenstein (1953; 98) admits that language does not follow strict rules, that linguistic vagueness makes sense, and that "there must be perfect order even in the vaguest sentence." According to (Pelletier and Berkeley, 1995), two practical issues have perpetually puzzled philosophers: (a) how to appropriately represent vague phenomena, and (b) how to appropriately reason with such representations.

Research on vagueness seems to fall into two main categories: *cognitive* (Hampton, 1998) and *linguistic* (Russell 1923). The first one is attributable to our feeble and limited cognitive system. The human cognitive system is capable of perceiving and representing only a vague "figure" of reality (Hampton, 1998). Linguistic vagueness, caused by vague terms/expressions, is probably the form most widely accepted by most scientists, while many* believe that it is the only acceptable form.

Vagueness, especially a linguistic one, is often confused with *ambiguity*, which can be lexical, syntactic, or semantic.† Words/expressions in language can be both vague and ambiguous. Vagueness refers to the inability to clearly understand the meaning of a concept in a context. "Large," "tall," "high," "dense," "rich" are common vague terms. On the other hand, ambiguity is the existence of more than one (but finite) context-dependent specific meaning. The words "bank" or "good" are common examples of ambiguous terms. Thus, ambiguity can be easily resolved by fixing the context (Prinz, 1998). When there is a lack of sharp conceptual boundaries, the notion of vagueness is employed to represent it. In this respect, ambiguity is not that important (Pletcher, 1995).

One main concern in the present context is about the implication of vagueness on the way geographic reality is partitioned (carved) into concepts. Because such partitionings are expressed by ontologies, such vagueness could be called *ontological vagueness*. As already discussed in Chapter 1, semantic heterogeneities are the result of different ways of carving geographic reality. If ontologies are the result of "carving Nature at its joints,"‡ then there are greater chances of finding common grounding and facing fewer problems during ontology/concept comparison. An important issue here is the discrepancy between language concepts and concepts in

---

* For instance, see defenders of van Fraassen's *supervaluation* regarding "all vagueness being a matter of linguistic indecision" (*Stanford Encyclopedia of Philosophy*, http://plato.stanford.edu/entries/vagueness/#5, accessed 1 May 2007).

† *Semantic ambiguity*, opposite to *lexical ambiguity*, is based on idiomatic expressions, which can have any number of non-agreed interpretations, thus resembling *vagueness*.

‡ A famous metaphor by Plato (*Phaedrus* 265d–266a) towards producing "objective" taxonomies (natural kinds).

Nature. Prinz (1998) attributes linguistic vagueness to a number of possible reasons, such as (a) a larger number of language (vocabulary) concepts compared to the number of natural joints, (b) a "random" carving of language not at Nature's boundaries (natural joints) but in between, or (c) a large number of natural joints which are, however, too "fine" compared to the coarser linguistic concepts.

So far, vagueness has been seen as an inherently representational issue, attributed to our thought system and language. This follows more or less Russell's position (1923) that vagueness does not exist in the "mind-independent world." This is contrary to another highly controversial type of vagueness supported by a number of philosophers—yet not the majority—called *ontic vagueness*. Not to be confused with *ontological vagueness* introduced above, which examines vagueness with respect to the cognitive and linguistic human systems, *ontic vagueness* (also known as *metaphysical vagueness*) supports that there is vagueness in the objects of the real world, irrespective of the human cognitive and linguistic systems. From the large number of those contributing to the debate, one can indicatively mention Russell (1923), Evans (1978), Lewis (1988), Williamson (1994), Sainsbury (1995), Heller (1996), Prinz (1998), and Colyvan (2001). Regardless of their acceptance, the three proposed kinds of vagueness create a *vagueness triangle* (Figure 2.2), which could be associated with the famous *meaning triangle* (see Chapter 6). That is, (a) *ontic vagueness* relates to reality/objects/referent, (b) *cognitive vagueness* relates to concepts (thought or reference), and finally (c) *linguistic vagueness* relates to symbolization.

Despite running the risk of adjudicating or coming across as arrogant, we shall attempt to take a practical stance in the issue of vagueness in geographic concepts and

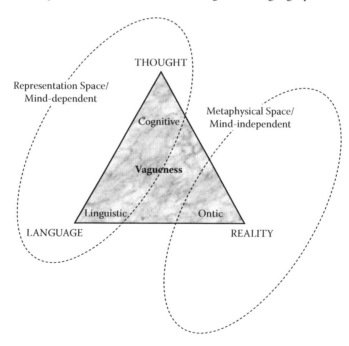

FIGURE 2.2   The vagueness triangle representing three types of vagueness met in the literature.

its effect on integration. It is quite interesting that the geospatial domain is used commonly as a case to demonstrate positions about vagueness. The well-known issues of precision, accuracy, scale, boundaries, generalization, temporality, etc. provide the right conditions for such a demonstration. It seems that geographic concepts suffer from mind-dependent vagueness at various levels of importance. It is important to deal with this representation vagueness inherent in geographic ontologies. Ontic vagueness is not considered here for the following two reasons.

First of all, it seems that Nature is as it is and does not use the representation mechanisms humans do. A "mountain" or a "forest" may appear vague with respect to their delineation (boundaries), but this is because of us humans deciding to define such concepts. Nature does not attempt to sort, name, or define "mountains" or "forests," much less their boundaries. The difficulty in defining the bounding surface of a "cloud" is a problem of the construct we create and not a problem of the "object" itself. We use "object" in quotes to indicate that objects, at least in the way we use them, are human (representational) constructs. Furthermore, even if we accepted the case of natural (and thus, objective, mind-independent) objects, any vagueness in their boundaries, parts, life, or motion does not imply that the identity of the objects is also vague. Vagueness is ontologically (and not ontically) important when it is based on the defining properties and relations of the concepts and not on anything else which is not semantically pertinent.

Second, when dealing with the problem of integration of existing (and imperfect) geographic ontologies, the available semantic information (a) is not always detailed or accurate, (b) is often of an unknown origin or ontology engineering process, and (c) has encompassed various imperfections in an unknown way. This makes any speculation about the portion of the ontic vagueness that may have crept into the final concept as presented to us academic and impractical.

Geographic ontologies that have resulted from or express different carvings of reality might be looked upon as layers of different semantic regions in the representation space. Comparing two such ontologies can be conceived as a semantic-layer overlay (similar to the classical polygon overlay in GIS). The main emphasis is put on the implication of vagueness on the identity of geographic concepts. As the determination of identity is a complex task relying mostly on defining semantic properties/ relations, considering vagueness becomes crucial as it affects the certainty of the outcome of concept/ontology comparison. From this perspective, spatial vagueness, which also includes the case of vague boundaries, is important if it affects the identity of geographic concepts.

From a long list of problems related to vagueness (e.g., Burrough and Frank, 1996; Couclelis 1996; Fisher, 2000; Bennett, 2001; Kulik, 2001; Varzi, 2001; Bittner and Smith, 2003) and without accepting ontic vagueness, cognitive and linguistic types of vagueness have an effect on geographic concepts (Kavouras and Tomai, 2001). Cognitive vagueness is inherent in cognitive categorization, such as Rosche's typicality structure (see Chapter 5), which also applies to geographic categories (Mark, Smith, and Tversky, 1999). The result is some semantic overlapping between concepts, which has to be dealt with. Linguistic vagueness is mostly caused by the existence of vague predicates in the definition of concepts. Characteristic examples are "a pond is a small lake," "a mountain is a high earth formation," and "a forest is the trees and plants in a large densely wooded area (WordNet)" (Kavouras and Tomai, 2001).

Normally, in proper geographic ontologies expressed as tree hierarchies, same-level concepts do not overlap semantically and are supposed to exhaust the domain of interest. When, however, concepts from different nonassociated (or standardized) ontologies are compared, semantic ovelappings are quite common. The overlapping of concepts/ontologies due to vagueness (cognitive, linguistic, or both) is a major issue in semantic integration and requires a systematic analysis in order to resolve it to the greatest possible extent. An approach based on *Semantic Factoring* (Chapter 8) can help to determine the extent of such overlappings and their decomposition to non-overlapping constituent subconcepts (Chapter 16). The process of clarifying vague predicates is also known as "sharpening of vagueness" (Gaifman, 2002; Tomai and Kavouras 2002); however, such sharpening is not always attainable (Collins and Varzi, 2000). The common use of qualitative linguistic terms (QLTs) in geographic concept definitions also relates to vagueness; their modeling proves useful in concept disambiguation (Tomai and Kavouras, 2005) (Chapter 17).

REFERENCES

Anderson, J. R., E. E. Hardy, J. T. Roach, and R. E. Witmer. 1976. *A land use and land cover classification system for use with remote sensor data.* Geological Survey Professional Paper 964. Washington, DC: U.S. Government Printing Office.

Bennett, B. 2001. What is a forest? On the vagueness of certain geographic concepts. *Topoi* 20(2): 189–201.

Bittner, T. and B. Smith. 2003. Vague reference and approximating judgments. *Spatial Cognition & Computation* 3(2/3): 137–156.

Bueno, O. and M. Colyvan. 2007. Just what is vagueness? Work in Progress, http://homepage. mac.com/mcolyvan/papers/jwiv.pdf (accessed 1 May 2007).

Burrough, P. A. and A. U. Frank, eds. 1996. *Geographic objects with indeterminate boundaries.* London: Taylor and Francis.

CARA (Consortium for Atlantic Regional Assessment). 2006. http://www.cara.psu.edu/land/ lu-primer/luprimer01.asp (accessed 1 July 2006).

CLC2000. 2000. European topic centre on terrestrial environment. Topic Centre of European Environment Agency. http://terrestrial.eionet.europa.eu/CLC2000 (accessed 1 May 2007).

Collins, J. and A. C. Varzi. 2000. Unsharpenable vagueness. *Philosophical Topics* 28: 1–10.

Colyvan, M. 2001. Russell on metaphysical vagueness. *Principia* 5(1/2): 87–98.

Couclelis, H. 1996. A typology of geographic entities with ill-defined boundaries. In *Geographic objects with indeterminate boundaries,* ed. P. A. Burrough and A. U. Frank, 45–56. London: Taylor & Francis.

Evans, G. 1978. Can there be vague objects? *Analysis* 38: 208.

Fisher, P. 2000. Sorites paradox and vague geographies. *Fuzzy Sets and Systems* 113: 7–18.

Gahegan, M., M. Takatsuka, and X. Dai. 2001. An exploration into the definition, operationalization and evaluation of geographical categories. http://www.geocomputation. org/2001/papers/gahegan.pdf (accessed 1 May 2007).

Gaifman, H. 2002. Vagueness, tolerance and contextual logic. (Sums works presented in several conferences during 1997–2001.) http://www.columbia.edu/~hg17/VTC-latest. pdf (accessed 1 May 2007).

Gaver, W. 1991. Technology affordances. In *Proc. ACM CHI 91 Human Factors in Computing Systems Conference*, ed. S. P. Robertson, G. M. Olson, and J. S. Olson, 79-84. New York: ACM Press.

Gibson, J. 1979. *The ecological approach to visual perception*. Mahwah, NJ: Lawrence Erlbaum Associates.

Grenon, P. and B. Smith. 2003. SNAP and SPAN: Towards dynamic spatial ontology. *Spatial Cognition and Computation* 4(1): 69–104.

Hampton, J. A. 1998. Similarity-based categorization and fuzziness of natural categories. *Cognition* 65: 137–165.

Hampton, J. A. 2007. Typicality, graded membership and vagueness. *Cognitive Science* 31, no. 3: 355–384.

Heller, M. 1996. Against metaphysical vagueness. *Philosophical Perspectives*, 10.

Jordan, T., M. Raubal, B. Gartrell, and M. Egenhofer. 1998. An affordance-based model of place in GIS. In *8th Int. Symposium on Spatial Data Handling*, SDH'98, Vancouver, Canada. http://citeseer.ist.psu.edu/jordan98affordancebased.html (accessed 1 May 2007).

Kavouras, M. and E. Tomai. 2001. Understanding and reducing ontological vagueness in the geographic realm. In *SVUG-01 Symposium*, Ogunquit, Maine. http://www.comp.leeds.ac.uk/qsr/vug/SVUG-01/position/kavourastomai.pdf (accessed 1 May 2007).

Keefe, R. 2000. *Theories of vagueness*. Cambridge, U.K.: Cambridge University Press.

Keefe, R. and P. Smith. 1997. Introduction: Theories of vagueness. In *Vagueness: A reader*, ed. R. Keefe and P. Smith, 1–57. Cambridge, MA: MIT Press.

Kuhn, W. 2002. Modeling the semantics of geographic categories through conceptual integration. In Proc. *Second International Conference on Geographic Information Science (GIScience 2002)*, Boulder, CO, September 25–28. Lecture Notes in Computer Science 2478: 108–118. Berlin: Springer-Verlag.

Kulik, L. 2001. A geometric theory of vague boundaries. In *Spatial information theory. Foundations of geographic information science*, ed. D. Montello, 44–59. Berlin: Springer.

Lewis, D. 1988. Vague identity: Evans misunderstood. *Analysis* 48: 128–130.

Mark, D., B. Smith, and B. Tversky. 1999. Ontology and geographic objects: An empirical study of cognitive categorization. In *Spatial Information Theory, Cognitive and Computational Foundations of Geographic Information Science*, ed. C. Freksa and D. M. Mark. Lecture Notes in Computer Science 1661: 283–298. Berlin: Springer.

Martin, P. 1995. Knowledge acquisition using documents, conceptual graphs and a semantically structured dictionary. In *Proc. KAW'95, Ninth Knowledge Acquisition for Knowledge-Based Systems Workshop*, ed. B. R. Gaines, University of Calgary, Banff, Canada.

McGrenere, J. and W. Ho. 2000. Affordances: Clarifying and evolving a concept. In *Proc. Graphics Interface 2000*, Montreal, Quebec, Canada, 179–186. Montreal: Canadian Human Computer Communications Society.

Myaeng, S. and A. Lopez-Lopez. 1992. Conceptual graph matching: A flexible algorithm and experiments. *Experimental and Theoretical Artificial Intelligence* 4: 107–126.

NCGIA. 1997. The state of the art of land use modeling. In *The Land Use Modeling Workshop*. http://www.ncgia.ucsb.edu/conf/landuse97/summary.html (accessed 1 May 2007).

Nobis, N. 2000. Vagueness, borderline cases and moral realism: Where's the incompatibility?!?. *Philosophical Writings* 14: 29–39.

Norman, D. A. 1988. *The design of everyday things*. New York: Doubleday.

Noy, N. F. and M. A. Musen. 2003. The PROMPT suite: Interactive tools for ontology merging and mapping. *International Journal of Human-Computer Studies* 59(6): 983–1024.

Pelletier, F. J. and I. Berkeley. 1995. Vagueness. In *Cambridge dictionary of philosophy*, ed. R. Audi, 825–827. Cambridge, U.K.: Cambridge University Press. http://www.sfu.ca/~jeffpell/papers/VagueDictionary.pdf (accessed 1 May 2007).

Pletcher, G. K. 1995. Subject: Re: 9.268 vagueness. Humanist Discussion Group, Vol. 9, No. 275. (47), Center for Electronic Texts in the Humanities (Princeton/Rutgers). http://lists. village.virginia.edu/lists_archive/Humanist/v09/0254.html (accessed 1 May 2007).

Prinz, J. 1998. Vagueness, language, and ontology. *Electronic Journal of Analytic Philosophy*, 6, no. 10. http://ejap.louisiana.edu/EJAP/1998/prinz98.html (accessed 1 May 2007).

Rahm, E. and P. A. Bernstein. 2001. A survey of approaches to automatic schema matching. *VLDB Journal* 10(4): 334–350. http://citeseer.ist.psu.edu/rahm01survey.html (accessed 1 May 2007).

Rodriguez, M. and M. Egenhofer. 2003. Determining semantic similarity among entity classes from different ontologies. *IEEE Transactions on Knowledge and Data Engineering* 15(2): 442–456.

Russell, B. 1923. Vagueness. *The Australasian Journal of Psychology and Philosophy* 1: 84–92. http://www.santafe.edu/~shalizi/Russell/vagueness/ (accessed 1 May 2007).

Sainsbury, R. M. 1995. *Paradoxes*. 2nd ed. Cambridge, U.K.: Cambridge University Press.

Sanford, D. 1975. Borderline logic. *American Philosophical Quarterly* 12: 29–40.

Smith, B. 1996. Mereotopology: A theory of parts and boundaries. *Data and Knowledge Engineering* 20: 287–304.

Sorensen, R. A. 1997. Vagueness. In *The Stanford encyclopedia of philosophy*. http://plato. stanford/edu/entries/ (accessed 1 May 2007).

Tomai, E. and M. Kavouras. 2002. "Sharpening" vagueness: Identifying, measuring and portraying its impact on geographic categorizations. In *2nd International Conference on Geographic Information Science*, Boulder, CO.

Tomai, E. and M. Kavouras. 2005. Qualitative linguistic terms and geographic concepts-quantifiers in definitions. *Transactions in GIS* 9(3): 277–290.

van Rees, R. 2003. Clarity in the usage of the terms ontology, taxonomy and classification. CIB73 2003 Conference Paper. http://vanrees.org/research/papers/cib2003.pdf (accessed 1 May 2007).

Varzi, A. C. 1996. Parts, wholes, and part–whole relations: The prospects of mereotopology. *Data and Knowledge Engineering* 20: 259–286.

Varzi, A. C. 2001. Vagueness, logic, and ontology, the dialogue. *Yearbooks for Philosophical Hermeneutics* 1: 135–154.

Werner, K. 2002. Modeling semantics through blendings. Technical Report, Institute for Geoinformatics, University of Muenster.

Williamson, T. 1994. *Vagueness*. London: Routledge.

Wittgenstein, L. 1953 (1967). *Philosophical investigations*. 3rd ed. Ed. by G. E. M. Anscombe and R. Rhees, trans by G. E. M. Anscombe. Oxford: Basil Blackwell.

Yang, L., M. Ball, V. C. Bhavsar, and H. Boley. 2005. Weighted partonomy-taxonomy trees with local similarity measures for semantic buyer-seller match-making. *Journal of Business and Technology* 1(1): 42–52.

3 Semantic Interoperability

3.1 NOTIONS AND PERSPECTIVES

The pragmatic need to understand and methodically deal with the meaningful communication and integration of information relates to what is known as *semantic interoperability*. This widely used notion can mean different things to different people. The first source of vagueness is attributed to the constituent *semantic*, which has already been discussed in the first two chapters. The second source of vagueness is caused by the all-pervading notion of "interoperability" or "being interoperable." The literature on interoperability is vast (Fileto and Medeiros, 2003), often with confusing views and numerous overlapping definitions. Therefore, there is no significant harm if we attempt our own version:

> Interoperability is the ability of "systems" or "products" to operate effectively and efficiently in conjunction, on the exchange and reuse of available resources, services, procedures, and information, according to the intended use of their providers, in order to fulfill the requirements of a specific task.

Notice first that "systems" do not refer strictly to technical systems (e.g. hardware/ software systems or products), but more generally to any organized field of human activity (such as organizations, programs, infrastructures, etc.). Therefore, it may also include interoperability of administration services, financial systems, political systems, legal frameworks, etc. The second requirement specifies the subject of cooperation of the two parties, namely, the exchange and reuse of resources in the context of a specific task. This cooperation, which involves communication, processing, and interpretation of the exchanged resources, has to be not only effective but also efficient. Specifying efficiency is another story; nevertheless, the two ends must get the feeling that their task is not hindered but sufficiently alleviated by the interoperation. A cumbersome communication of two systems does not comply with the notion of being interoperable. Another issue is that of the "intended use," similar to another commonly employed term of "proper use," implying that interoperability should not allow misuse.

This definition may be more or less accurate but is too general to be of any practical use. According to the book context, what is of interest here is the technical issue of *interoperability of geo-ontologies*. Notice that "technical" should not be confused with "explication" issues, which are not our focus. It just implies a systematic approach of practical value, which deals with theoretical and conceptual issues but aims at some palpable results. Furthermore, in order to fulfill the "according to intended use" requirement, it is essential that geoinformation be unambiguously understood by users (humans or computer programs), so that it is meaningfully dealt with. Understanding involves the study of *meanings*, which is *semantics*. The semantics of a geospatial domain are best elucidated by a geo-ontology. This narrows our interoperability focus to *semantic interoperability between geo-ontologies*.

Semantic interoperability addresses the ultimate level of interoperation—the only level not necessitating knowledge of the semantics involved. There are elementary levels of nonsemantic communication, which are currently well established. Bishr (1998) for instance, presents an order of six interoperability levels. The order signifies the level of sophistication or advancement in information communication, moving from architectural to modeling issues. Level 1 supports network protocol communication with a distant system (e.g., via telnet), but requires knowledge of the operation system used. Level 2 supports interoperation with a distant system independently of its operating system (e.g., via ftp). It is required, however, to know everything about the files transferred. Level 3 supports data communication with the source files using a query language and an interface of the source. The user still needs to know what is in the source data. Level 4 supports a database management system (DBMS)-type communication allowing the use of the local query mechanism, and not that of the source (as in level 3), as long as the model and meaning (semantics) of the source information are known. Level 5 of interoperability does not require knowledge of the model of the source but allows a connection to it via a single general model. In order to know, however, exactly what the data represents in reality, it requires knowledge of the data semantics. Finally, the last level (6) facilitates communication with the source without needing to know data semantics as these accompany the data and provide a self-explanation. The above interoperability levels are just indicatory. Interoperation expresses a communication between a user and a source, indicating what can be automatically supported and what one needs to know. Considering the degree of maturity and development, the first five levels of communication are already supported by sound technical solutions. The last one, based on semantics, is the focal point of enormous research worldwide, GIScience being no exception.

Once information and resources are interoperable, it means that they can be integrated; in fact, they probably need to be integrated in order to provide holistic and consistent answers to questions. If *integration* is based on semantics, then we talk about *semantic integration*. Interoperability is not exhausted with (i.e., does not always engage in) integration. It also involves other means of intelligent communication (Doerr, 2004) such as querying, extraction, transformation, etc. Therefore, *integration* is a subset of *interoperability* and *semantic integration* is a subset of *semantic interoperability*. This raises a notable difference. In addition to the interoperability requirement for heterogeneous systems to communicate and exchange information without prior knowledge of their associated semantics, *semantic integration* further necessitates the reconciliation of mismatches, conflicts, heterogeneities, etc.

Hence, interoperability and especially integration aim at dealing with existing heterogeneities between information that is communicated and associated. If information were homogeneous, there would be no need for interoperability. The commonest and also traditional way to avoid heterogeneity is to impose standards (DIFFUSE PROJECT, 2001). Standards can be considered as an interoperability approach in the sense that they support exchange and reuse of information. It is well known though, that standards have limited effectiveness. Rapid technological change, independent development of systems and databases, compliance cost, inferior performance of general standardized versions when compared to tailor-made ones, lack of universally accepted standards, and cost of adaptation constitute important reasons for systems, especially legacy ones, to resist change unless there is a good reason/motive for the opposite. In addition, existing standards—especially old ones—address mainly explication and not semantic matters (Bittner, Donnelly, and Winter, 2005).

Interoperability goes beyond standards. Instead of enforcing a common view of the world, it accepts variability as something unavoidable, yet useful, and attempts to establish advanced "bridges" or "mappings" between the associated ontologies. In its simplest form, also existing before standards, it provides "translators" between two heterogeneous systems. In its advanced form, interoperability allows for a consistent integrated view of the world, which can be broken down to different complementary views. Building, however, this consistent integrated view between heterogeneous systems is not simple to achieve and cannot be done magically. It requires (a) identification of all existing heterogeneities, (b) an analysis of their importance and priority, and (c) a systematic strategy to resolve/reconcile them.

In principle, semantic aspects of interoperability and integration, being a matter of necessity rather than a matter of choice, are of great value. The opposite would certainly lead to all sorts of problems. Actually, it does not make sense to be intentionally contented with nonsemantic interoperability or integration, which is equivalent to mixing "apples and oranges." When other interoperability tasks are deployed to resolve certain nonsemantic heterogeneities (e.g., explication differences), it is implicitly assumed that there are no semantic differences in the associated information, or that they have been resolved at a prior stage. Even in nonsemantically aware applications, for example, when dealing with a single well-known domain, interpreters somewhat always consider semantic aspects as inherent in the application. The difference is that in the past an explicit, systematic, and conscious consideration of semantics was commonly neglected.

3.2 SEMANTIC HETEROGENEITIES

Identifying heterogeneities (also called conflicts or mismatches) is crucial to integration. Identifying semantic heterogeneities is crucial to ontology integration. There have been numerous attempts to classify heterogeneities (see also Chapter 14). An exhaustive review of these would be inescapably incomplete, of limited practicality, and outside the scope of the book. More important is to attempt to identify the critical issues and provide sufficient explanation for them. The probably most common view, based on a wide consensus, classifies heterogeneities into three types: *syntactic, schematic, and semantic.* This classification obviously has its origins in linguistics, which find direct application here since natural language is a very important means of expressing and communicating geographic knowledge. It, nevertheless, also applies (metaphorically) to other formal representations of knowledge (formal languages, databases, maps, etc.). In short, whatever the representation, it normally uses a syntax and a schema (structure) of relations. Semantics provides meaning by associating the *representing* to what is *represented* in the real world. This meaning cannot be provided by syntax and schema.

Syntactic heterogeneity may, for instance, be called the difference between a relational model and an object-oriented model in two database management systems (DBMS). The difference between a Triangulated Irregular Network (TIN) and a Digital Elevation Model (DEM) structure of the same topographic relief, or the difference between a raster and a quadtree structure of the same matrix also count as syntactic differences. Differences in conceptual models regarding relations between entities, attributes, and entities and attributes count as schematic heterogeneity. For example, in one model there may be different entities corresponding to what, in the other model, is expressed as a single entity with various values of a specific attribute. An explicit topological relation in one model that is implicit in the other or is expressed differently are also considered as schematic heterogeneities.

Semantic heterogeneity is caused by different conceptualizations of the same real world entities/phenomena. Naming heterogeneities (i.e., synonyms and homonyms) are the simplest case of semantic heterogeneity. Similarly to Doerr's typology (2001) concerning semantic heterogeneities between different thesauri, it could be assumed (Figure 3.1) that semantic heterogeneities between different ontologies are caused by:

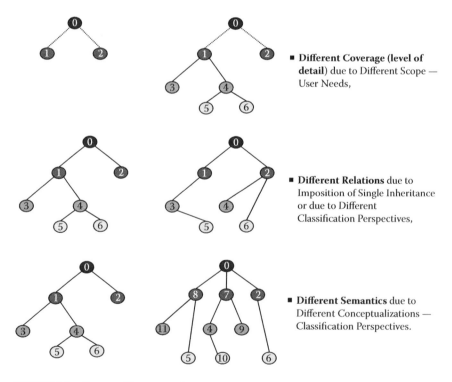

- **Different Coverage (level of detail)** due to Different Scope — User Needs,

- **Different Relations** due to Imposition of Single Inheritance or due to Different Classification Perspectives,

- **Different Semantics** due to Different Conceptualizations — Classification Perspectives.

FIGURE 3.1   Causes of heterogeneity.

1. Different coverage (level of detail) due to different scope–user needs. For example, one ontology stops at general concepts of agricultural use (e.g., cultivating cereals), whereas the other uses additional subconcepts to provide more detail on the type of agriculture (e.g., cultivating wheat). Notice that the general concepts are the same in both ontologies. The new, more detailed concepts provide additional levels in the tree hierarchy.
2. Different relations often due to imposition of a single inheritance or due to different classification perspectives. For example, the ontology of the taxation office considers a doctor's office and a drugstore as private businesses, whereas the ontology of the statistical service or ministry of health considers both as nodes of the health care service system. Notice that the subconcepts in both ontologies are the same. They are just classified differently due to the different perspectives.
3. Different semantics due to different conceptualizations–classification aspects. These are the result of different partitions of reality.

The first cause of semantic heterogeneity does not induce conflicting but complementary views of the same domain, and, therefore, it does not obstruct the integration process. The second cause can be overcome by the permission of multiple inheritances in the integrated ontology. The third cause of semantic heterogeneity, that is, different semantics due to different conceptualizations of similar concepts, is most difficult to identify and tackle. Some examples of cases are very instructive:

Case 1: Concept LAND PARCEL in ontology A divides into two subconcepts, URBAN (land parcel) and RURAL (land parcel), while in ontology B it divides into BUILD-UP (land parcel) and NON-BUILD-UP (land parcel).
Case 2: Concept TERRAIN is further classified in one ontology as MOUNTAINOUS (terrain) and PLANE (terrain), while in the other as MOUNTAINOUS (terrain), SEMI-MOUNTAINOUS (terrain) and PLANE (terrain).
Case 3: Concept CANAL in ontology A includes TRANSPORTATION (canal) and IRRIGATION (canal), while in ontology B it includes IRRIGATION (canal) and DRAINAGE (canal).

Ontological mismatches have also been addressed by several other frameworks, as explained in Chapter 14. An issue that remains unclear, however, is the validity of the widely purported view that schematic information does not relate to semantic information. The practical question posed here is: Can schematic differences be excluded from semantic comparison? This inevitably leads to more critical questions: Are schematic (structural) relations part of a concept's meaning? Can the same concept possess different superconcepts and subconcepts in different ontologies? So far the answer to these is assumed to be yes. One the other hand, there are definitions of concepts in which such relations to superconcepts/subconcepts are stated or implied, turning such information essential (i.e., semantically important) to the concept's identity. Kashyap and Sheth (1996), among others, have attempted to reconcile the two perspectives of semantic and schematic differences.

3.3 GEOSPATIAL INTEROPERABILITY EFFORTS

Looking into the geospatial domain, out of the numerous approaches reported in the literature, one can identify a few indicative. Vckovski (1998) and Goodchild et al. (1999) stress the importance of GI interoperability. Mark (1993) was one of the first to discuss the inadequacy of existing standards/formats for a meaningful (semantically aware) data transfer. The importance of semantic aspects is also addressed by many others, for example: Frank (1986), Frank and Raubal (1998), Nyerges (1989a; 1989b), Shepherd (1991), Townshend (1991), Ram and Park (1997), Yuan (1997), Devogele, Parent, and Spaccapietra, (1998), Gahegan (1999), Harvey (1999), Sheth (1999), Uitermark et al. (1999), Fabrikant and Buttenfield (2001), Worboys and Duckham (2002), and Bittner, Donnelly, and Winter, (2005). Furthermore, Bishr (1997, 1998) and Bishr et al. (1999) distinguish three types of interoperability and provide examples for all these types. They place emphasis on the importance of the semantic type, for it is the one creating most sharing problems among "geospatial information communities" (GICs) (Bishr, 1997, 1998; Bishr et al., 1999). They advocate the use of a "semantic translator" to share transparently the views of the GICs. Kuhn (2002, 2003) focuses on semantic interoperability via his work on semantic reference systems and cognitive semantics. Rodriguez and Egenhofer (2003) calculate semantic similarity using other features such as attributes, parts, and functions. This approach is suitable for comparing categories when both categories have such complete and detailed descriptions.

In the context of the web, which is characterized by vast amounts of somewhat similar but not always relevant information, providing semantics proves critical in avoiding information overload. Furthermore, introducing *space* as a means of coherence and semantic relevance of most information inevitably leads to the development of a geospatial semantic web proposed by Egenhofer (2002). Duckham and Worboys (2005) have also presented an extensional approach to information integration ("fusion"), also known as *information flow*, based on database instances, rather than on intensional information of concept definitions in ontologies. Hakimpour and Timpf (2002) present an approach to schema integration based on formal definitions in Description Logic. Kavouras and Kokla (2002) present an approach based on Formal Concept Analysis (FCA) for the integration of heterogeneous geographic ontologies. Semantic information (relations and properties) is extracted from definitions using Natural Language Processing (NLP) techniques. The issue of semantic granularity in information integration has been addressed by Fonseca et al. (2002). An integration approach introducing the notion of spatial concept lattices of multiscale/context information has also been presented by Kokla and Kavouras (1999).

3.4 CORE ISSUES

Although interoperability in the geospatial domain has been identified as a hot research area for over a decade, there are only very few practical results to demonstrate. Robust techniques are only available at the explication level to deal mostly with syntactic differences. At the semantic level, despite the important steps made forward (see previous section), there are many problems that remain unclear, making, thus, their solution harder to attain. From the variety of semantic interoperability issues, it is important to identify those pertaining to the scope of this book.

- Any available methods for semantic interoperability should also provide an adequate solution to the difficult issue of geo-ontology integration.
- The methods should support dealing with existing bodies of knowledge; knowledge that often resides in legacy systems and suffers by various imperfections. Only when what characterizes existing knowledge is realized, will it be possible to successfully deal with the issue of developing new ontologies (ontology engineering).
- Because semantic integration is to be implemented in a computer environment, it is necessary to provide an adequate formalization of the representations involved, by making use of the appropriate conceptual structures.
- The success of semantic integration relies on the ability to identify and, subsequently, resolve any existing heterogeneities.
- In order to improve efficiency but more importantly systematize and objectify the integration process, it is important that user involvement is kept to the absolutely critical one.

These key interoperability issues are addressed in the remainder of the book.

REFERENCES

Bishr, Y. 1997. Semantic aspects of interoperable GIS. Ph.D. dissertation, Enschede, The Netherlands.

Bishr, Y. 1998. Overcoming the semantic and other barriers to GIS interoperability. *International Journal of Geographical Information Science* 12(4): 299–314.

Bishr, A., H. Pundt, W. Kuhn, and M. Radwan. 1999. Probing the concept of information communities: A first step toward semantic interoperability. In *Interoperating geographic information systems*, ed. M. F. Goodchild, M. Egenhofer, R. Fegeas, and C. Kottman, 55–69. Boston, MA: Kluwer Academic Publishers.

Bittner, T., M. Donnelly, and S. Winter. 2005. Ontology and semantic interoperability. In *Large-scale 3D data integration: Problems and challenges*, ed. D. Prosperi and S. Zlatanova, 139–160. Boca Raton, FL: CRC Press.

Devogele, T., C. Parent, and S. Spaccapietra. 1998. On spatial database integration. *International Journal of Geographic Information Science* 4: 335–352.

DIFFUSE PROJECT. 2001. Geographic Data Exchange Standards. http://www.dcc.ac.uk/diffuse/?s=-1&ct=25 (accessed 1 May 2007).

Doert, M. 2001. Semantic problems of thesaurus mapping. *Journal of Digital Information*, 1(8). http://journals.tdl.org/jodi/article/view/jodi-35/32 (accessed 1 May 2007).

Doerr, M. 2004. Semantic interoperability: Theoretical considerations. Technical Report 345, Institute of Computer Science, Foundation for Research and Technology–Hellas, Science and Technology Park of Crete, Crete, Greece.

Duckham, M. and M. F. Worboys. 2005. An algebraic approach to automated geospatial information fusion. *International Journal of Geographic Information Science* 19(5): 537–558.

Egenhofer, M. 2002. Toward the semantic geospatial web. In *10th ACM International Symposium on Advances in Geographic Information Systems*, ed. A. Voisard and S. C. Chen. 1-4. New York: ACM Press.

Fabrikant, S. I. and B. P. Buttenfield. 2001. Formalizing semantic spaces for information access. *Annals of the Association of American Geographers* 91(2): 263–280.

Fileto, R. and C. B. Medeiros. 2003. A survey on information systems interoperability. TR-IC-03-30, icunicamp. http://www.dcc.unicamp.br/ic-tr-ftp/2003/03-30.ps.gz (accessed 1 May 2007).

Fonseca, F., M. Egenhofer, P. Agouris, and G. Câmara. 2002. Using ontologies for integrated geographic information systems. *Transactions in Geographic Information Systems* 6(3): 231–257.

Frank, A. 1986. Integrating mechanisms for storage and retrieval of land data. *Surveying and Mapping* 46(2): 107–121.

Frank, A. and M. Raubal. 1998. Specifications for interoperability: Formalizing image schemata for geographic space. In Proc. *8th International Symposium on Spatial Data Handling, SDH'98*, Vancouver, Canada.

Gahegan, M. 1999. Characterizing the semantic content of geographic data, models, and systems. In *Interoperating geographic information systems*, ed. M. Goodchild, M. Egenhofer, R. Fegeas, and C. Kottman, 71–84. Norwell, MA: Kluwer Academic Publishers.

Goodchild, M. F., M. Egenhofer, R. Fegeas, and C. Kottman, eds. 1999. *Interoperating geographic information systems*. Norwell, MA: Kluwer Academic Publishers.

Hakimpour, F. and S. Timpf. 2002. A step towards geodata integration using formal ontologies. In *5th AGILE Conference on Geographic Information Science*, ed. M. Ruiz, M. Gould, and J. Ramon, 5. Universitat de les Illes Balears, Palma de Mallorca, Spain.

Harvey, F. 1999. Designing for interoperability: Overcoming semantic differences. In *Interoperating geographic information systems,* ed. M. F. Goodchild, M. J. Egenhofer, R. Fegeas, and C. A. Kottman, 85-98. Norwell, MA: Kluwer Academic Publishing.

Kashyap, V. and A. Sheth. 1996. Semantic and schematic similarities between database objects: A context-based approach. *VLDB Journal* 5: 276–304.

Kavouras, M. and M. Kokla. 2002. A method for the formalization and integration of geographical categorizations. *International Journal of Geographical Information Science* 16(5): 439–453.

Kokla, M. and M. Kavouras. 1999. Spatial concept lattices: An integration method in model generalization. *Cartographic Perspectives* 34(Fall): 5–19.

Kuhn, W. 2002. Modeling the semantics of geographic categories through conceptual integration. In *Proc. Second International Conference on Geographic Information Science*, ed. M. Egenhofer and D. M. Mark. GIScience 2002, Boulder, CO.

Kuhn, W. 2003. Semantic reference systems. *International Journal of Geographic Information Science* 17(5): 405–409.

Mark, D. 1993. Toward a theoretical framework for geographic entity types. In *Spatial information theory*, ed. A. Frank and I. Campari, 270-283. Berlin: Springer-Verlag.

Nyerges, T. 1989a. Schema integration for the development of GIS databases. *International Journal of Geographical Information Systems* 3(2): 153–183.

Nyerges, T. 1989b. Information integration for multipurpose land information systems. *URISA Journal* 1(1): 27–38.

Ram, S. and J. Park. 1996. Modeling spatial and temporal semantics in a large heterogeneous GIS database environment. *Proc. 2nd Americas Conference on Information Systems (AIS '96)*, 16-18, Phoenix, Arizona,.

Rodríguez, A. and M. Egenhofer. 2003. Determining semantic similarity among entity classes from different ontologies. *IEEE Transactions on Knowledge and Data Engineering* 15(2): 442–456.

Shepherd, I. D. H. 1991. Information integration and GIS. In *Geographical information systems: Principles and applications*, ed. D. Maguire, M. Goodchild, and D. Rhind, 337-360. London: Longman.

Sheth, A. 1999. Interoperability and spatial information theory. In *Interoperating geographic information systems*, ed. M. Goodchild, M. Egenhofer, R. Fegeas, and C. Kottman, 5–29. Dordrecht, The Netherlands: Kluwer.

Townshend, J. R. G. 1991. Environmental data bases. In *Geographical information systems: Principles and applications*, ed. D. Maguire, M. Goodchild, and D. Rhind, 201-216. London: Longman.

Uitermark, H., P. V. Oosterom, N. Mars, and M. Molenaar. 1999. Ontology-based geographic dataset integration. In *Proc. Workshop on Spatio-Temporal Database Management*, 60-79, Edinburgh, Scotland.

Vckovski, A. 1998. *Interoperable and distributed processing in GIS*. London. Taylor & Francis.

Worboys, M. F. and M. Duckham. 2002. Integrating spatio-thematic information. In *Geographic Information Science*, ed. M. J. Egenhofer and D. M. Mark. Lecture Notes in Computer Science 2478: 346–361. Berlin: Springer.

Yuan, M. 1997. Development of a global conceptual schema for interoperable geographic information. *The International Specialist Meeting on Interoperating Geographic Information Systems, sponsored by NSF and OpenGIS*. December 5–6, Santa Barbara, CA. http://ncgia.ucsb.edu/conf/interop97/program/papers/yuan/yuan.html (accessed 1 May 2007).

Part 2

Theoretical Foundations

4 Ontologies

4.1 ONTOLOGY IN PHILOSOPHY

Ontology as a branch of *metaphysics* is the science of being; it investigates the kinds of objects (abstract and concrete, existent and nonexistent, real and ideal, independent and dependent), and their in-between relations. *Metaphysics* is the pure part of philosophy, in contrast to the applied. The word originates from *Metaphysica,* the title of Aristotle's treatise on the subject, from Greek (ta) meta (ta) physika, literally, the (works) after the physical (works) (*Merriam-Webster's Collegiate Dictionary,* 2003). According to Kant (1997), the word refers to the science beyond the boundaries of nature, that is, metaphysics is the philosophy of nature so far as it is based on a priori principles. Physics is the philosophy of nature so far as it is based on principles from experience (Kant, 1790–1791?). Bolzano (1810) compares mathematics and metaphysics as the two main parts of a priori knowledge; mathematics investigates the conditions under which existence of things is possible, whereas metaphysics is concerned to prove a priori which things are real.

Metaphysics studies the fundamental nature of reality and being. However, there is no agreement in literature as to the divisions of metaphysics. According to Peirce (1932), metaphysics has three subdivisions: (a) *general metaphysics* or *ontology,* (b) *psychical* or *religious metaphysics,* which deals with issues of God, freedom, and immortality, and (c) *physical metaphysics,* which studies the issues of time, space, matter, etc. Some divide metaphysics into two main subdivisions: *ontology* and *cosmology* (Cocchiarella, 2004); whereas others also include *philosophical theology* (*The Columbia Encyclopedia,* 2004), or *epistemology* (*Merriam-Webster's Collegiate Dictionary,* 2003). *Epistemology* studies the nature and sources of knowledge (*The American Heritage New Dictionary of Cultural Literacy,* 2007) in contrast to ontology, which studies reality regardless of our knowledge about it.

Ontology studies the nature and categories of being and, thus, surveys general notions such as essence, existence, properties, modes, necessity, place, time, change, life, etc. According to Sommers (cited in Englebretsen, 1990), ontology is the science of categories, since it seeks an answer to such questions as: what categories are, how they are specified, and how they are related to each other; Plato's theory of forms, Aristotle's Categories, and Russell's theory of types are ontological essays in providing an answer to these questions.

One of the greatest debates in philosophy emanates from the seventeenth and eighteenth centuries and concerns reality and what can be known about it (Lowe, 2001). The first group of philosophers (originating from Plato and Aristotle) believes that we can know reality to some extent. According to the opposite view (led by Kant), we cannot know anything about reality, since "it is in itself." For the first group, ontology is considered as the "science of being," whereas for the second, ontology is the "science of our thought about being" (Lowe, 2001).

Ontological theories are formulated from the perception of the world (Hartmann, 1949) through a continuous and toilful process of examination and improvement on the basis of scientific research. These methods are called *fallibilistic*, since they do not declare to reach the truth, although their aim is to discover the truth behind reality (Smith, 2003). During the twentieth century, the methods used by ontologists to derive their theories are supplemented by formal tools and theories (e.g., category theory, set theory, mereology, topology), which, together with formal logic and formal semantics, allow the rigorous definition and validation of ontological theories (Smith, 2003). In order to make explicit what kinds of objects compose the real world, it is important to state the *ontological commitment* of a theory. Quine (1961) was the first to introduce this notion in philosophy, with his famous *criterion of ontological commitment*: "to be is to be the value of a bound variable." For Quine, ontology is the study of the theories of natural sciences through the specification of the ontological commitments underlying them.

Different schools of philosophy deal with the classification of objects in the world in different ways. *Substantialists* consider that ontology is focused on substance, whereas *fluxists* regard it as based on process or function or continuous fields of variation (Smith, 2003). *Adequatists* search the kinds of objects at all levels of granularity (from the microphysical to the cosmological); *reductionists*, on the other hand, believe that every area of reality can be analyzed into main components and focus their research on these (Smith, 2003).

Bunge (1999) classifies ontology into *general* and *special* (or *regional*). Whereas *general ontology* deals with the study of all existents, focusing on general concepts such as space, time, event, etc., *special ontologies* focus on the study of one kind of thing or process (e.g., social ontology examines sociological concepts, biological ontology examines biological objects, etc.). Regardless of being general or special, ontology operates either in a *speculative* or in a *scientific* manner (Bunge, 1999). Examples of speculative ontologies are those defined by Leibniz, Whitehead, Wittgenstein, and Heidegger. Chemical, biological, and medical ontologies are examples of ontologies that operate in a scientific manner.

Perzanowski (1990) distinguishes ontology into three parts: (a) *ontics*, (b) *ontomethodology*, and (c) *ontologic*. *Ontics* studies ontological problems and notions, and how these are classified, analyzed, and distinguished. *Ontomethodology* deals with ontological methods, whereas *ontologic* is concerned with the logic, the laws and base pertaining to the ontic realm, as well as with ontological relations.

The first attempt to organize reality into a system of categories was undertaken by Aristotle in his work *Categories*. Ten basic categories for classifying anything were presented: *Substance* (ousia), *Quality* (poion), *Quantity* (poson), *Relation* (pros ti), *Activity* (poiein), *Passivity* (paschein), *Having* (echein), *Situatedness* (keisthai), *Spatiality* (pou), and *Temporality* (pote). At the level of everyday human experience, Aristotle defined substances and accidents as the basic constituents of reality (Smith, 1998). *Substances* have properties and are subject to changes, which are called *accidents*. Substances (e.g., human beings, animals, trees, cars, etc.) exist on their own, in contrast to accidents (e.g., discussions, phone calls, songs, thirst, etc.),

TABLE 4.1
Kant's Organization of Aristotle's Categories

Quantity	Quality	Relation	Modality
Unity	Reality	Substantiality and inherence	Possibility and impossibility
Plurality	Negation	Causality and dependence	Being and non-being
Totality	Limitation	Reciprocity	Necessity and contingency

whose existence depends on substances. Substances form collectives such as a family, a class of students, a government, etc. Accidents, on the other hand, are classified into *relational* and *nonrelational* with regard to their dependence on a single or many substances.

Kant (cited in Weber, 1908 and Sowa, 2000) in the "Critique of Pure Reason" made an attempt to organize Aristotle's categories by introducing a framework consisting of four groups corresponding to the fundamental categories of (a) quantity, (b) quality, (c) relation, and (d) modality. The symmetry of this table (Table 4.1), resulting from its triadic structure, originates from a profound law, since the third concept in the triad emanates from a combination of the second and the first concept (Sowa, 2000).

Kant's triadic patterns raised confusion and induced research among the following German philosophers on the profound principle immanent in them (Sowa, 2000). Peirce (cited in Burch, 2006) identified three primary concepts: *firstness*, *secondness*, and *thirdness*. *First* indicates the mode of being independent of anything else. *Second* indicates the mode of being relative to something else. *Third* defines the mode of being, which relates the first and the second, that is, the concept of mediation. According to Sowa (2000), firstness is defined by a feature inherent in something (e.g., woman), secondness defines a relation with something else (e.g., mother), and thirdness defines a mediation, which leads to the association of two or more entities (e.g., motherhood).

Whitehead (Simons, 1998; Sowa, 2000; George, 2006) formulated eight categories of existence. The first three correspond to physical categories, the second three correspond to abstract categories, and the last two categories generate combinations:

1. An *actual entity*, also called actual occasion, represents a real thing existing in the world and being independent of anything else.
2. A *prehension* represents an internal relation between two actual entities.
3. A *nexus* represents the connection or grouping of actual entities, which are interrelated by their prehension of each other.
4. An *eternal object* is a "pure potential for the specific determination of fact"; it becomes available through prehensions from actual entities (e.g., sourness, hardness, etc.).
5. A *proposition* is the relation between an actual entity or group of entities and an eternal object.
6. A *subjective form* represents the way in which an actual entity prehends another; examples of subjective forms are emotions, valuations, purposes, etc.

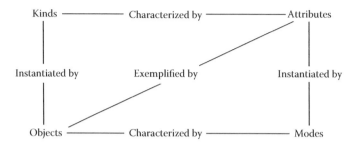

FIGURE 4.1 Lowe's four-category ontology. (From Lowe, E. J. 2001. Recent advances in metaphysics. In *Proc. 2nd International Conference on Formal Ontology in Information Systems*, Ogunquit, Maine. New York: ACM Press. http://www.cs.vassar.edu/~weltyc/fois/fois-2001/keynote/. With permission.)

7. A *multiplicity* is a grouping of actual entities that share a common characteristic.
8. A *contrast* is a "mode of synthesis of entities in one prehension."

More recently, Lowe (2001) introduced a four-category ontology, which provides a metaphysical basis for natural science. It considers three ontological categories to be fundamental: (a) objects or individual substances, (b) universals (i.e., properties or else "repeatable" entities ascribed to different objects at several times and places), and (c) tropes or modes (i.e., properties or "nonrepeatable" entities, ascribed to a specific object). Universals are further classified into: (i) kinds of objects and (ii) properties of objects (property-universals or attributes). The four-category ontology is shown in Figure 4.1.

4.2   FORMAL ONTOLOGY

The philosopher Edmund Husserl in his work Logical Investigations has introduced *formal ontology* in order to account for categories such as object, property, relation, relatum, manifold, part, whole, state of affairs, existence, etc. (Smith and Woodruff Smith, 1995).

The term *formal ontology* has been studied from two different research directions: an *analytic* one mainly inspired by the work of Cocchiarella and a *phenomenological* one inspired by the work of Brentano and Husserl (Guarino, 1995; Poli, 2003). The first one analyzes formal ontology within the framework of formal logic, whereas the second focuses on issues of mereology, topology, and dependence.

According to Cocchiarella (1991, 640; 2004), formal ontology is a discipline that combines the philosophical analyses and principles of ontology with the formal methods of mathematical logic. Formal ontology may be defined as "the systematic, formal, axiomatic development of the logic of all forms and modes of being" (Cocchiarella, 1991, 2004). According to Guarino (1995), this definition is rather pregnant, since it considers both senses of the adjective "formal": (a) rigorous, and (b) corresponding to the forms of being. Guarino (1995) defines formal ontology as the theory of a priori distinctions among the entities of the world (physical objects, events, processes, quantities, etc.) and among the meta-categories used to model the world (concepts, properties, qualities, states, etc.).

Poli (2003) draws the distinction between *descriptive, formal,* and *formalized ontology,* supporting that they are different levels of theory construction. *Descriptive ontology* studies the objects existing in the world, dependent or independent, real or ideal, either generally or under a particular domain. *Formal ontology* categorizes and organizes this information and is called "formal" according to Husserl, because it investigates pure categories of reality such as thing, process, matter, whole, part, and number. *Formalized ontology* undertakes the formalization of this information using a specific formalism. A set of formal tools supporting ontological research, such as formal logic, algebra, category theory, set theory, mereology, and topology, have been developed during the twentieth century. *Mereology* deals with the notion of parthood or overlap, whereas *topology* deals with the notion of wholeness or connection. However, formal ontology needs to be supplemented by other tools in order to account for important ontological considerations; morphology, qualitative geometry, kinematics, and dynamics are some of them (Varzi, 1998).

Smith (1998), based on Husserl, identifies the basic concepts of formal ontology focusing on the mesoscopic level, that is, the level of everyday human experience. The combination of mereology and topology, that is, *mereotopology,* in contrast to set theory, is identified as a basic formal tool to account for the objects, which constitute the realm of everyday experience and expand in space forming a "boundary-continuum structure" (Smith, 1998, 20). In this way, the world is perceived as consisting not of elements or abstract properties, as in set theory, but of discernible objects connected with different mereological and topological relations, as well as relations of dependence. Mereotopology, dependence, and separability are used as basic ontological tools to define the relations between objects (substances, accidents, parts, wholes, and boundaries).

An important ontological distinction originating from Whitehead concerning the identification of entities is that between *continuants* and *occurrents* (Muller, 1998; Sowa, 2000; Grenon and Smith, 2004; Galton, 2006). *Continuants* are entities that remain stable through time (they endure, thus also called *endurants*); they have spatial and not temporal parts. *Occurrents* on the other hand are processes or events that are temporally restricted and have temporal parts (they perdure, thus called *perdurants*). The distinction between continuants and occurrents is meaningful only within the framework of a specific time scale and level of detail (Sowa, 2000). A coastline may be defined as an occurrent on a minute time scale and a fine level of detail, due to sea waves, but for a long-standing time scale and a coarser level of detail a coastline may be considered as a continuant. Correspondingly, a city whose area and population are changing may be considered a continuant or an occurrent, dependent on the choice of time scale and level of detail.

4.3 THE PROBLEM OF UNIVERSALS

Another important philosophical distinction, which originated two thousand years ago, is that between *universals* and *particulars. Universals* are the category of mind independent entities, which represent the properties or qualities of single objects, called *individuals*, or *particulars*, or *instances* (MacLeod and Rubenstein, 2006). *Individuals* are nonrepeatable, that is, they exist in only one place at a time, in contrast to universals, which are repeatable. However, individuals may have properties or qualities that change in the course of time. Examples are the objects of everyday experience, such as cars, trees, tables, etc. These examples of individuals are material, that is, they occupy a spatiotemporal region. Other types of individuals include immaterial ones (e.g., holes, spirits, souls, angels), and individuals outside space and time (e.g., numbers and God) (MacLeod and Rubenstein, 2006).

A longstanding debate among metaphysicians concerns the nature of universals and whether they may be employed to account for similarity among individuals. Questions of whether there are general truths or commonality in reality and whether universal knowledge of individual entities is feasible relate to what is called the *problem of universals* (Klima, 2004). There are three competing views on the problem of universals that are represented by *realists*, *nominalists*, and *conceptualists* (Klein and Smith, 2005) (Figure 4.2). *Realists* believe that there are general truths and commonality in reality and it is through the existence of universals that similarity relations among particulars may be identified. These similarity relations exist independently of the agents that perceive them. Universals are multi-exemplifiable through various particulars. Universals according to the realist position are called *general concepts* or *general terms* (Klein and Smith, 2005). There are different variations of realism, for example, *extreme realism* originating from Plato or *strong realism* originating from Aristotle.

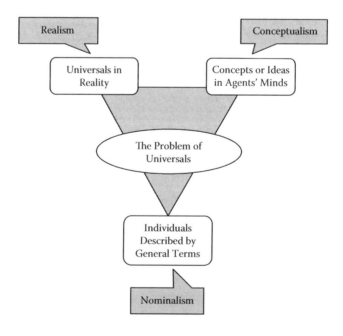

FIGURE 4.2   The problem of universals.

Nominalists believe that only individuals exist; universals do not exist, either in reality, or in the agents' minds. They believe that there are not general concepts, only general terms referring to ad hoc collections of objects. For them, the problem of universals can be resolved by taking into account only the nature and relations among individuals (MacLeod and Rubenstein, 2006). Nominalism has a series of variants: *predicate nominalism, resemblance nominalism,* and *trope nominalism* are three of them. For *predicate nominalism,* commonality in nature is explained relatively to individuals and linguistic expressions. Two individuals are similar because the same predicate applies to both of them. For example, the predicate "mouse" applies to both Mickey Mouse and Minnie Mouse. Therefore, according to predicate nominalism, Mickey and Minnie are both mice, since the predicate (i.e., the linguistic expression) holds for both of them. The criticism against predicate nominalism is that it does not solve the problem of universals, since it does not give an account of the similarity between individuals, for example, the reason why the predicate "mouse" holds for both Mickey and Minnie.

For *resemblance nominalism,* similarity is explained on the basis of resemblance relations among the members of a set (MacLeod and Rubenstein, 2006). Sets are formed by individuals that are similar according to a property. For example, resemblance nominalism accounts for an individual being tall by the fact that it is a member of the set of tall entities. Similarity in the property of being tall is explicated by the fact that all tall individuals belong to the same set. However, resemblance nominalism has weak points. First of all, it cannot give an explanation for the case of the same set being described by two distinct properties, which are always coexistent (e.g., having a heart and a kidney). In that case, either the same set is built twice, or the two properties are not distinguishable. Other problems of resemblance nominalism consist in the method of set construction and the resemblance relation (MacLeod and Rubenstein, 2006).

Trope nominalism is based on the notion of *tropes.* A *trope* is a particular property instance and exhibits qualitative simplicity in contrast to the qualitative complexity of common particulars (MacLeod and Rubenstein, 2006). A particular is considered as a collection of tropes, which ascribe to the particular specific qualities. Similarity between two particulars is explained on the basis of the similarity between the tropes these particulars consist of; the first particular has a trope that is qualitatively identical but numerically discrete from the trope of the second particular. For example, the green color of the grass is numerically discrete but qualitatively identical to the green color of the tree. However, although tropes are embraced in order to account for similarity between particulars, trope nominalism has fallen short of explaining the similarity between tropes. Therefore, it seems that the problem of similarity is shifted to another level.

Conceptualists believe that there are no universals in reality but in the minds of the agents that perceive them. For them, qualitative similarity is identified relatively to concepts or ideas. These concepts are not arbitrarily created but signify the similarities among particulars (*The Columbia Encyclopedia,* 2004). For example, the concept "mouse" represents the similarity between Mickey and Minnie Mouse. The criticism against conceptualism is that it cannot account for proper in contrast to improper concept application (MacLeod and Rubenstein, 2006).

4.4 ONTOLOGY IN COMPUTER SCIENCE

During the last decade, ontologies have also been the center of research in the computer science community and in different research fields, such as knowledge representation and engineering, information integration, information retrieval and extraction, etc. Research on ontologies emanated from the need to overcome conceptual inconsistencies and provide a solid basis for information exchange and reuse among different information sources, especially due to the development of information systems and the World Wide Web. According to computer science, an ontology is "an explicit specification of a conceptualization" (Gruber, 1995, 908), that is, a formal definition of the concepts and relations describing a domain of reality. This specification should be neutral and computationally tractable in order to facilitate its exchange and reuse for different applications (Smith and Mark, 2001).

Although ontology is a central research field both for philosophy and computer science, ontological notions, research topics, as well as basic assumptions, differ a lot between these two research fields for a number of reasons (Guarino, 1998; Smith, 2003). First, for philosophy, ontology is a theoretical endeavor independent of the language used to represent it, whereas for computer science an ontology is an engineering artifact dependent on a specific vocabulary and computational environment. Second, ontology from the philosophical point of view is searching for the truth behind reality, whereas for computer scientists the truth is not a primary concern, since they are foremost interested in designing ontologies for specific needs pursuing functional criteria and constraints. Third, philosophy focuses on entities, properties, relations, actions, etc., whereas information science focuses primarily on languages, and generally representational and computational issues. In order to differentiate the two notions of ontology, Guarino (1998) adopts the term *conceptualization* for the philosophical perspective and the term *ontology* for the computer science perspective. According to that perception, a conceptualization is language-independent, whereas an ontology is language-dependent.

The computer science's vision takes a *functional view* of ontology as an engineering artifact dependent on specific application needs and on the perspective of a group of agents. However, in order to facilitate knowledge sharing and reuse, knowledge engineering shifted from the "functional view" of knowledge acquisition and organization to the *modeling view*, which considers that knowledge should be related, not to the goals and perspectives of specific agents, but to an objective reality (Clancey, 1993; Guarino, 1995). The association of a knowledge base to reality provides a solid common basis for communication and interaction among different agents for diverse tasks.

Ontologies exhibit distinctive characteristics both from the methodological and the architectural points of view (Guarino, 1998). From the methodological point of view, they are the center of an interdisciplinary endeavor, incorporating research results from different disciplines such as philosophy and linguistics. From the architectural view, they play a primary role in information systems' components, that is, application programs, information resources, and user interfaces (Guarino, 1998).

An *ontological commitment*, from the computer science perspective, is the specification of the meaning of the terms used to describe a particular domain of reality. According to Quine's criterion "to be is to be the value of a variable" (Quine, 1961), ontological commitments are specified by types of variables in the knowledge representation (Sowa, 2000). In order to ensure the consistent explication of the semantics of a domain and achieve information sharing across domains, Guarino, Carrara, and Giaretta (1994) present a formal definition of ontological commitment, which represents a specific conceptualization. The formalization of the ontological commitment of a logical language implies the specification of the meaning of its vocabulary by the restriction of its models, the description of the modeling primitives, and the a priori relations used (Guarino, Carrara, and Giaretta).

Ontologies may take various forms, but they always consist of a vocabulary of terms and the description of their meaning, usually through definitions, as well as a specification of their between relations (Uschold and Gruninger, 1996). Depending on the level of formality, some ontologies may also include axioms which impose rules and constrain the values of concepts. The concepts constituting an ontology may be classified into two types (Stevens, Goble, and Bechhofer, 2000):

- *Primitive concepts* determining class membership with necessary conditions only.
- *Defined concepts* determining class membership with necessary and sufficient conditions.

The relations between the concepts of an ontology are also classified into two types (Stevens, Goble, and Bechhofer, 2000):

- *Taxonomical relations*, which relate concepts according to subsumption hierarchies; the most common taxonomical relations are the specialization relations (is a kind of) and the partitive relations (is a part of).
- *Associative relations*, which relate concepts across hierarchical structures and describe the properties, functions, and processes of a concept.

In the literature, different classifications of ontology types exist taking into account different criteria.

Uschold (1996) identifies three key dimensions along which ontologies vary: formality, purpose, and subject matter. According to the degree of formality in specifying the meaning of a vocabulary, ontologies are classified as:

- *Highly informal*, when the meaning of terms is loosely expressed in natural language.
- *Structurally informal*, when the meaning is expressed in natural language, but in a restricted and structured way.
- *Semiformal*, when the meaning is expressed in an artificial formal language (e.g., Ontolingua).
- *Rigorously formal*, when the meaning is expressed in a formal language with formal semantics, theorems, and proofs.

According to the purpose for which an ontology is designed, Uschold (1996) identified three main categories:

- Communication between people.
- Interoperability among systems.
- Systems engineering benefits, and specifically: reusability, knowledge acquisition, reliability, and specification.

Finally, according to the subject matter, ontologies are classified as:

- *Domain ontologies*, which refer to subjects such as medicine or geography treated separately from the problems or tasks that may result due to the subject.
- *Task*, *method*, or *problem-solving ontologies*, which refer to the subject of problem solving.
- *Representation ontologies*, which are relevant to knowledge representation languages and *meta-ontologies* provide the metadata about ontologies.

Van Heijst, Schrieber, and Wielinga, (1997) classify ontology types along two dimensions: (a) the amount and type of structure of the conceptualization, and (b) the subject of the conceptualization. Regarding the first dimension, three ontology types are identified:

- *Terminological ontologies* define the terms that represent a domain (e.g., lexicons).
- *Information ontologies* define the structure of a database (e.g., database schemata).
- *Knowledge modelling ontologies* define conceptualization of the knowledge, often focusing on a specific use of the knowledge.

Regarding the second dimension, that is, the subject of the conceptualization, van Heijst, Schrieber, and Wielinga, (1997) identify four ontology types:

- *Generic ontologies* define general concepts such as entity, property, relation, process, etc. that are commonly used across different domains.
- *Domain ontologies* describe concepts applying to specific domains. Usually the concepts described by domain ontologies are specializations of those defined by generic ontologies.
- *Application ontologies* define the concepts essential for the description of a specific application.
- *Representation ontologies* set the framework for knowledge representation formalisms; they provide the primitives used by generic and domain ontologies to describe reality, without making any assertions about reality themselves.

Guarino (1998) distinguishes ontologies according to their level of generality as follows:

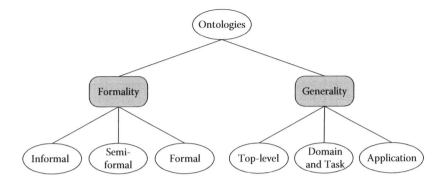

FIGURE 4.3   Distinction of ontology types according to formality and generality.

- *Top-level ontologies* define general concepts such as entity, process, action, property, relation, space, time, etc., which provide a general consideration of reality independently of any specific application domain.
- *Domain* and *task ontologies* define the concepts, which describe a domain (e.g., biology, chemistry, geography, etc.), or a general task (e.g., way-finding), respectively. Domain and task ontologies specialize the concepts defined by the top-level ontology.
- *Application ontologies* define concepts relative to a particular domain and a task, and they may specialize both respective ontologies. An example of an application ontology is a real estate ontology, which defines concepts regarding both the geography-cadastre domain and the land property buying and selling task ontology.

Gruber (2003, 2004) also classifies ontologies according to the level of formality of the specification as: (a) informal, (b) semiformal, and (c) formal, noting, however, that the distinction is somehow fuzzy, because ontologies usually incorporate both formal and informal parts; formal parts support automated processing and analysis (e.g., axioms), whereas informal parts support human understanding (e.g., dictionary definitions). Semiformal ontologies, which are the most common case, basically include informal parts but some formal parts as well, and are very useful especially for information integration purposes. The main reason for that is that semiformal ontologies are more flexible and adaptable to real-world applications, since they may incorporate partial information (Sheth and Ramakrishnan, 2003).

Figure 4.3 illustrates the distinction of ontology types according to two criteria: formality and generality.

4.5   TOP-LEVEL ONTOLOGIES

As already mentioned, a *top-level ontology* explicates fundamental notions such as space, time, entity, property, quantity, relation, process, identity, etc. These notions are domain-independent; therefore, their unified definition may provide the framework for the definition of ontological categories, properties, and relations and for the evolvement of domain ontologies. A top-level ontology may constitute the "common neutral backbone" (Smith, 2003) for the conjunction of domain ontologies. Philosophical inquiry into fundamental concepts is the necessary basis to relate more specialized ontologies and share knowledge across different domains (Sowa, 2000). In that way, domain-specific concepts are defined based on the well-founded semantics of the top-level concepts, hence, facilitating information exchange and reuse across different domains.

However, there is no single universally agreed top-level ontology, and some believe that the different attempts to develop a top-level ontology may shift the interoperability problems to a higher level. On the other hand, defenders of the top-level ontology notion assert that although there may be no universal top-level ontology to adhere to, the advantages of adopting one for large communities of users are important and cannot be overlooked. Gangemi et al. (2001) maintain that the ability to map a top-level ontology into large knowledge bases (e.g., WordNet) presents important advantages, among which is the association of the top-level concepts with language and cognition. Having this requirement in mind, they present a methodology for defining primitive relations and formal properties in order to illuminate common ontological distinctions.

Several initiatives in developing a top-level ontology exist. Sowa (2001) developed a top-level ontology based on three fundamental philosophical distinctions or dimensions for subdividing the universal type or entity: (a) *Physical* or *Abstract*, (b) *Independent*, *Relative*, or *Mediating*, and (c) *Continuant* or *Occurrent*. The first distinction is relevant to Heraclitus's distinction between *physis* and *logos*; *Physical* refers to anything consisting of matter or energy, whereas *Abstract* refers to information structures. The second distinction emanates from Peirce's trichotomy (firstness, secondness, thirdness). *Independent* is the category of actual entities (objects and processes) that exist independently of anything else. *Relative* is the category of directed relations or reactions between two entities. *Mediating* is the category of connection of two or more actual entities. The third distinction emanates from Whitehead's classification of entities which possess stable characteristics over some period of time (*continuants*) in contradiction to entities, which constantly change and can only be identified in some spatiotemporal region (*occurrents*). The combinations among these three distinctions result in a lattice of twelve categories (Figure 4.4). For example, *Object* is the category resulting from the combination Independent – Physical – Continuant (IPC); it represents a physical entity, whose identity is maintained stable over time. *Process* is the category resulting from the combination Independent – Physical – Occurrent (IPO); it represents a physical entity which has no stable characteristics over a period of time. *Juncture* is the category resulting from the combination Relative – Physical – Continuant (RPC); it represents a prehension functioning as a continuant over a time period, for example, a node, where three lines intersect.

The IEEE Standard Upper Ontology Working Group (SUO WG, 2003) is involved in the development of a standard for an upper ontology able to address a variety of domains and to support applications such as data interoperability, information search and retrieval, automated inferencing, and natural language processing. The standard specifies the structure and a set of general concepts, in order to provide the framework to build more specialized domain ontologies. There are three ontologies on

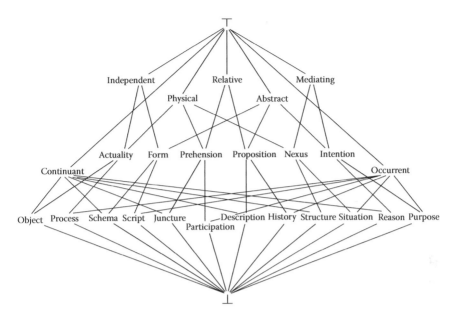

FIGURE 4.4 Sowa's top-level ontology. (From Sowa, J. F. 2001. Top-Level Categories. http://www.jfsowa.com/ontology/toplevel.htm. With permission.)

which the SUO WG is working: (a) the IFF Foundation Ontology, (b) the Suggested Upper Merged Ontology (SUMO), and (c) the Open Cyc ontology. The SUO WG also aims at the development of a *lattice of theories* based on the above ontologies.

The Information Flow Framework (IFF) is "a descriptive category metatheory... that provides an important practical application of category theory to knowledge representation, knowledge maintenance and semantic web" (Kent, 2006, 3). IFF's design was influenced by Category Theory, Information Flow (Chapter 10) and Formal Concept Analysis (Chapter 8). IFF's development began in the framework of the standardization activity of the IEEE Standard Upper Ontology (SUO) and represents the structural aspect of the SUO (Kent, 2005). IFF's purpose is to make possible the interoperability between distributed systems like database applications and knowledge bases and to provide a principled methodology for ontology sharing, manipulating, relating, partitioning, composing, and discussing (Kent, 2003).

The Suggested Upper Merged Ontology (SUMO) (Niles and Pease, 2001; Pease, Niles, and Li, 2002; Pease and Niles, 2002) is also developed under the IEEE SUO WG initiative. It provides the framework for middle-level and domain ontologies. It includes approximately four thousand assertions and one thousand concepts. It deals with structural concepts (e.g., instance, subclass), general types of objects and processes, abstractions (set theory, attributes, and relations), numbers and measures, temporal concepts, parts and wholes, basic semiotic relations, and notions such as agency and intentionality.

The Upper Cyc Ontology (Cycorp, 2002) is the core of the Cyc Knowledge Base (KB), which aims at the formalized representation of basic commonsense knowledge. The Upper Cyc Ontology includes three thousand terms, which describe the most general concepts of reality (collections, individuals, predicates, relations, and functions); their selection has been tested on the basis of logical deduction of tens of millions of examples. The Upper Cyc Ontology is designed according to two criteria: universality and articulation. Universality is the ability of the ontology to correctly embody every concept regardless of its context and granularity. Articulation is the design of the ontology according to necessary and sufficient distinctions, that is, there is theoretical and pragmatic basis for the choices made in order to provide sufficient grounding for applications such as knowledge sharing, natural language disambiguation, semantic database integration, semantic information retrieval, etc. The aim of the Upper Cyc Ontology is to provide the foundation for incorporating more specific ontologies, which may represent different contexts or "microtheories."

Basic Formal Ontology (BFO) (Grenon, 2003a) is a bi-categorial ontology, which embodies and modulates both a three-dimensionalist and a four-dimensionalist view. BFO comprises two complementary ontologies: a SNAP ontology of endurants in order to represent static views of the world, and a SPAN ontology of perdurants in order to account for processes developing through time. Each ontology consists in a partition of reality into categories or universals.

Descriptive Ontology for Linguistic and Cognitive Engineering (DOLCE) is the first module of the WonderWeb Foundational Ontologies Library (WFOL) (Masolo et al., 2003). DOLCE focuses on the representation of ontological categories at the mesoscopic level (Smith, 1995), constituting the basis of natural language and commonsense. DOLCE follows a "multiplicative approach," that is, entities can occupy the same space-time because of incompatible essential properties assigned to them. DOLCE is an ontology of particulars further specified by the distinction between endurants and perdurants.

WordNet (Fellbaum, 1998) is a lexical reference system developed by the Cognitive Science Laboratory at Princeton University. WordNet is organized as a semantic network based on psycholinguistic theories of human lexical memory. Although it is not designed to be an ontology, it is used as an upper ontology by a number of applications, especially for Natural Language Processing. It consists of synonym sets, which organize English nouns, verbs, adjectives, and adverbs. Each synonym set corresponds to a lexical concept. Synonym sets are linked with different types of relations such as hyponymy and meronymy.

Mikrokosmos ontology (Mahesh and Nirenburg, 1996; Mahesh, 1996) is part of the Mikrokosmos knowledge-based machine translation system, developed by the Computer Research Laboratory, New Mexico State University. It aims at the development of a concept repository for interlingual text meaning representation for various input languages. It contains knowledge about concepts organized in a hierarchy. The concepts included in the ontology describe things, events, and relations. These concepts are mainly related with IS-A relations, although other semantic relations (referred to as relational slots) are also identified (e.g., IS-A, PART-OF, MANUFAC-TURED-BY, etc.). For each relation, constraints are specified in order to define the appropriate concepts for its domain and its range. Literal slots (e.g., COLOR) are also assigned to concepts. The development of the Mikrokosmos ontology is based on ten commitments—principles: (a) broad coverage, (b) rich properties and interconnections, (c) ease of understanding, searching, and browsing, (d) NLP oriented, (e) economy/cost effectiveness/tractability, (f) language independence, (g) absence of unconnected terms, (h) taxonomic organization and inheritance, (i) intermediate-level grain size, and (j) equal status of all properties.

The Generalized Upper Model 2.0 (Bateman, Henschel, and Rinaldi, 1995), originally developed as the Penman Upper Model (Bateman et al., 1989), is currently maintained by the Bremen Ontology Research Group. It is a general task- and domain-independent linguistic ontology aiming at information organization for expression in natural language. It gives an account on how the grammar and/or the semantics of a specific natural language categorize reality. The Generalized Upper Model 3.0 (GUM 3.0) (Bateman et al., 2006a, 2006b) is a linguistically driven ontology, which acts as a mediator between natural language and different applications. It is developed by the Collaborative Research Center on Spatial Cognition at the Bremen University, and hence emphasis is put on spatial concepts in order to enable the use of the ontology in spatiolinguistic applications. The GUM 3.0 is structured in two hierarchies: the concept hierarchy and the relation hierarchy. The concept hierarchy includes all concepts under the general category *GUM-Thing*. GUM-Thing has three subconcepts: (a) *Configuration* is an activity or a state of affairs, (b) *Element* is an independent object or conceptual item, which may participate in a configuration, and (c) *MultiConfiguration* is a complex entity, which relates configurations. The concept hierarchy includes concepts that represent locations, spatial modalities and activities, and static and dynamic spatial configurations. The relation hierarchy is organized under the most abstract concept *GUM-Relation* and includes spatial relations, such as Locatum, Relatum, HasSpatialRelation, etc.

SENSUS (Knight and Luk, 1994) is a terminological ontology consisting of a 90,000-node taxonomy. It aims at functioning as a framework for placing additional information in order to facilitate semantic understanding of texts for machine translation, summarization, and information retrieval. It is an extension of WordNet, also based on the Penman Upper Model (Bateman et al., 1989). Each node in the SENSUS ontology corresponds to one concept, that is, one specific sense of a word. Concepts are linked through IS-A relations, although multiple inheritance links are allowed, resulting in a non-strict hierarchy. However, other types of semantic relations are also included, such as part-of, synonym, antonym, etc. The root of the hierarchy is the concept THING, further classified as OBJECT, PROCESS, and QUALITY. Furthermore, SENSUS includes cross-ontology alignment algorithms, in order to link existing ontologies such as Cyc, and Mikrokosmos. The Omega Ontology (Philpot, Fleischman, and Hovy, 2003; Philpot, Hovy, and Pantel, 2005) is Sensus's descendant, with 120,000 nodes resulting from the synthesis of WordNet and Mikrokosmos. Omega includes only a few semantic relations between concepts and no formal concept definitions in order to be adaptable to various applications.

4.6 ONTOLOGY IN THE GEOSPATIAL DOMAIN

Ontology has been widely recognized as a priority research theme for the geospatial domain (Smith and Mark, 1998; Fonseca et al., 2000). The two different directions in ontological research, that is, philosophical and computer science, also influenced GIScience. For that reason, ontological research in the geospatial domain embodies both a philosophical as well as a computer science perspective. The philosophical perspective studies the constituents of geographic reality, that is, seeks to define the fundamental geospatial concepts, processes, and relations, as well as the theories governing them. The computer science perspective deals with issues relative to the rigorous description and formalization of geospatial categories and relations in order to facilitate human understanding and interaction with geographic information, information standardization and integration for effective information exchange and reuse.

A particularity of the geographic domain is that it is based on geographic reality, which exists independently of human cognition. This independence results in a closer interconnection of the philosophical and the information science perspectives in the geographic domain than in other domains (Mark et al., 2004). Smith and Mark (2003) maintain that the philosophical and computer science ontological perspectives may be reconciled into a single framework by exploiting Horton's (1982) distinction between *primary* and *secondary theory*. *Primary theory* refers to the commonsense knowledge stored by human beings in order to function in their everyday lives; primary theory concerns entities, processes, and phenomena directly perceived by human beings, since they are related to the mesoscopic level (Smith and Mark, 2001). The geographic domain is part of the primary theory of the commonsense world (Smith and Mark, 2001). Due to our common biological and psychological mechanisms, primary theory is shared by all societies and cultures. *Secondary theories* consist in the knowledge used to explain entities, processes, and phenomena failed to be explained by primary theory; these are usually either too big or too large to be directly observed, or hidden from the observer, for example, spirits, ghosts, etc. Secondary theories differ considerably among societies and cultures.

Ontological research should be carried in a systematic, integrated way taking into account conceptual, representational, logical, and implementation issues (Mark et al., 2004). However, the generation of a common reference ontology, even for specialized domains such as the geospatial one, is a tremendous task, which can only be accomplished by making a settlement between neutrality and expressive power (Mark et al., 2004). Agarwal (2005) argues that it is debatable whether a unified approach to geospatial ontology may exist, taking into account the interdisciplinary nature of geographic information research, and the different conceptualizations and terminology used for the same geospatial concepts.

An often proposed solution to overcome this problem is the association of different geographic ontologies with a top-level ontology. A top-level ontology describes general concepts such as space, time, quantity, quality, object, property, relation, event, process, etc. (see Section 4.5), and may provide a solid framework for the association of more specialized domain ontologies (Guarino 1998; Sowa 2000; Kokla and Kavouras, 2001; Mark et al., 2004). In that way, the definition of the more specialized domain concepts is based on the rigorously defined top-level concepts. This solution reinforces the interconnection of domain ontologies that refer to different contexts under the umbrella of the top-level ontology permitting information integration and transition among contexts.

Top-level concepts such as space, time, and motion are central to ontological investigations (Galton, 1997; Casati and Varzi, 1997; Casati, Smith, and Varzi, 1998; Muller, 1998). According to Casati, Smith, and Varzi, (1998), geographic entities exhibit a balance between empirical and ontological issues. Questions relative to the hypostasis of geographic entities and their dependence on spatial regions and boundaries are essential for the foundation of geospatial ontology and geographic knowledge representation. More specifically, there are five basic categories of ontological issues relative to geographic entities (Casati, Smith, and Varzi, 1998; Muller, 1998):

1. Regions of space and things in space, that is, whether and how a theory of geospatial ontology and geographic knowledge representation should incorporate as primitives things (objects and events) occurring in space, as well as spatial regions.
2. Absolutist versus relational theories of space, that is, whether space exists independently of spatial objects and relations functioning as a container in which things happen, or whether only spatial objects exist and are spatially related to each other.
3. Types of spatial entities, that is, the difference between material entities that occupy a specific spatial location and immaterial entities that share their spatial location with other entities (e.g., processes, events, holes, and shadows).
4. A view of reality considering space and time to be different dimensions (the world is represented as successions of snapshots) vs. a view of reality following a unified space-time approach (concerned with changes and processes).
5. Boundaries and vagueness, that is, how boundaries mark individual entities, and other issues relative to the nature of boundaries (e.g., fiat and bona fide boundaries, fuzziness and indeterminacy, etc.).

The primary issue in developing an ontology of the geospatial domain is the notion of space itself. Space may be conceptualized from different perspectives. Distinctions between *physical* and *commonsense space*, *absolute* and *relative space*, *global* and *local space* (Vieu, 1997) are all relevant to the definition of a geospatial ontology. An *absolute space* is a container whose existence is independent of the objects located in it; it consists of spatial entities, and a location function, in order to assign objects to their place. *Relative space*, on the other hand, consists of spatial relations over non-purely spatial entities; these may be material in case of physical space or mental in case of cognitive space. *Global space* defines a location for each spatial entity in a reference framework so as the relative position to other spatial entities is fully specified. *Local space* defines the location of each spatial entity with respect to explicit spatial relations with other spatial entities (Vieu, 1997).

Casati, Smith, and Varzi (1998) identify general issues concerning the nature of geographic entities, as well as the notion of boundaries and propose the use of mereology, location, and topology as the basic theoretical tools for dealing with these issues. Furthermore, they propose the enrichment of classical geography with the inclusion of nonclassical geographies in order to account for complicated nonstandard cases, for example nonspatial geographic entities, existence of gaps between spatial regions, or the opposite case, where more than one geographic entities are attached to the same spatial region, etc. Following the distinction between physical and human geography, they specify two types of geographic entities: (a) *physical* (e.g., mountains, rivers, etc.), and (b) *socioeconomic* (land parcels, cities, countries, etc.). Furthermore, two different types of boundaries for geospatial entities are distinguished: (a) *bona fide boundaries*, which result from spatial discontinuities and physical prominences (e.g., mountains, rivers, islands, etc.), and (b) *fiat boundaries*, which are the result of human delimitation (e.g., the boundaries between central European countries). According to the boundary type, corresponding entity types are specified: *bona fide entities* are delineated by bona fide boundaries, whereas *fiat entities* are delineated by fiat boundaries.

Entities existing in the geospatial realm are sometimes more complicated and difficult to explicate and formalize than objects of everyday experience such as organisms (e.g., human, animals, etc.) and artifacts (e.g., cars, tables, etc.) (Mark, Smith, and Tversky, 1999). Objects of everyday experience have determinate boundaries that distinguish them from neighbor objects; furthermore, they belong to categories that are clearly differentiated from neighboring categories. Geospatial entities on the other hand, do not always have determinate boundaries, nor do they belong to clear-cut categories. Smith and Mark (2003) and Mark and Smith (2004) explore mountains as a representative example of complicated geospatial entities. Mountains, like hills, valleys, islands, lakes, etc., are categories resulting from the elevation variations in the Earth's surface, as well as the covering of certain surfaces with water. These categories are derived by object-based views of reality; properties (e.g., size, shape, cover, etc.) and relations between objects (e.g., higher than, closer to, etc.) are also assigned to them. However, because of their graded or vague boundaries, mountains are represented by Earth scientists according to the field-based views of the Earth's surface and not according to the object-based view. Therefore, an attempt to develop an ultimate ontology of the geospatial domain, which would serve different application needs and uses, should take into account both the commonsense object-based conceptualizations and the scientific field-based conceptualizations used to model runoff and erosion (Mark and Smith, 2004). Peuquet, Smith, and Brogaard (1998) explore the ontology of geographic phenomena with regard to the notion of fields. The ontological questions addressed are how fields may be embodied in the ontology of the geospatial domain, when they should be preferred to objects, how they should be formalized, what operations are possible, etc.

Another issue related to spatiotemporal representation is the continuants–occurrents opposition, that is, objects enduring through time vs. objects that are constrained by time and have temporal parts (e.g., states, processes, events, etc.) (Muller, 1998). Bittner and Smith (2003) present an ontological theory for space and time that incorporates both continuants and occurrents. Reitsma and Bittner (2003) explore the notion of scale within the context of the continuant–occurrent distinction. Scale is considered as the level of granularity in the observation of reality, that is, it differs from the traditional cartographic point of view. Mereology, theory of granular partitions, and hierarchy theory are applied in order to clarify the notion of scale for the ontologies of continuants and occurrents. Grenon (2003b) also adopts the distinction between continuants (enduring entities) and occurrents (perduring entities) in order to develop a formal framework for spatiotemporal reality and to explicate issues pertaining to the existence of entities in time, the changes they undergo, and the processes in which they are involved. The primitives of which the framework consists are: existents, enduring and perduring entities, mereological features, ontologies, and their constituents. Furthermore, the framework identifies three types of relations: (a) *intra-ontological,* that is, relations between existents which are constituents of the same ontology, (b) *trans-ontological,* that is, relations between existents which are constituents of different ontologies, and (c) *meta-ontological,* that is, relations between ontologies or between an ontology and an existent.

Change, process, and motion are also central notions for the development of a geospatial ontology. Muller (1998) presents a theory of spatiotemporal entities in order to account for a theory of motion. This theory considers that space is constituted of extended spatial regions, that is, regions of a space-time primitive, in contrast to theories that regard space and time to be different dimensions. Temporal relations apply on the spatiotemporal regions and not on time extents. Grenon and Smith (2004) propose a modular ontology incorporating both snapshot views of the world and a spatiotemporal ontology of change and process in order to account for the dynamic nature of reality. The ontology proposed relies on the concepts of *realism, perspectivalism, fallibilism,* and *adequatism. Realism* declares that reality exists independently of humans' conceptualizations and representations of it. *Perspectivalism* supports that there exist alternative plausible views of reality. However, realism imposes a restriction on perspectivalism in that not all views of reality are equally appropriate in representing reality. Scientific inquiry and examination may provide the criteria for selecting the appropriate views for representing reality. Nevertheless, since scientific inquiry is a process subject to change over time, the *fallibilistic* approach holds that scientific theories undergo continuous evolution and change. *Adequatism* asserts that there is no single view of reality incarnating all other, but there exist different views, evenly genuine in relation to different domains, perspectives, granularity, etc.

An ultimate ontology of some domain can be developed on the basis of ontologies grounded on good conceptualizations, that is, those formed by scientific research but also those cultivated by commonsense reasoning and cognition (Smith and Mark, 2001). The endeavor of developing an ultimate ontology of some domain and consequently, of the geographic one too, has fallen down because such attempts "ignored the real world of flesh-and-blood objects in which we all live and focused instead on closed world models" (Smith and Mark, 2001, 594). In an attempt to discern good commonsense conceptualizations from bad ones, with the long-term aim of developing an ontology of geospatial categories, they performed a series of experiments in order to determine the fundamental geospatial concepts shared by all humans independently of their geographic education and identify the way nonexperts perceive geographic reality (Smith and Mark, 2001; Mark, Skupin, and Smith, 2001). The experiments revealed a difference in the meaning ascribed to terms such as "geographic feature" and "geographic object" between experts and nonexperts. The term "geographic feature" evoked natural and not artificial items such as mountain, river, and ocean in contrast to "geographic object," which elicited small, manipulable items, such as map and globe. "A geographic concept" induced examples with low coherence, probably due to the fact that the noun "concept" refers to abstract things. On the other hand, the term "something that could be portrayed on a map" evoked items either artificial, or with fiat boundaries, such as city, road, country, and state. The study is based on the assumption that the concepts that comprise geographic reality as conceived by nonexperts constitute a good conceptualization, that is, a conceptualization transparent to reality. According to this assumption, concepts such as rivers, seas, mountains, cities exist in reality and have properties that are perceived by commonsense experience.

In the geospatial domain, ontologies are usually considered to be part of an abstraction framework evolving at different levels. In order to be able to embody different ontological approaches in a unified way, Frank (2001, 2003) proposed that a geospatial ontology be built in five coordinated tiers. Tier 0 of the ontology is the tier of the physical reality, which can be perceived as a four-dimensional space-time continuum. Tier 1 of the ontology represents the reality as observed by agents, like human beings. The assumption made is that agents observe properties at a point in space at the time "now." Tier 2 of the ontology deals with the formation of objects with respect to uniform properties. Tier 3 represents social reality; at this tier, language names result from social processes and facts, while relationships between them are created by social rules. Tier 4 of the ontology represents the knowledge of reality possessed by cognitive agents. This knowledge is incomplete and partial, but it is used to derive facts and make decisions.

Fonseca, Egenhofer, and Agouris (2002) and Fonseca et al. (2002) specify an abstraction model evolving in five "universes" in order to account for the different characteristics of geographic phenomena: (1) the *physical universe* refers to geographic phenomena occurring in the physical world, (2) the *cognitive universe* represents the conceptualization of geographic phenomena by the human cognitive system, (3) the *logical universe* represents the formalization of the conceptualizations using formal structures, (4) the *representation universe* embodies the distinctive characteristics of the geospatial domain, for example, field or object representations, and (5) the *implementation universe* codifies the representation universe into computer language and data structures. Two types of ontologies are supported within this abstraction model, giving rise to the notion of ontology-driven Geographic Information Systems, that is, Geographic Information Systems that use ontologies as an active component. The phenomenological domain ontology specifies the characteristics and dimensions of geographic concepts and is part of the representation universe, whereas the application domain ontology describes particular subjects and tasks and is part of the logical universe.

Geographical space is considered not only to consist of geographical entities, but also to incorporate the activities occurring in them (Câmara et al., 2000; Kuhn, 2001; Raubal, 2001; Soon and Kuhn, 2004). Relative to activities is the notion of *affordances* (Gibson, 1986), which is invented in order to describe what the environment provides to living beings. Câmara et al. (2000), taking into account Santos' (1997) view of geographical space as a system of entities and actions, propose the notion of *action-driven ontologies*. This notion intends to capture the dynamic nature of geographic entities, which change over time, as well as the users' intentions concerning geographic entities. Kuhn (2001) argues that ontologies of the geographic domain should focus on human activities in space in order to increase their usefulness and usability, and proposes a method to develop ontologies, which relies on activity theory and the notion of affordances. The basic idea of the method is to use natural language texts describing human activities in order to capture the activities relevant to a particular domain; verbs and verbal expressions correspond to actions, whereas nouns correspond to objects. The derived actions are then hierarchically organized in an action hierarchy. In order to develop task-oriented geographic ontologies, Soon and Kuhn (2004) present an approach for the formalization of user actions using Formal Concept Analysis (Chapter 8) and Entailment Theory (Fellbaum, 1990). Formal Concept Analysis is used to construct an action lattice, which is further complemented with entailment relations, that is, troponymy, proper inclusion, backward presupposition, and causation.

Raubal (2001) follows an ecological approach to model the ontology and epistemology of an agent-based wayfinding system. The distinction between ontology and epistemology is crucial in agent-based systems, since the knowledge of an agent about the environment is dependent on personal characteristics (e.g., experience, the specific task, etc.). At the ontological level, the way-finding environment is perceived as consisting of a *medium, substances,* and *surfaces.* At the epistemological level, the central notion is that of affordances, which are further classified into *physical, social-institutional,* and *mental.* Timpf (2002) develops and compares two ontological models for wayfinding: the ontology of the traveler and the ontology of the public transportation system. These ontologies are analyzed on the basis of the framework provided by activity theory (Kaptelinin, Nardi, and Macauley, 1999; Nardi, 1996), according to which wayfinding is structured at three levels of granularity: (1) activity, (2) action, and (3) operations. Based on the analysis, the comparison process leads to the conclusion that the two ontologies overlap.

Following an information science perspective, Fonseca et al. (2002) explore the notion of semantic granularity to account for the different generalization levels in the conceptualization and representation of geographic concepts. Hence, the development of a Geographic Information System that incorporates ontologies at different generality levels, where low-level ontologies are based on higher-level ones, allows for changes of semantic granularity supporting the generalization and specialization operations (Fonseca et al., 2002). Kokla and Kavouras (2001) propose a method for the fusion of geographic domain ontologies with a top-level ontology in order to support information exchange. The method is based on the *principle of complementarity* (Bohr, 1934), which states that different views reflect different aspects of reality, congruent with different contexts. The principle of complementarity may be implemented by the association of different, complementary ontologies, but, according to Sowa (2000), only when the contexts are explicitly distinguished. Work by Kokla and Kavouras (2001) is based on the assumption that geographic contexts consist of concepts, essential properties and semantic relations and on the formation of these contexts using Formal Concept Analysis (Chapter 8).

Fonseca, Câmara, and Monteiro (2006) introduce a framework for measuring the interoperability between two geographic ontologies in order to specify the degree to which these ontologies are interoperable. The framework uses a measure based on the model management algebra for metadata (Bernstein, 2003). Within this framework, an ontology is considered to be a model, which applies to a specific application domain and consists of a single hierarchy of concepts associated through specialization relations. Generally, model representation considers that concepts may be related to other concepts through similarity, generalization, and subsumption.

Interoperability between two ontologies is measured using a "match" operator that calculates the similarity and generalization relations between the concepts of the two ontologies.

REFERENCES

Agarwal, P. 2005. Ontological considerations in GIScience, *International Journal of Geographical Information Science (IJGIS)* 19(5): 501–535.

Bateman, J., R. Henschel, and F. Rinaldi. 1995. *Generalized upper model 2.0.* Technical Report, GMD, Institut für Integrierte Publikations und Informationssysteme, Darmstadt, Germany.

Bateman, J., R. Kasper, J. Moore, and R. Whitney. 1989. *A general organization of knowledge for natural language processing: The Penman upper model.* Technical Report, Information Sciences Institute, Marina del Rey, California.

Bateman, J., S. Farrar, J. Hois, and R. Ross. 2006a. *The generalized upper model 3.0: Documentation.* Internal Report SFB/TR8, Collaborative Research Center for Spatial Cognition, University of Bremen, Germany.

Bateman, J., J. Hois, R. Ross, and S. Farrar. 2006b. *The generalized upper model 3.0: Space components.* Internal Report SFB/TR8, Collaborative Research Center for Spatial Cognition, University of Bremen, Germany.

Bernstein, P. 2003. Applying model management to classical meta data problems. In *Proc. Conference on Innovative Database Research*, 209–220. CIDR, Asilomar, CA.

Bittner, T. and B. Smith. 2003. *Formal ontologies for space and time.* Technical Report, IFOMIS, University of Leipzig.

Bohr, N. 1934. *Atomic theory and the description of nature.* Cambridge: Cambridge University Press.

Bolzano, B. 1810. *Contributions to a better-grounded presentation of mathematics,* Translated by William Ewald, 1996. In *From Kant to Hilbert: A source book in the foundations of mathematics*, vol. I, 183. Oxford: Clarendon Press. From *"What is Ontology?" Definitions by leading philosophers from Christian Wolff to Edmund Husserl.* http://www.formalontology.it/section_4.htm (accessed 1 May 2007).

Bunge, M. 1999. *Dictionary of philosophy.* Amherst, NY: Prometheus Books.

Burch, R. 2006. Charles Sanders Peirce. In *The Stanford encyclopedia of philosophy*, Fall Edition, ed. E. N. Zalta. http://plato.stanford.edu/archives/fall2006/entries/peirce/ (accessed 1 May 2007).

Câmara, G., A. Monteiro, J. Paiva, and R. Souza. 2000. Action-driven ontologies of the geographical space. In *Proc. 1st International Conference on Geographical Information Science (GIScience2000)*, Savannah, GA, October 28–31.

Casati, R. and A. Varzi. 1997. Spatial entities. In *Spatial and Temporal Reasoning*, ed. O. Stock, 73–96. Dordrecht, The Netherlands: Kluwer Academic Publishers.

Casati, R., B. Smith, and A. Varzi. 1998. Ontological tools for geographic representation. In *Proc. First International Conference of Formal Ontology in Information Systems (FOIS'98)*, Trento, Italy, ed. N. Guarino, 77–85. Amsterdam: IOS Press.

Clancey, W. J. 1993. The knowledge level reinterpreted: Modeling socio-technical systems. *International Journal of Intelligent Systems* 8: 33–49.

Cocchiarella, N. 1991. Formal ontology. In *Handbook of Metaphysics and Ontology*, ed. H. Burkhardt and B. Smith, 640–647. Munich: Philosophia Verlag.

Cocchiarella, N. 2004. *Elements of formal ontology.* Lecture Notes from an Intensive Course on Formal Ontology, Rome, Lateran University.

Cycorp, Inc. 2002. *OpenCyc selected vocabulary and upper ontology.* http://www.cyc.com/cycdoc/vocab/vocab-toc.html (accessed 31 October 2006).

Englebretsen, G. 1990. A reintroduction to Sommers' tree theory. In *Essays on the philosophy of Fred Sommers,* 1-32. Lewiston, NY: The Edwin Mellen Press. Also available in *Definitions of Ontology From Nicolai Hartmann to Current Days.* http://www. formalontology.it/section_4_two.htm (accessed 2 May 2007).

Fellbaum, C. 1990. English verbs as a semantic net. *International Journal of Lexicography.* 3: 270–301.

Fellbaum, C., ed. 1998. *WordNet: An electronic lexical database.* Cambridge, MA: MIT Press.

Fonseca, F., G. Câmara, and A. M. Monteiro. 2006. A framework for measuring the interoperability of geo-ontologies. *Spatial Cognition and Computation* 6(4): 307–329.

Fonseca, F. M. Egenhofer, and P. Agouris. 2002. Using ontologies for integrated geographic information systems. *Transactions in GIS* 6(3): 231–257.

Fonseca, F., M. Egenhofer, C. Davis, and K. Borges. 2000. Ontologies and knowledge sharing in urban GIS. *CEUS Computer, Environment and Urban Systems,* 24(3): 232–251.

Fonseca, F., M. Egenhofer, C. Davis, and G. Câmara. 2002. Semantic granularity in ontology-driven geographic information systems. *AMAI Annals of Mathematics and Artificial Intelligence Special Issue on Spatial and Temporal Granularity* 36(1-2): 121–151.

Frank, A. U. 2001. Tiers of ontology and consistency constraints in geographic information systems. *International Journal of Geographical Information Science* 15(7): 667–678.

Frank, A. U. 2003. Ontology for spatio-temporal databases. In *Spatiotemporal Databases: The Chorochronos Approach*, ed. M. Koubarakis et al., 9-78. Lecture Notes in Computer Science. Berlin: Springer-Verlag.

Galton, A. 1997. Space, time and movement. In *Spatial and Temporal Reasoning*, ed. O. Stock, 321–352. Dordrecht, The Netherlands: Kluwer Academic Publishers.

Galton, A. 2006. Processes as continuants. In Proc. *Thirteenth International Symposium on Temporal Representation and Reasoning (TIME'06)*, Budapest, Hungary, June 15–17, 2006, 15–17.

Gangemi, A., N. Guarino, C. Masolo, and A. Oltramari. 2001. Understanding top-level ontological distinctions. In *Proc. of IJCAI 2001 Workshop on Ontologies and Information Sharing*, Seattle, WA: August 4–5, ed. A. Gomez Perez, M. Grunninger, H. Stuckenschmidt, and M. Ushold, 26–33. Seattle, WA: AAAI Press.

George, U. X. 2006. Whitehead: *Defending Brentano against the criticisms of heretical disciples.* http://libspirit.blogspot.com/2006/10/whitehead-defending-brentano-against. html (accessed 26 October 2006).

Gibson, J. J. 1986. *The ecological approach to visual perception.* Hillsdale, NJ: Lawrence Erlbaum.

Grenon, P. 2003a. *BFO in a nutshell: A bi-categorial axiomatization of BFO and comparison with DOLCE.* Technical Report, ISSN 1611-4019, Institute for Formal Ontology and Medical Information Science (IFOMIS), University of Leipzig, Germany.

Grenon, P. 2003b. The formal ontology of spatio-temporal reality and its formalization. In *Foundations and applications of spatio-temporal reasoning (FASTR)*, ed. H. W. Guesguen, D. Mitra, and J. Renz, 27-34. AAAI Spring Symposium Technical Report Series. Menlo Park, CA: AAAI Press.

Grenon, P. and B. Smith. 2004. SNAP and SPAN: Towards dynamic spatial ontology. *Spatial Cognition and Computation* 4(1): 69–104.

Gruber, T. R. 1995. Toward principles for the design of ontologies used for knowledge sharing. *International Journal of Human-Computer Studies* 43(5-6): 907–928.

Gruber, T. 2003. It is what it does: The pragmatics of ontology. Invited talk at Sharing the Knowledge, International CIDOC CRM Symposium, Washington, DC, March 26–27.

Gruber, T. 2004. Every ontology is a treaty, Interview for Semantic Web and Information Systems, SIG of the Association for Information Systems. *SIGSEMIS Bulletin* 1(4).

Guarino, N. 1995. Formal ontology, conceptual analysis and knowledge representation. *International Journal of Human and Computer Studies* Special Issue. *The Role of Formal Ontology in Information Technology* 43(5/6): 625–640.

Guarino, N. 1998. Formal ontology and information systems. In *Proc. 1st International Conference on Formal Ontologies in Information Systems (FOIS'98)*, Trento, Italy, ed. N. Guarino, 3-15. Amsterdam: IOS Press.

Guarino, N., M. Carrara, and P. Giaretta. 1994. Formalizing ontological commitments. In *Proc. 12th National Conference on Artificial Intelligence*, Seattle, WA, July 31–August 4, 1994, Vol. 1, 560–567.

Hartmann, N. 1949 [1953]. *New ways of ontology*, trans. R. C. Kuhn. Chicago: Henry Regnery Company.

Horton, R. 1982. Tradition and modernity revisited. In *Rationality and relativism*, ed. M. Hollis and S. Luke, 201–260. Oxford: Blackwell.

Kant, I. 1997. *Lectures on metaphysics, Part 5. Metaphysik L2*, trans. and ed. K. Ameriks and S. Naragon. Cambridge: Cambridge University Press. From *"What is ontology?" Definitions by leading philosophers from Christian Wolff to Edmund Husserl.* http://www.formalontology.it/section_4.htm (accessed 1 May 2007).

Kaptelinin, V., B. Nardi, and C. Macauley. 1999. The activity checklist: A tool for representing the "space" of context. *Interactions* 6(4): 27–39.

Kent, R. E. 2003. The IFF foundation for ontological knowledge organization. In *Knowledge Organization and Classification in International Information Retrieval, Cataloging and Classification Quarterly.* Binghamton, NY: The Haworth Press Inc.

Kent, R. E. 2005. Semantic integration in the information flow framework. In *Proc. Dagstuhl Seminar 04391 Semantic Interoperability and Integration*, ed. Y. Kalfoglou, M. Schlorlemmer, A. Sheth, S. Staab, and M. Uschold. Internationales Begegnung sund Forschungs zentrum (IBFI), Dagstuhl, Germany. http://drops.dagstuhl.de/opus/frontdoor.php?source_opus=41 (accessed 2 May 2007).

Kent, R. E. 2006. The Information Flow Framework: New architecture. In International Category Theory Conference, CT 2006, White Point, Nova Scotia, June 25–July 1. http://www.mscs.dal.ca/~selinger/ct2006/slides/CT06-Kent.pdf (accessed 1 May 2007).

Klein, G. and B. Smith. 2005. Concept systems and ontologies: Recommendations based on discussions between realist philosophers and ISO/CEN experts concerning the standards addressing "concepts" and related terms. CEN/TC 251/WGII, Terminology and Knowledge Bases, WGII/N05-19. http://ontology.buffalo.edu/concepts/ConceptsandOntologies.pdf (accessed 1 May 2007).

Klima, G. 2004. The medieval problem of universals. In *The Stanford Encyclopedia of Philosophy*, Winter edition, ed. E. N. Zalta. http://plato.stanford.edu/archives/win2004/entries/universals-medieval/ (accessed 1 May 2007).

Knight, K. and S. K. Luk. 1994. Building a large-scale knowledge base for machine translation. In *Proc. 12th National Conference on Artificial Intelligence (AAAI'94)*, Seattle, WA, 773–778.

Kokla, M. and M. Kavouras. 2001. Fusion of top-level and geographical domain ontologies based on context formation and complementarity. *International Journal of Geographical Information Science* 15(7): 679–687.

Kuhn, W. 2001. Ontologies in support of activities in geographical space. *International Journal of Geographical Information Science* 15(7): 613–631.

Lowe, E. J. 2001. Recent advances in metaphysics. In *Proc. 2nd International Conference on Formal Ontology in Information Systems, FOIS 2001*, Ogunquit, Maine, October 17–19. New York: ACM Press. http://www.cs.vassar.edu/~weltyc/fois/fois-2001/keynote/ (accessed 1 May 2007).

MacLeod, M. C. and E. M. Rubenstein. 2006. Universals. In *Internet encyclopedia of philosophy*. http://www.iep.utm.edu/u/universa.htm#H5 (accessed 18 October 2006).

Mahesh, K. 1996. *Ontology development for machine translation: Ideology and methodology*. Technical Report MCCS 96-292. Computing Research Laboratory, New Mexico State University, Las Cruces, NM.

Mahesh, K. and S. Nirenburg. 1996. Meaning representation for knowledge sharing in practical machine translation. In *Proc. of the FLAIRS-96 track on Information Interchange*, May 19–22, AI Research Symposium, Key West, FL.

Mark, D. M. and B. Smith. 2004. A science of topography: Bridging the qualitative-quantitative divide. In *Geographic Information Science and Mountain Geomorphology*, ed. M. P. Bishop and J. Shroder, 75-100. Chichester, England: Springer-Praxis.

Mark, D. M., A. Skupin, and B. Smith. 2001. Features, objects, and other things: Ontological distinctions in the geographic domain. In *Spatial information theory: Foundations of geographic information science*, ed. D. R. Montello, 488–502. Lecture Notes in Computer Science 2205. Berlin: Springer-Verlag.

Mark, D. M., B. Smith, and B. Tversky. 1999. Ontology and geographic objects: An empirical study of cognitive categorization. In *Spatial Information Theory, Cognitive and Computational Foundations of Geographic Information Science*, ed. C. Freksa and D. M. Mark. Lecture Notes in Computer Science 1661: 283-298.

Mark, D. M., B. Smith, M. Egenhofer, and S. C. Hirtle. 2004. Ontological foundations for geographic information science. In *Research Challenges in Geographic Information Science*, ed. R. McMaster and L. Usery, 335–350. Boca Raton, FL: CRC Press.

Masolo, C., S. Borgo, A. Gangemi, N. Guarino, A. Oltramari, and L. Schneider. 2003. WonderWeb, Deliverable D17, The WonderWeb Library of Foundational Ontologies, Preliminary Report, Version 2.1, ISTC-CNR.

Merriam-Webster's Collegiate Dictionary, 11th ed. 2003. Springfield, MA: Merriam-Webster Inc.

Muller, P. 1998. Space-time as a primitive for space and motion. In *Proc. First International Conference in Formal Ontology in Information Systems (FOIS'98)*, Trento, Italy, ed. N. Guarino, 63-76. Amsterdam: IOS Press.

Nardi, B. A. 1996. *Context and consciousness: Activity theory and human computer interaction*. Cambridge, MA: MIT Press.

Niles, I. and A. Pease. 2001. Origins of the standard upper merged ontology: A proposal for the IEEE standard upper ontology. Working Notes, IJCAI-2001 IEEE Standard Upper Ontology Workshop, Seattle, WA.

Pease, A. and I. Niles. 2002. IEEE standard upper ontology: A progress report. *Knowledge Engineering Review, Special Issue on Ontologies and Agents* 17: 65–70.

Pease, A., I. Niles, and J. Li. 2002. The suggested upper merged ontology: A large ontology for the semantic web and its applications. Working Notes, AAAI-2002 Workshop on Ontologies and the Semantic Web, Edmonton, Alberta, Canada.

Peirce, C. S. 1932. Collected papers of Charles Sanders Peirce. vol. 1: *Principles of Philosophy*, ed. C. Hartshorne and P. Weiss. Cambridge, MA: Harvard University Press.

Perzanowski, J. 1990. Ontologies and ontologics. In *Logic Counts*, ed. E. Zarnecka-Bialy, 23–24. Dordrecht: Kluwer.

Peuquet, D., B. Smith, and B. Brogaard. 1998. *The ontology of fields: Report of a specialist meeting held under the auspices of the Varenius project: Panel on computational implementations of geographic concepts*. Technical Report, National Center for Geographic Information and Analysis, Bar Harbor, ME.

Philpot, A. G., M. Fleischman, and E. H. Hovy. 2003. Semi-automatic construction of a general purpose ontology. Invited Paper, International Lisp Conference, October 12–15, New York.

Philpot, A., E. H. Hovy, and P. Pantel. 2005. The Omega ontology. In *Proc. ONTOLEX Workshop at the International Conference on Natural Language Processing (IJCNLP)*, Jeju Island, Korea.

Poli, R. 2003. Descriptive, formal and formalized ontologies. In *Husserl's Logical Investigations Reconsidered*, ed. D. Fisette, 183–210. Dordrecht: Kluwer Academic Publishers.

Quine, W. O. 1961. *From a logical point of view: Nine logico-philosophical essays.* Cambridge, MA: Harvard University Press.

Raubal, M. 2001. Ontology and epistemology for agent-based wayfinding simulation. *International Journal of Geographical Information Science* 15(7): 653–665.

Reitsma, F. and T. Bittner. 2003. Scale in object and process ontologies. In *Spatial information theory: Foundations of geographic information science*, ed. W. Kuhn, M. F. Worboys, and S. Timpf, 13–30. COSIT03, Ittingen, Switzerland. Lecture Notes in Computer Science, 2825. Berlin: Springer.

Santos, M. 1997. *The nature of space.* São Paulo: HUCITEC. [Published in Portuguese as *A Natureza do Espaço.*]

Sheth, A. and C. Ramakrishnan. 2003. Semantic (Web) technology in action: Ontology driven information systems for search, integration and analysis. *IEEE Data Engineering Bulletin, Special issue on Making the Semantic Web Real*, ed. U. Dayal, H. Kuno, and K. Wilkinson, 40–48.

Simons, P. 1998. Metaphysical systematics: A lesson from Whitehead. *Erkenntnis* 48(2-3): 377–393.

Smith, B. 1995. Formal ontology, common sense and cognitive science. *International Journal of Human-Computer Studies* 43: 641–667.

Smith, B. 1998. Basic concepts of formal ontology. In *Proc. First International Conference on Formal Ontology in Information Systems (FOIS'98)*, Trento, Italy, ed. N. Guarino, 19–28. Amsterdam: IOS Press.

Smith, B. 2003. Ontology. In *Blackwell Guide to the Philosophy of Computing and Information*, ed. L. Floridi, 155–166. Blackwell: Oxford, preprint version available at http://ontology.buffalo.edu/smith/articles/ontology_pic.pdf (accessed 10 September 2007).

Smith, B. and D. Mark. 1998. Ontology and geographic kinds. In *Proc. 8th International Symposium on Spatial Data Handling (SDH'98)*, ed. T. K. Poiker and N. Chrisman, 308–320. Vancouver, Canada: International Geographical Union.

Smith, B. and D. Mark. 2001. Geographical categories: An ontological investigation. *International Journal of Geographical Information Science* 15(7): 591–612.

Smith, B. and D. M. Mark. 2003. Do mountains exist? Towards an ontology of landforms. *Environment and Planning* 30(3): 411–427.

Smith, B. and D. Woodruff Smith, eds. 1995. *The Cambridge Companion to Husserl.* Cambridge, UK: Cambridge University Press.

Soon, K. and W. Kuhn. 2004. Formalizing user actions for ontologies. In *Proc. GIScience 2004*, ed. M. J. Egenhofer, C. Freksa, and H. J. Miller, 299–312. Lecture Notes in Computer Science 3234. Berlin: Springer-Verlag.

Sowa, J. F. 2000. *Knowledge representation: Logical, philosophical and computational foundations.* Pacific Grove, CA: Brooks Cole Publishing.

Sowa, J. F. 2001. Top-level categories. http://www.jfsowa.com/ontology/toplevel.htm (accessed 1 May 2007).

Stevens, R., C. A. Goble, and S. Bechhofer. 2000. Ontology-based knowledge representation for bioinformatics. *Journal Briefings in Bioinformatics* 1(4): 398–414.

SUO WG. 2003. IEEE P1600.1 Standard Upper Ontology Working Group Home Page. http://suo.ieee.org/index.html (accessed 30 October 2006).

The American Heritage® New Dictionary of Cultural Literacy, Third Edition. Epistemology. American Psychological Association (APA) from Dictionary.com website: http://dictionary.reference.com/browse/epistemology (accessed 1 May, 2007).

The Columbia Encyclopedia. 2004. Conceptualism. In *The Columbia Encyclopedia*, 6th ed. New York: Columbia University Press. http://www.bartleby.com/65/co/conceptu.html (accessed 1 May 2007).

Timpf, S. 2002. Ontologies of wayfinding: A traveler's perspective. *Networks and Spatial Economics* 2(1): 9–33.

Uschold, M. 1996. Building ontologies: Towards a unified methodology. Presented at 16th Annual Conference of the British Computer Society Specialist Group on Expert Systems, Cambridge, December 16–18.

Uschold, M. and M. Gruninger. 1996. Ontologies: Principles, methods and applications. *Knowledge Engineering Review* 11: 93–155.

Van Heijst, C., A. Schrieber, and B. Wielinga. 1997. Using explicit ontologies in KBS development. *International Journal of Human-Computer Studies* 45: 183–292.

Varzi, A. 1998. Basic problems of mereotopology. In *Proc. of the First International Conference on Formal Ontology in Information Systems (FOIS'98)*, Trento, Italy, ed. N. Guarino, 29–38. Amsterdam: IOS Press.

Vieu, L. 1997. Spatial representation and reasoning in AI. In *Spatial and Temporal Reasoning*, ed. O. Stock, 5–41. Dordrecht: Kluwer.

Weber, A. 1908. *History of philosophy*, trans. F. Thilly. New York: Charles Scribner's Sons. http://www.class.uidaho.edu/mickelsen/ToC/Weber%20ToC.htm (accessed 1 May 2007).

5 Concepts

5.1 ONTOLOGY OF CONCEPTS

Concepts are one of the most significant notions for the study of thought and language. They are "the most fundamental constructs in theories of the mind" (Laurence and Margolis, 1999, 3); they are at the root of cognitive abilities such as categorization and understanding and they constitute the primary building blocks of ontologies and linguistic systems. Therefore, it is reasonable that they have attracted the intense interest of many scientific fields, such as philosophy, linguistics, cognitive science, conceptual analysis, etc. During the centuries of studying concepts, different theories have been formulated regarding the ontological status of concepts, their nature, their structure, their relation to natural language, etc. (Laurence and Margolis, 1999; Margolis and Laurence, 2006).

In the literature, in certain contexts, the terms *category* and *concept* are used interchangeably. However, the two terms are not exactly equivalent, and, what is worse, it seems that they do not have universally accepted definitions in all contexts. In philosophy, the term *concept* is traced back to Plato and is highly relevant to the problem of universals (see Section 4.3). The three main views on this problem, that is, *realism*, *conceptualism*, and *nominalism*, reflect the way the term concept is used (Klein and Smith, 2005) (Figure 5.1). According to realists, concepts refer to universals, that is, mind-independent entities, which represent the properties of individual instances. According to conceptualists, concepts refer to mind-dependent entities, whereas, following nominalists, they are regarded as general terms in a controlled language (Klein and Smith, 2005).

Under a cognitive perspective, a concept is usually considered to be a mental representation of a category (Laurence and Margolis, 1999; Medin and Rips, 2005). It includes the conceptual information associated with the category. A *category* connects similar entities according to some properties, characteristics, functions, roles, and so forth that differentiate these entities from other entities in the world. A concept distinguishes instances that fall under a specific category. For example, the concept "building" (i.e., a structure with roof and walls) helps to identify and categorize different instances (e.g., the parliament, the Empire State, the university building, my apartment building, etc.) as falling under the category "building." However, there are other approaches to the notion of concepts; concepts are thought of as abilities of cognitive agents or even abstract objects that constitute propositions and are identified with Fregean senses (Margolis and Laurence, 2006) (Figure 5.2).

In ontology and knowledge representation literature, the notion of concept arouses confusion and misunderstanding due to the lack of an unambiguous definition. However, embracing a cognitive approach according to which concepts are

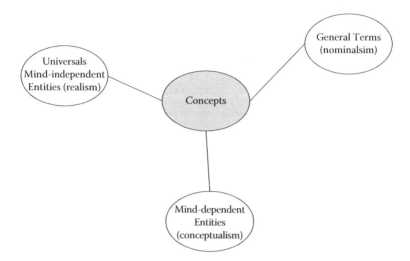

FIGURE 5.1   Different notions of concepts according to philosophy.

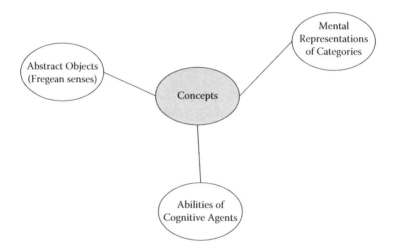

FIGURE 5.2   Different notions of concepts according to cognitive science.

mental representations formed by human cognition independently of reality is not appropriate for ontologies of natural sciences (B. Smith, 2004). Klein and Smith (2005) define a concept as "the meaning of a term as agreed upon by a group of responsible persons" to emphasize the agreement of a group of people on the meaning of a term in contrast to individual conceptions (i.e., cognitive representations). This notion of a concept contributes to the effective understanding and communication of information among different individuals or groups of individuals.

For the purposes of our work, we consider that the terms "category" and "concept" are not equivalent. A category is a set of entities (i.e., instances) bound by a unifying principle, such as a set of distinguishing properties, roles, or functions. A concept embodies all knowledge, which describes the meaning of a category (i.e., its intension). For example, the category "river" comprises all individual rivers in the world, such as Nile, Mississippi, Missouri, Euphrates, Rhine, Volga, Danube, Congo, etc., whereas the concept "river" describes the knowledge associated to the category, that is, that it refers to a large, natural stream of water, which forms meanders and usually has tributaries, that large rivers are navigable, etc. Other terms that are used alternatively to "category" are "class," "kind," and "type"; however, these terms are associated to disparate scientific fields acquiring different meanings, such as biology, mathematics, object-oriented programming, and classification theory. Thus, they are not used here to avoid misinterpretations.

We hold that concepts represent the meaning of categories of entities existing in reality and more specifically, entities existing in the geospatial realm. We do not consider concepts to be cognitive artifacts (a conception which may be the focus of geographic cognition), since we focus on entities existing in reality, that is, the spatiotemporal continuum (tangible or abstract, continuants or occurrents) and not on conceptualizations of these entities. This consists in a realist conception traced back to Aristotle also embraced by contemporary philosophers like Lowe (2001) and B. Smith (2004). According to this approach, ontologies are grounded on reality and their underlying concepts are regarded as universals, whose instances correspond to particulars. This grounding on reality provides a common solid basis for communication and common understanding during ontology engineering, comparison, and integration. Only under this realist view is it reasonable to seek the development of ontologies (top-level, domain, and application) with wide acceptance and adoption since ontologies representing individual (not shared) conceptualizations somehow miss the objective of their existence.

5.2 THEORIES OF CONCEPTS

As mentioned in the previous section, the study of concepts has led to the formulation of different theories concerning their nature, their structure, their relation to the mind and language. The term *theories of concepts* primarily concerns the structure of concepts (Margolis and Laurence, 2006). Five main theories exist (Laurence and Margolis, 1999): (1) Classical, (2) Probabilistic, (3) Theory-based, (4) Neoclassical, and (5) Conceptual Atomism (Table 5.1).

TABLE 5.1
Concept Representation According to the Theories of Concepts

Theory	Concept Representation
Classical	Necessary and sufficient conditions
Probabilistic	Properties that instances tend to possess
Theory-based	Mental theories
Neoclassical	Partial definitions
Conceptual atomism	Primitive, innate, with no structure

5.2.1   THE CLASSICAL THEORY OF CONCEPTS

The *Classical theory* of concepts is traced back to Aristotle and claims that concepts exhibit definitional structure, that is, there is a set of necessary and sufficient conditions, which sort out exclusively the instances that fall under the extension of a category. These conditions usually take the form of defining features (i.e., characteristics, properties). An instance is said to belong to the extension of a concept in case it has all the defining features that describe the concept.

According to this perspective, categories have clear boundaries and are defined based on the set of necessary and sufficient properties shared by all members of the category. Therefore, there are not "better" examples of a category; all members of a category are equally representative.

However, the ability of definitions to properly describe concepts as well as their psychological validity has been questioned; hence Classical theory's adherence to definitions has been criticized. Especially the criticism originating from cognitive science is based (a) on the difficulty in finding necessary and sufficient conditions for concepts, and (b) on empirical evidence revealing typicality effects as well as conceptual fuzziness in the structure of categories; phenomena which are not dealt with by the Classical theory.

5.2.2   PROBABILISTIC THEORIES OF CONCEPTS

In contrast to the Classical theory, according to which concepts are thought to have definitional structure, *Probabilistic theories* hold that concepts exhibit probabilistic structure. An instance belongs to a category in case it has an adequate number of properties that its members tend to possess (Laurence and Margolis, 1999; Margolis and Laurence, 2006). Wittgenstein (1953) was the first to object to the Classical Theory's persistence in necessary and sufficient conditions with his famous example of the category "game." According to Wittgenstein, the category "game" is not described by a set of necessary and sufficient conditions, but by a series of features, which, however, are not ascribed to all its instances. There are games of chance (e.g., card games), games of skill (e.g., chess), games that require physical activity (e.g., football), games that involve competition among players (e.g., tennis), solitary games (e.g., patience), and others played only for amusement. Driven by this example, Wittgenstein used the term *family resemblances* to express categories characterized by similarities and relations among the different instances.

During the 1970s, extensive empirical research signaled a retreat from the Classical Theory's adherence to definitions and provided evidence for the Probabilistic Theories of Concepts. According to these theories, a concept is described by a set of distinguishing features in accordance with the classical view; however, these do not set necessary conditions for the concept's application. Thus, an instance may belong to a concept's extension if it exhibits a sufficient number of features that describe the concept. Furthermore, the probabilistic view allows the comparison of the significance of the distinguishing features since some of them may be more important for the description of the concept and, therefore, are weighted higher than the rest. Probabilistic theories consider two models for concept representation: *prototypes* and *exemplars*. A *prototype* is "a summary representation that is a central tendency of all of the experienced members of a category" (Nosofsky and Zaki, 2002, 924). The *exemplar* model, on the other hand, considers concepts to be represented as a collection of the individual members, or otherwise exemplars, of a category (Nosofsky and Zaki, 2002)).

Rosch (1978) was among the first to question the validity and relevance of the Classical theory to categorization and to provide empirical evidence for the existence of typicality effects. She formulated two basic principles for the formation of categories.

1. *Cognitive economy*: category systems aim at offering maximum information requiring minimum cognitive effort and hence, categories represent the way the world is perceived as closely as possible.
2. *Perceived world structure*: the world is perceived as structured and not as arbitrary or unpredictable information.

According to Rosch (1978), category systems are depicted as extending along two dimensions: the *vertical dimension* refers to the categories' level of abstraction, whereas the *horizontal dimension* refers to the partition of categories at the same level of abstraction. The principles of categorization affect both dimensions. For the vertical dimension they entail that all levels of abstraction are not equally important, but among them there is a level, called the *basic level*, which is the most inclusive level representing the perceived world structure. Rosch's experiments revealed that categories at the basic level exhibit commonalities in attributes, motor movements, shape, as well as identifiable shapes. For the horizontal dimension, the principles of categorization entail that categories should be as distinct and unambiguous as possible; this is accomplished by defining categories with regard to clear cases of category membership, that is, prototypes. Prototypical category members appear to have more properties in common with other category members and fewer properties in common with members of contrasting categories.

Properties can be used for the description of prototypes, since they are considered to be cognitively important for categorization (E. E. Smith et al., 1988). E. E. Smith et al. (1988) propose the representation of prototypes with properties consisting of attribute-value pairs. For example, the prototype "basketball ball" could be described with attribute-value pairs such as shape: round; size: big; texture: rough; weight: heavy; and color: orange. Lakoff (1987) interprets prototype effects in terms of cognitive models. Cognitive models are tools for knowledge representation and organization from a cognitive perspective. Different types of cognitive models exist: propositional, image-schematic, metaphoric, metonymic, and symbolic (Lakoff, 1986). E. Smith and Medin (1981) consider that categories are represented in terms of exemplars; an exemplar may be either an instance or a subset of the concept. In case an exemplar is an instance, then it is represented with reference to the properties it exhibits. In case the exemplar is a subset of the concept, then it is represented with reference to either other exemplars or to a series of distinguishing properties (E. Smith and Medin, 1981).

Although Probabilistic theories have been formulated to deal with the problems of the Classical theories, they also raise many problems themselves. Probabilistic theories cannot account for the fact that some concepts lack prototypes. Furthermore, prototypes or exemplars cannot always specify the correct extension of a category either because they are not so rich as to include all instances or because they are too rich and include inappropriate instances (Margolis and Laurence, 1999).

Another criticism of the Probabilistic Theories is that the existence of typicality effects in people's judgments does not necessarily manifest prototypical (i.e., graded) structure in concepts. Besides, typicality effects have induced some misleading interpretations regarding the structure of concepts (Rosch, 1978; Lakoff, 1987). As Armstrong, Gleitman, and Gleitman (1999, 247) point out, "prototype descriptions apply to an organization of 'exemplariness' rather than to an organization of class membership." According to them, graded responses to exemplars do not necessarily manifest a family resemblance in the structure of a category, but may be related to what is called the "identification function," that is, a function used for quick categorizations of things. Armstrong, Gleitman, and Gleitman (1999) distinguish between exemplariness, which contributes to the *identification function*, and *sense*, which constitutes a systematic description of a category. The notion of *sense* is parallel to what others have called *conceptual core* and is contrasted to the notion of *real essence*, which consists in the scientific description of a category (Armstrong, Gleitman, and Gleitman, 1999). The real essence is considered to be close to reality, whereas the sense or conceptual core may be prone to errors.

Furthermore, although empirical results on prototypical effects show that prototypes play a role in concept representation, processing, and learning, prototypes do not consist in mental representations of categories either in the form of an abstraction or in the form of an exemplar (i.e., a particular example). A prototype is a term to describe human judgments about the degree of prototypicality. Especially for natural language categories, the claim that there is a single entity that stands for the prototype of a category is misleading. Prototypes do not constitute a model of representation, processing, or learning (Rosch, 1978).

Rey (1999) discusses unclear cases and typicality judgments in relation to the metaphysical and epistemological functions that concepts are called to perform (see Section 5.3). According to this perspective, although people may be slightly hesitant to epistemologically decide for certain unclear cases, this does not imply that these cases are also metaphysically unclear. Rey (1999) draws a distinction between two types of unclear cases: metaphysically unclear cases, like "euglena," and epistemologically unclear cases, like "penguin" and "tomato." Epistemologically unclear cases, that is, those for which subjects may not give definite and immediate answers, do not reflect ambiguity or even lack of metaphysical defining conditions for the concept. Rey (1999) stresses the assumption that many natural kinds do indeed have definitions, although competent users may be ignorant of them. This is because sound definitions are provided by the corresponding scientific field that deals with the natural kind. He introduces the *hypothesis of external definitions*, which states that "the correct definition of a concept is provided by the optimal account of it, which need not be known by the concept's competent users" (293). Although there is a differentiation in the metaphysical and epistemological functions of concepts, these seem to coincide in the core of the concept (Rey, 1999).

According to Wierzbicka (1996), the Classical and the Probabilistic approaches to human categorization are not two opponent theories. On the contrary, this "contrasting" view has proven to be worthless in semantic research; therefore, instead of deciding on one approach against the other, a synthesis of these approaches is proposed. The weakness of necessary and sufficient features to account for certain concepts such as "bird," "game," and "bachelor" is due to the focus on physical features ignoring mental ones. Prototypes may be rather helpful in semantic research if they are cautiously dealt with and accompanied by verbal definitions, instead of being their substitute. Wierzbicka (1996) also observes that the categories "bird" and "furniture" used in Rosch's experiments are not comparable because they belong to different semantic and grammatical categories: "bird" is a taxonomic concept (i.e., "a kind of" vertebrate) expressed by a count noun, whereas "furniture" is a collective concept (i.e., "a heterogeneous collection of things of different kinds"), expressed by a mass noun.

It is now recognized (Lloyd, Patton, and Cammack, 1996; Mark et al., 1999; Smith and Mark, 1999; Mennis, Peuquet, and Qian, 2000; Kokla and Kavouras, 2002; Mennis, 2003; Usery, 2003) that certain aspects of the representation of geographic information may avail themselves of the conformance to geographic cognition and basic level categorization. Tversky and Hemenway (1983) concluded to analogous results with Rosch regarding the existence of the basic level in the categorization of environments and spatial settings either indoor (e.g., school, home) or outdoor (e.g., beach, mountain, city), based on common appearance and performance of common activities. Lloyd, Patton, and Cammack (1996) provided empirical evidence for the existence of basic-level geographic categories based on characteristics, parts, and activities associated with geographic categories. Basic-level categories such as country, region, state, city, and neighborhood are more distinct and informative than categories at the superordinate and subordinate levels.

5.2.3   THEORY-BASED THEORIES OF CONCEPTS

Theory-based theories of concepts are so called because they define a concept's identity and structure not according to a set of distinguishing features, like Classical and Probabilistic theories, but according to the concept's role within a mental theory. They respect the human tendency to categorize not based on the observation of a set of perceptible features, but on account of being aware of the possession of an *essence*, that is, "an appropriate internal structure or some other hidden property" (Margolis and Laurence, 1999, 45). The perceptible features are the result of the concept's possession of the essence. For example, the perceptible features of a particular tree (e.g., the roots, the trunk, the branches, and the leaves) are the result of its having the essence of a tree. It seems that people have intuitions that certain feature correlations are relevant and may be associated whereas others are not. Thus, they tend to focus on the relevant feature correlations, ignoring any others. Keil (cited in Smith and Mark, 2001, 599) compares these intuitions to "intuitive theories of how things in a domain work and why they have the structure they do." Murphy and Medin (1985) assert that "representations of concepts are best thought of as theoretical knowledge or, at least, as embedded in knowledge that embodies a theory about the world." Embedding theories constrain the properties used for the representation of concepts. Subjects come up with properties "particularly salient and diagnostic" when describing a concept, instead of mentioning everything about it (Murphy and Medin, 1999). They cite Miller and Johnson-Laird (1976), who compared theory-based attributes, which describe the concept's core with perceptually salient attributes, used in the identification procedure.

People's belief about the existence of an essence that designates existence, properties, identity conditions, and category membership is called *psychological essentialism* (Medin and Rips, 2005). However, essence is considered to be a nonexperts' conception of natural kinds, rather than an innate characteristic of natural kinds themselves (Medin and Rips, 2005). *Sortalism* is a version of essentialism, which holds that the essence is called to fulfill a deeper function than to determine category membership and this is to define an object's identity. Categories, which satisfy these identity conditions, are called *sortals*.

Theory-based theories of concepts are also related to the differences between experts' and novices' conceptual structures. It seems that experts' concepts are more specific and distinct than those of novices. This is attributed to the fact that "experts have better developed theories about the domain than do novices" (Murphy and Medin, 1999, 443). However, although novices may not have such developed theories as experts do, they seem to believe that there are defining properties for concepts, probably incited from naive theories. This implies a connection between theoretical and conceptual knowledge. Defining properties are those centrally located in our knowledge, whereas characteristic properties are located at the periphery of our knowledge (Murphy and Medin, 1999).

The problem with the theory-based view is that people may have an inadequate or incorrect mental theory about a concept but still possess the concept. Furthermore, people's difficulty to describe the hidden essence of a concept questions the Theory-based Theory's ability to account for categorization of new instances.

5.2.4 NEOCLASSICAL THEORIES OF CONCEPTS

Neoclassical theories of concepts provide a rethink of Classical theory especially under philosophical and linguistic contexts, properly adapted to accommodate the late findings of cognitive science. According to the Neoclassical theories, concepts have partial definitions that describe the conditions that must be satisfied by the instances falling under their extensions (Margolis and Laurence, 1999). According to Jackendoff (1983), these conditions are distinguished into three types: (a) *necessary conditions*, (b) *centrality conditions*, which represent variable attributes, for example, color, and (c) *typicality conditions*, which represent features typical but subject to exceptions.

A problem generated by partial definitions is that they cannot contribute to the determination of a concept's extension. However, in case a partial definition is transformed into a full definition, then Neoclassical theory is charged with all the problems of the Classical theory. In order to overcome this problem, Jackendoff (1989) proposed the elaboration of a concepts' representation with spatial information (geometric and topological properties) organized as a three-dimensional model structure (Marr, 1982), based on the idea that "knowing the meaning of a word that denotes a physical object involves in part knowing what such an object looks like" (Jackendoff, 1993, 431).

5.2.5   CONCEPTUAL ATOMISM

In contrast to the previously analyzed theories, Conceptual Atomism holds that concepts are mental representations with no structure (Fodor et al., 1980; Fodor, 1990). According to this theory, a concept is described with reference to causal relations with things in the world. Furthermore, the content of a concept is independent from its representation.

The existence of primitive concepts with no structure usually implies that such concepts are innate. However, this belief poses difficulties concerning mental processes such as concept acquisition and categorization. It is hard to explain how concepts such as "galaxy" and "carburetor" are primitive and innate (Margolis and Laurence, 1999), and how they are learned independently of simpler concepts, like the previous theories suppose. Wierzbicka (1996) contradicts Fodor's claim of concepts' innateness and conceptual primitiveness by presenting counterexamples of words (e.g., the word "paint" in English and some close equivalents in Polish and Russian), which are language specific, and therefore, not universally conceived.

5.3   FUNCTIONS OF CONCEPTS

Concepts are vital for many mental processes such as perception, learning, and decision making, and they are called on to fulfill different functions (Rey, 1999; Smith, 1988; Davidsson, 1996; Medin and Rips, 2005; Swoyer, 2006) (Figure 5.3). Medin and Rips (2005) specify two main functions of concepts: (a) *categorization* and (b) *communication. Categorization* is the ability to specify whether an entity is a member of a category. This function incites further subfunctions, such as understanding, prediction, learning, and reasoning. *Communication*, on the other hand, presupposes the existence of comparable concepts among humans and also incorporates the ability to combine concepts in order to form new ones (Medin and Rips, 2005).

Swoyer (2006) identifies three main functions of concepts: (a) *classification*, (b) *perception*, and (c) *inference. Classification* is used to "sort the world into types of things ..., events ..., stuff ..., qualities ..." (2006, 137). *Perception* interrelates the mind and the world and contributes to the understanding and interpretation of things seen and heard, that is, we see different sorts of things (e.g., houses, trees, people) instead of shapes and hear different words and sentences in known languages instead

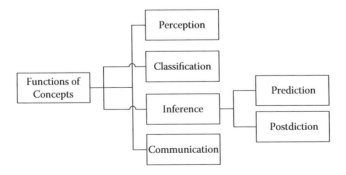

FIGURE 5.3   Functions of concepts.

of sounds. *Inference* refers to making *predictions* or *postdictions*. An example of a prediction is that extensive sunbathing will cause sunburn. An example of a postdiction is that the sight of black smoke leads to the conclusion that something is on fire, and therefore, it is dangerous to approach.

Rey (1999) specifies the following functions that concepts are called on to fulfill: (a) *stability functions*, (b) *linguistic functions*, (c) *metaphysical functions*, and (d) *epistemological functions*. *Stability functions* are fundamental for conceptual capability and comparison of cognitive states either within an agent (intrapersonal) or among agents (interpersonal). *Linguistic functions* refer to the representation of the meaning of concepts using linguistic terms. *Metaphysical functions* provide the foundation for assertions regarding the true existence, nature, and classification of things. *Epistemological functions* on the other hand, refer to the way an agent categorizes things into kinds.

5.4 KINDS OF CONCEPTS

There are different ways to distinguish types of concepts. Concepts may be classified according to their content (moral, social, logical, abstract, aesthetic, etc.), their structure (e.g., sortal, quality, relational, etc.), the way they are related to each other (e.g., via hierarchical relations), or whether they are based on similarity (e.g., green) or rules (e.g., prime numbers) (Swoyer, 2006).

Medin, Lynch, and Solomon (2000) provide a classification of the kinds of concepts (both object and nonobject) based on three criteria: *structure, processing*, and *content. Structure* is examined with regard to the features that describe the concepts and the relations among the features. *Processing* refers to the concepts' development and maintenance procedures (e.g., bottom-up or top-down). Structure and process are interrelated and are sometimes studied in combination. *Content* refers to the domains to which the concepts apply (e.g., naive physics, naive geography). Based on structure, Medin, Lynch, and Solomon (2000) draw the following distinctions among kinds of concepts: (a) nouns versus verbs, (b) count nouns (e.g. chair) versus mass nouns (e.g., furniture), (c) isolated (e.g., man) versus interrelated concepts (e.g., grandfather), (d) objects versus mental events, (e) artifacts versus natural kinds, (f) abstract concepts, (g) basic level versus subordinate and superordinate concepts, and (h) hierarchies versus paradigms (i.e., social categories based on age, gender, race, etc.). Based on processing, Medin, Lynch, and Solomon (2000) distinguish the following kinds of concepts: (a) common taxonomic versus goal-derived categories, (b) social information processing and individuation, and (c) stereotypes, subtypes, and subgroups. Finally, different contents derived from different domains may also distinguish kinds of concepts (Medin, Lynch, and Solomon, 2000).

5.5 PROPERTIES AND RELATIONS

5.5.1 PROPERTIES

Research on properties and the study of their existence and nature originate since the provenance of philosophical inquiries (Denkel, 1996). *Properties* are the "attributes or qualities or features or characteristics of things" (Swoyer, 2000). They are embraced in order to account for the existence of a phenomenon or the truth of a sentence, and they "have an important role to play in explaining such epistemological phenomena as our ability to recognize and categorize things in the world around us" (Swoyer, 2000). They are contrasted with individual things, also called *particulars*; properties are considered to be *instantiated* or *exemplified* by individuals, but the opposite is not possible.

Properties constitute a controversial philosophical issue regarding their existence, the kinds of properties that exist, their nature, the functions they are called to fulfill, etc. The subject of properties is closely related to the philosophical distinction between *universals* and *particulars* (see Section 4.3). Those who are in favor of the view that properties are universals believe that different instances may exemplify the same property, that is, two distinct pastures may have the same green color. A contrary view advocates that properties are particulars, that is, although two properties may seem similar (as in the case of the two pastures' green color), they are not the same, but they constitute numerically distinct individuals. The three competing views on the *problem of universals* (i.e., realism, nominalism, and conceptualism) are also relevant to the discussion on properties.

Properties are employed by philosophers to fulfill certain explanatory roles (Swoyer, 2000). A major one is *qualitative similarity*, that is, the resemblance of certain entities with respect to a property or a set of properties. For example, plants are alike on account of the possession of certain characteristics, that is, roots, a stem, and leaves. Another role is the *recognition and classification of new instances*. A new instance is recognized in terms of a property or a set of properties already encountered in known instances. *Explanation of the meaning of terms* and consequently, of the things these terms refer to is another role that properties are called to fulfill. Apart from philosophy, semantics is also an area where properties play an important role, although there are also many different views on their existence and usefulness. Probably, the most important role that properties are called to play in semantics is to explicate the meaning of natural languages.

Although philosophy has a great interest in the nature and existence of properties, there are controversial examples of the *kinds of properties* that may exist. Moreover, properties, as used in the philosophical sense, do not necessarily coincide with *attributes* but rather refer to the property of "being something," for example, the property of "being a lake." Different kinds of properties are proposed in the literature according to different perspectives (Bigelow 2000; Guarino and Welty 2000a, 2000b, Swoyer, 2000). Some of these are directly pertinent to geographic concepts and are presented below in the rest of this section (Figure 5.4).

Particularizing properties designate the identity of the object and determine the principles by which it can be distinguished from other objects. Particularizing properties (also called *sortal properties* by Strawson, 1959) allow us to count objects, that is, "to divide the world up into a definite number of things," and thus, they specify *counting principles*, or *principles on identity* (Swoyer, 2000). An example of a particularizing property is that of "being a tree" because it is readily possible to count trees. Particularizing properties are contrasted with *characterizing properties* and *mass properties*. *Characterizing properties* (e.g., high) although they also refer to

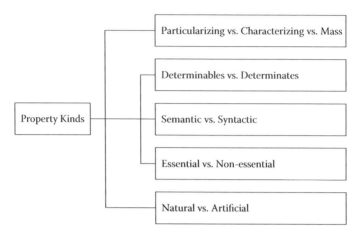

FIGURE 5.4   Some distinctions of property kinds.

individuals, do not partition the world into a specific number of entities. *Mass prop-erties* (e.g., water, cement, and furniture) do not refer to individuals but to stuff, and, like characterizing properties, they do not partition the world into a specific number of things.

Another distinction, which, however, is a relative one, is that between *determin-ables* and *determinates*. *Determinables* are general properties (e.g., color, shape), while *determinates* are more specific versions of these (e.g., red, triangular). Deter-minables and determinates can be used to model properties and their values.

Bigelow (2000) distinguishes between *semantic* and *syntactic* properties under the framework of referential semantics. *Referential semantics* accounts for the way in which the properties of compound expressions rely on the properties of the con-stituent words. According to Bigelow, *syntactic properties* are considered intrinsic, physical, and natural and are perceived through the senses, whereas *semantic* ones are relational, latent, and result from social conventions. He cites (and partly agrees with) Devitt (1996) in that semantic properties (at least primarily) play an explana-tory role in describing reality and predicting behavior.

Essential or *rigid properties* are the properties that are indispensable to the existence of individuals. For example, a riverbed may be considered as an essential property of a river. Guarino and Welty (2000a) present a formal ontology of proper-ties based on the metaproperties of *identity, unity, essence,* and *dependence. Identity* refers to the question of how an instance of a category is distinguished from all other instances of the same category by using a prominent, individual property of the instance, called *essential property. Unity* refers to the question of how the parts of an instance are distinguished from anything else surrounding it through a unify-ing relation, which connects all the parts of the instance. These metaproperties are used to analyze and develop a taxonomic structure of unary properties such as type, category, role, attribution, etc.

Natural kind properties (e.g., river) are important properties that "carve nature at its natural joints" and are contrasted with *artificial properties* (e.g., road) (Swoyer, 2000). *Qualities* (e.g., distance, volume, area, speed, and temperature) as well as *roles* (e.g., land use versus land cover, hotel versus building) are also considered as kinds of properties.

As already mentioned in the Introductory Part, this book holds a realistic metaphysical view. Properties are considered to be universals; thus, they constitute a different kind of entity from individuals. Properties are located in different places (i.e., exemplified by different entities) at the same time since, unlike individuals, they have no spatial parts. Properties are very important for the definition of geographic concepts and consequently, for the comparison and integration of geographic ontologies, the computation of semantic similarity, information retrieval, etc. In the geographic domain, importance mainly lies in essential properties, which ensure the semantic and univocal definition of geographic categories. Nevertheless, it is not always easy to determine essential properties for geographic concepts. Thus, the question arises whether essential properties are the only way to create identity or a certain combination of nonessential properties can still provide identity to geographic concepts. For example, the properties of being a "wetland," an "arable land," "flat," "flooded," and containing "irrigation channels," although they are individually nonessential, may in combination quite confidently identify the concept "rice field." According to Elder (1998), "any distinctive or peculiar property, essential to an individual object must go together with yet other properties, which likewise distinguish that object with other similar ones."

5.5.2   RELATIONS

Relations are also important conceptual units that contribute to meaning attribution (Bean, Green, and Myaeng, 2002). Similarly to concepts and properties, they are subject to interdisciplinary research (philosophy, cognitive science, linguistics, information science, etc.) because they play a central role in many aspects of human thought and interaction with the world, such as knowledge organization, representation, and reasoning. Relations, although they are polyadic, in contrast to common properties, which are monadic, raise the same philosophical questions and, thus, they are usually considered to be a kind of property (Swoyer, 2000).

Storey (1993) identifies seven kinds of semantic relations based on Landis, Herrmann, and Chaffin (1987), Winston, Chaffin, and Herrmann (1987), and Chaffin, Herrmann, and Winston (1988) (Figure 5.5):

1. *Inclusion,* the relation that holds between an entity type that contains other entity types. Inclusion is further classified into three subtypes (Winston, Chaffin, and Herrmann, 1987):
   - *Class,* that is, the subtype/supertype relation that applies to natural kinds, artifacts, states, and activities (Chaffin, Herrmann, and Winston, 1988).
   - *Meronymic,* that is, the part-whole relation.
   - *Topological,* that is, "situations where one object is surrounded by another but is not part of the thing that surrounds it," (Storey, 1993, 467) for example, the ship is-in the harbor.
2. *Possession,* the relation that holds when an entity type owns another entity type.

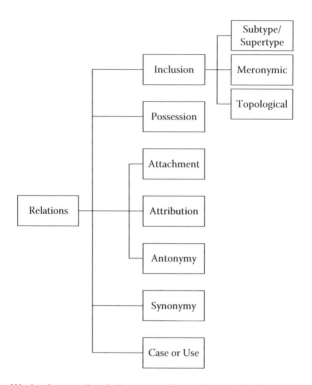

FIGURE 5.5   Kinds of semantic relations according to Storey (1993).

3. *Attachment,* the relation that holds between entity types that are "connected or joined."
4. *Attribution,* the relation between an entity type or object and its attributes.
5. *Antonymy,* the relation between opposite entities, attributes, or relations (e.g., shallow-deep, hot-cold).
6. *Synonymy,* the relation that describes the same or nearly the same entities, attributes, or relations (e.g., stream and watercourse).
7. *Case* (or *use*), the relation that applies to things that are typically true in the real world and may describe the actions of agents, the instruments, and objects involved in these actions.

Bean, Green, and Myaeng (2002) identify three kinds of relations from a linguistic point of view: (a) equivalence (or identity) relations, (b) hierarchical (or subsumption) relations, and (c) association relations. The hierarchical relation is further distinguished into: (i) hyponymy relation, which associates entities or else nouns; (ii) troponymy, which associates processes or else verbs; and (iii) meronymy, which deals with wholes and parts of entities (Bean, Green, and Myaeng, 2002).

The *hyponymy/hypernymy (inclusion, is-a, subtype/supertype, kind of)* relation is the subordination/superordination relation. The *hyponym* inherits all the characteristics of the more generic concept (*hypernym* or *hyperonym*) and adds at least one characteristic that distinguishes it from the generic concept and the other hyponyms of that superordinate, for example, a "river" is a kind of "stream." The hyponymy relation most commonly and naturally applies to nouns. *Taxonymy* is a variety of hyponymy; although hyponymy is illustrated by *an X is a Y*, taxonymy is elucidated by *an X is a kind/type of Y*, which is more discriminating than hyponymy and represents "the "vertical" relation in a taxonomy" (Cruse, 2002). Categories may be further divided into hyponyms, regardless of their hierarchical level, according to different modes. Three main modes of hyponymy are specified: (a) *natural kind mode*, (b) *nominal kind mode*, and (c) *functional mode*. *Natural kinds* (e.g., chemical elements or biological species) are groupings of entities based on properties characterized by a natural law. A single distinguishing feature or even a small set of features cannot describe the relation of a natural kind to its hyperonym. The exact opposite stands for *nominal kinds*, that is, the relation to their hyperonym can be represented with a single feature, as in the case of "mare" relative to its hyperonym "horse" (Cruse, 2002). The *functional mode* relies on the *function* of the primary feature for the specification of the meaning of the hyperonym, as, for example, in the cases "gun-weapon" or "hammer-tool" (Cruse, 2002).

Troponymy is the main relation for the association of verbs in a semantic network (Fellbaum, 1990). Concepts expressed by verbs are hierarchically related but not with the is-a or kind-of relation, as in the case of concepts expressed by nouns. For example, although it is eligible to say that "a fir is a tree" (or a kind of tree), it is not that legitimate to say that "to walk is to move" (or walking is moving). The hierarchical relation among verbs comprises other subrelations such as *manner, function*, and *result* and forms wide and shallow hierarchies, which consist mostly of three or four levels.

Meronymy/holonymy or *partonomy* is the part-whole relation (e.g., "midstream," "ford," and "meander" are parts of "stream"). It is another fundamental relation both for ontology and for cognition. *Parts* pervade every entity in the world, physical or abstract, spatial or temporal; they are also identified both for concepts and for individual instances of a concept (Pribbenow, 2002). Parts and part-whole relations have raised many problems such as the hypostasis of holes, (temporal) change, and identity, etc. Like the association of entities by the is-a relation forms taxonomies, the association of entities by the meronymic relation creates *partonomies*. Taxonomies are hierarchically organized into three levels: the superordinate level, the basic level, and the subordinate level (see Chapter 5). Partonomies are also hierarchically structured and comprise three levels: (i) the whole object, (ii) the parts, and (iii) the subparts. Four dimensions contribute to the distinction among meronymic relations (Winston, Chaffin, and Herrmann, 1987; Chaffin, Herrmann, and Winston, 1988; Storey, 1993).

- *Function*: when parts are located in a particular position in order to play a specific role.
- *Separability*: when parts may be physically dissociated from the whole (Winston, Chaffin, and Herrmann, 1987).
- *Homogeneity*: similarity of the parts to each other and to the whole.
- *Contemporaneity*: simultaneity in the occurrence of the parts.

Meronymic relations are further classified into different types. Winston, Chaffin, and Herrmann (1987) distinguish six *types of meronymic relations*: (1) *component-integral object* (e.g., sail-sailing boat, engine-automobile), (2) *member-collection* (e.g., tree-forest), (3) *portion-mass* (e.g., scoop-ice-cream bucket), (4) *stuff-object* (e.g., clay-pottery) (5) *feature-activity* (e.g., take-off and landing-air traveling), and (6) *place-area* (e.g., bedroom-house, Central Park-New York).

Gerstl and Pribbenow (1995) distinguish two categories of part-whole relations according to whether they depend or not on the "compositional structure of the whole." The first category is designated by an "inherent compositional structure" and is further distinguished into three subcategories of part-whole relation.

- A *mass* is an entity with no compositional structure and thus is treated as homogeneous, for example, a quantity of sugar in a pack.
- A *collection* is an entity with uniform compositional structure. The element-collection relation represents the case where all parts have the same relation to the whole with reference to specific properties (e.g., the trees of a forest or the bricks of a brick wall).
- On the other hand, a *complex* has a heterogeneous compositional structure; the parts or components of the whole are arranged differently in space and time or perform a different function. For example, an airport is comprised by runways, terminals, and a control tower, each of which performs a specific task. The component-complex relation also holds for events or activities, which are further distinguished into subevents or subactivities, respectively. For example, the activity of air traveling can be further analyzed into subactivities such as arrival at the departure airport, check-in, boarding, take-off, flight, landing, arrival at the destination airport, etc.

The second category is designated by "independence of the compositional structure of the whole" and is further distinguished into two subcategories of part-whole relation.

- The partition of the whole into *segments* depends on an external scheme, which is generally a spatial one (e.g., the end of the highway).
- The partition of the whole into *portions* depends on internal attributes, for example, the land parcels of a city with commercial land use.

Apart from hyponymy and meronymy, which are the most widely studied, other relations are also important to philosophy, psychology, linguistics, cognition, and knowledge representation.

Roles describe the relations (static or dynamic) among entities and more specifically, the properties that the entities engaged in the relation should possess (Allwood, 1999). Roles are further classified into the following basic types: (1) cause–motive–reason–origin, (2) result–function–product–effect, (3) direction–purpose–goal, (4) need, (5) object–material, (6) agent, (7) potential, (8) resource, (9) patient–other participants, (10) instrument, (11) manner–organization, and (12) surrounding (Allwood, 1999). *Thematic roles* or *case relations* associate a process with its participants (Sowa, 2000). According to this perspective, *thematic roles* are further divided by two distinctions, which specify the role that the participants play in a process: *determinant* or *immanent* and *source* or *product*. A *determinant* participant specifies the direction of a process either as the *initiator* (i.e., from the start of the process) or as the *goal* (i.e., from the end of the process). An *immanent* participant, although present during the process, does not play an active role in it. A *source* participant participates at the start of the process, in contrast to a *product* participant who participates at the end of the process. This twofold distinction, if represented in a lattice, results in four categories: (a) *initiator* (i.e., the generation of a change or state), (b) *resource* (i.e., matter or substrate), (c) *goal* (i.e., purpose), and (d) *essence* (i.e., the true substance). These four categories are chosen by Sowa (2000) to reflect Aristotle's four causes (*aitia*), that is, *efficient cause, material cause, final cause*, and *formal cause*, correspondingly. This distinction is used to further classify thematic roles such as agent, instrument, matter, patient, theme, etc.

The *cause-effect relation* is a relation pertinent to categorization and concept formation. Khoo, Chan, and Niu (2002) focus on the facets of the cause-effect relation and examine the conditions under which this relation occurs. Different types of cause-effect relation may be identified. For example, *general causation* refers to the cause-effect relation between two *types of events*; whereas *singular* or *local causation* refers to the cause-effect relation between two *particular events*. The cause-effect relation is expressed in text in different ways; for example causal links such as "hence" and "because of," or the causative adverb "fatally" explicitly express the cause-effect relation.

Relations are central to many research fields regarding ontology engineering, development of knowledge bases, information retrieval, extraction, and visualization.

The Unified Medical Language System (UMLS) (McCray and Bodenreider, 2002; USNLM, 2007) is a project pursued by the U.S. National Library of Medicine (NLM) that aims at the development of knowledge sources and tools for the biomedical domain. The semantic constituents of UMLS are the Metathesaurus and the Semantic Network. The Metathesaurus includes vocabularies from the biomedical domain, whereas the Semantic Network manages the heterogeneity caused by the different vocabularies by providing a semantically coherent structure and the general categories to which Metathesaurus concepts are assigned. The UMLS Semantic Network is a semantic network consisting of nodes, which correspond to 134 semantic types or categories and 54 links, which, in their turn, correspond to the relations between the 134 semantic types (McCray and Bodenreider, 2002; USNLM, 2007). The Semantic Network distinguishes two main categories for semantic types, (a) entities and (b) events, and two main categories for relations, (a) is-a and (b) nonhierarchical associative relations. Entities are further categorized into physical objects and conceptual entities, whereas events include the subtypes of activities and phenomena or processes. Nonhierarchical associative relations are further distinguished into physical, spatial, functional, temporal, and conceptual relations. Semantic types apply to the biomedical domain, whereas relations may be implemented to other domains as well.

WordNet (Miller et al., 1993) is an online lexical database, which represents lexical concepts as synonym sets falling under one of the following five categories: nouns, verbs, adjectives, and adverbs. WordNet uses synonymy, antonymy, hyponymy, and meronymy as the main semantic relations for the association of synonym sets. Morphological relations between word forms are also represented. WordNet has motivated research concerning semantic relations for natural language processing and information retrieval. Pennacchiotti and Pantel (2006) describe two algorithms for the automatic attachment of binary semantic relations (such as part-of and cause-effect) to WordNet. Hearst (1998) proposes a method for automatically discovering semantic relations through lexico-syntactic patterns in large text corpora.

From the knowledge representation perspective, semantic relations are also considered to be an issue of primary importance especially for ontology development, comparison, and integration. Hovy (2002) presents an approach to compare semantic relations and their underlying ontologies based on a hierarchy of features whose top level is comprised of *form*, *content*, and *usage*. Guarino and Welty (2002) use the notions of identity, unity, and essence, which ontologically constrain the subsumption relation in order to reason about its ontological foundation. Jouis (2002) proposed a specification system for the association of logical properties with relations. This system is based on the following set of primitives: (a) types, (b) relations, and (c) properties. Relations are described by their functional type, their algebraic properties, and their relations to other entities within the same context.

Relations also play a central role in information retrieval and extraction. Evens (2002) reports on the progress in using thesaurus relations for information retrieval. She cites several studies that propelled the use of lexical relations such as hyponymy, synonymy, antonymy, meronymy, functional, attributive, spatial, etc. and explores their usefulness in information retrieval. Khoo and Myaeng (2002) focus on the automatic identification of semantic relations in text for different purposes, such as information retrieval (e.g., relation matching), information extraction, construction of relational thesauri, and other natural language processing applications. In proportion to the computation of feature-based similarity, Turney (2006) computes relational similarity, that is, similarity between relations. The method used is called Latent Relational Analysis (LRA) and may be applied in fields such as information extraction, word sense disambiguation, and information retrieval. Relational similarity specifies the analogy between word pairs, while attributional similarity defines similarity between words. Hetzler (2002) provides examples of implementing visualization techniques for the representation and exploration of various relations, such as hierarchical, temporal, cause/effect, similarity, etc., in order to ease understanding and communication.

REFERENCES

Allwood, J. 1999. Semantics as meaning determination with semantic-epistemic operations. In *Cognitive Semantics: Meaning and Cognition*, ed. J. Allwood and P. Gärdenfors, 1–17. Amsterdam: John Benjamins Publishing.

Armstrong, S., L. Gleitman, and H. Gleitman. 1999. What some concepts might not be. In *Concepts: Core readings*, ed. E. Margolis and S. Laurence, 225–259. Cambridge, MA: MIT Press (originally published in 1983, *Cognition* 13: 263–308).

Bean, C. A., R. Green, and S. H. Myaeng. 2002. Introduction. In *The semantics of relationships: An interdisciplinary perspective*, ed. R. Green, C. A. Bean, and S. H. Myaeng. Dordrecht: Kluwer Academic Publishers, vii–xvi.

Bigelow, J. 2000. *Semantic properties*. Victoria, Australia: Monash University. wysiwyg://body.11/http:www.arts.m...il/department/bigelow/semantic.html (accessed 10 June 2005).

Chaffin, R., D. J. Herrmann, and M. Winston. 1988. An empirical taxonomy of part-whole relations: Effects of part-whole type on relation identification. *Language and Cognitive Processes* 3(1): 17–48.

Cruse, D. A. 2002. Hyponymy and its varieties. In *The semantics of relationships: An inter-disciplinary perspective*, ed. R. Green, C. A. Bean, and S. H. Myaeng, 3–21. Dordrecht: Kluwer Academic Publishers.

Davidsson, P. 1996. Autonomous agents and the concept of concepts. Ph.D. thesis, Department of Computer Science, Lund University, Sweden.

Denkel, A. 1996. *Object and property*. Cambridge, UK: Cambridge University Press.

Devitt, M. 1996. *Coming to our senses: A naturalistic program for semantic localism*, Cambridge, MA: Cambridge University Press.

Elder, C. 1998. Essential properties and coinciding objects. *Philosophy and Phenomenological Research* 58: 317–331.

Evens, M. 2002. Thesaural relations in information retrieval. In *The semantics of relationships: An interdisciplinary perspective*, ed. R. Green, C. A. Bean, and S. H. Myaeng, 143–160. Dordrecht: Kluwer Academic Publishers.

Fellbaum, C. 1990. English verbs as a semantic net. *International Journal of Lexicography* 3: 270–301.

Fodor, J. 1990. Information and representation. In *Information, language, and cognition*, ed. P. Hanson, 175–190. Vancouver Studies in Cognitive Science, Vol. 1. Vancouver: University of British Columbia Press.

Fodor, J., M. Garett, E. Walker, and C. Parkes. 1980. Against definitions. *Cognition* 8: 263–367.

Gerstl, P. and S. Pribbenow. 1995. Midwinters, end games and body parts: A classification of part-whole relations. *International Journal of Human-Computer Studies* 43: 865–889.

Guarino, N. and C. Welty. 2000a. A formal ontology of properties. In *Proc. EKAW-2000: The 12th International Conference on Knowledge Engineering and Knowledge Management*, ed. R. Dieng and O. Corby, 97–112. Lecture Notes on Computer Science, Vol. 1937. Berlin: Springer Verlag.

Guarino, N. and C. Welty. 2000b. Ontological analysis of taxonomic relationships. In *Proc. ER-2000: The 19th International Conference on Conceptual Modeling*, ed. A. Laender and V. Storey, 210–224. Lecture Notes on Computer Science, Vol. 1920. Berlin: Springer-Verlag. http://www.cs.vassar.edu/faculty/welty/papers/er2000/LADSEB05-2000.pdf (accessed 1 May 2007).

Guarino, N. and C. Welty. 2002. Identity and subsumption. In *The semantics of relationships: An interdisciplinary perspective*, ed. R. Green, C. A. Bean, and S. H. Myaeng, 111–126. Dordrecht: Kluwer Academic Publishers.

Hearst, M. A. 1998. Automated discovery of WordNet relations. In *WordNet: an electronic lexical database and some of its applications*, ed. C. Fellbaum, 131–151. Cambridge, MA: MIT Press.

Hetzler, B. 2002. Visual analysis and exploration of relationships. In *The semantics of relationships: An interdisciplinary perspective*, ed. R. Green, C. A. Bean, and S. H. Myaeng, 199–217. Dordrecht: Kluwer Academic Publishers.

Hovy, E. 2002. Comparing sets of semantic relations in ontologies. In *The semantics of relationships: An interdisciplinary perspective*, ed. R. Green, C. A. Bean, and S. H. Myaeng, 91–110. Dordrecht: Kluwer Academic Publishers.

Jackendoff, R. 1983. *Semantics and cognition*. Cambridge, MA: MIT Press.

Jackendoff, R. 1989. What is a concept, that a person may grasp it? *Mind and Language* 4: 68–102.

Jackendoff, R. 1993. On Beyond Zebra: The Relation of Linguistic and Visual Information. In *Noam Chomsky: Critical assessments*, ed. C.-P. Otero, 417–442. London: Routledge (originally published in 1987, *Cognition* 26: 89–114).

Jouis, C. 2002. Logic of relationships. In *The semantics of relationships: An interdisciplinary perspective*, ed. R. Green, C. A. Bean, and S. H. Myaeng, 127–140. Dordrecht: Kluwer Academic Publishers.

Khoo, C. and S. H. Myaeng. 2002. Identifying semantic relations in text for information retrieval and information extraction. In *The semantics of relationships: An interdisciplinary perspective*, ed. R. Green, C. A. Bean, and S. H. Myaeng, 161–180. Dordrecht: Kluwer Academic Publishers.

Khoo, C. S. Chan, and Y. Niu. 2002. The many facets of the cause-effect relation. In *The semantics of relationships: An interdisciplinary perspective*, ed. R. Green, C. A. Bean, and S. H. Myaeng, 51–70. Dordrecht: Kluwer Academic Publishers..

Klein, G. O. and B. Smith. 2005. Concept systems and ontologies, Recommendations based on discussions between realist philosophers and ISO/CEN experts concerning the standards addressing "concepts" and related terms, CEN/TC 251/WGII Terminology and Knowledge Bases, WGII/N05-19. http://ontology.buffalo.edu/concepts/ConceptsandOntologies.pdf (accessed 6/3/2007).

Kokla, M. and M. Kavouras. 2002. Theories of concepts in resolving semantic heterogeneities. Paper presented at the 5th AGILE Conference on Geographic Information Science, Palma, Mallorca, Spain, April 25–27. http://ontogeo.ntua.gr/publications/agile2002_kokla.pdf (accessed 1 May 2007).

Lakoff, G. 1986. *Women, fire, and dangerous things: What categories tell us about the nature of thought*. Chicago: University of Chicago Press.

Lakoff, G. 1987. Cognitive models and prototype theory. In *Concepts and conceptual development: Ecological and intellectual factors in categorization*, 63–100, ed. U. Neisser, Cambridge, UK: Cambridge University Press.

Landis, T. Y., D. J. Herrmann, and R. Chaffin. 1987. Development differences in the comprehension of semantic relations. *Zeitschrift für Psychologie*, 195(2): 129–139.

Laurence, S. and E. Margolis. 1999. Concepts and cognitive science. In *Concepts: Core readings*, ed. E. Margolis and S. Laurence, 3–81.Cambridge, MA: MIT Press.

Lloyd, R., D. Patton, and R. Cammack. 1996. Basic-level geographic categories. *Professional Geographer* 48(2): 181–194.

Lowe, E. J. 2001. Recent advances in metaphysics. In *Proc. 2nd International Conference on Formal Ontology in Information Systems*, FOIS 2001, Ogunquit, ME, October 17–19. New York: ACM Press. http://www.cs.vassar.edu/~weltyc/fois/fois-2001/keynote/ (accessed 1 May 2007).

Margolis, E. and S. Laurence. 2006. Concepts. In *Stanford Encyclopedia of Philosophy*, Metaphysics Research Lab, CSLI, Stanford University, Stanford, CA. http://plato.stanford.edu/entries/concepts/ (accessed 1 May 2007).

Mark, D. M., C. Freksa, S. C. Hirtle, R. Lloyd, and B. Tversky. 1999. Cognitive models of geographical space. *International Journal of Geographical Information Science* 13(8): 747–774.

Marr, D. 1982. *Vision: A computational investigation into the human representation and processing of visual information*. New York: W. H. Freeman and Company.

McCray, A. and O. Bodenreider. 2002. A conceptual framework for the biomedical domain. In *The semantics of relationships: An interdisciplinary perspective*, ed. R. Green, C. A. Bean, and S. H. Myaeng, 181–198. Dordrecht: Kluwer Academic Publishers.

Medin, D. L. and L. J. Rips. 2005. Concepts and categories: Memory, meaning, and metaphysics, concepts and categorization. In *The Cambridge Handbook of Thinking and Reasoning,* ed. K. J. Holyoak and R. G. Morrison, 37–72. Cambridge, UK: Cambridge University Press.

Medin, D. L., E. B. Lynch, and K. O. Solomon. 2000. Are there kinds of concepts? *Annual Review of Psychology* 51: 121–147.

Mennis, J. L. 2003. Derivation and implementation of a semantic GIS data model informed by principles of cognition. *Computers, Environment, and Urban Systems* 27: 455–479.

Mennis, J. L., D. J. Peuquet, and L. Qian. 2000. A conceptual framework for incorporating cognitive principles into geographic database representation. *International Journal of Geographic Information Science* 14(6): 501–520.

Miller, G. A. and P. N. Johnson-Laird. 1976. *Language and Perception.* Cambridge, MA: Harvard University Press.

Miller, G. A., R. Beckwith, C. Fellbaum, D. Gross, and K. Miller. 1993. *Introduction to Word-Net: An on-line lexical database.* ftp://ftp.cogsci.princeton.edu/pub/wordnet/5papers. pdf (accessed 1 May 2007).

Murphy, G. and D. Medin. 1999. The role of theories in conceptual coherence. In *Concepts: Core readings,* 225–259, ed. E. Margolis and S. Laurence. Cambridge, MA: MIT Press (originally published in 1985, *Psychological Review* 92: 289–316).

Nosofsky, R. M. and S. R. Zaki. 2002. Exemplar and prototype models revisited: response strategies, selective attention, and stimulus generalization. *Journal of Experimental Psychology: Learning, Memory, and Cognition* 28(5): 924–940.

Pennacchiotti, M. and P. Pantel. 2006. Ontologizing semantic relations. In *Proc. Conference on Computational Linguistics/Association for Computational Linguistics COLING/ ACL-06,* Sydney, Australia, 793–800.

Pribbenow, S. 2002. Meronymic relationships: From classical mereology to complex part-whole relations. In *The semantics of relationships: An interdisciplinary perspective,* ed. R. Green, C. A. Bean, and S. H. Myaeng, 35–50. Dordrecht: Kluwer Academic Publishers.

Rey, G. 1999. Concepts and stereotypes. In *Concepts: Core readings,* ed. E. Margolis and S. Laurence, 279–299. Cambridge, MA: MIT Press (originally published in 1983, *Cognition* 15: 237–262).

Rosch, E. 1978. Principles of categorization, In *Cognition and categorization,* 27–48, ed. E. Rosch and B. Lloyd. Hillsdale, NJ: Lawrence Erlbaum.

Smith, B. 2004. Beyond concepts: Ontology as reality representation. In *Proc. International Conference on Formal Ontology and Information Systems (FOIS 2004),* November 4–6, 2004, ed. A. Varzi and L. Vieu, Turin.

Smith, B. and D. Mark. 1999. Ontology with human subjects testing: An empirical investigation of geographic categories. *American Journal of Economics and Sociology* 58: 245–272.

Smith, B. and D. Mark. 2001. Geographical categories: An ontological investigation. *International Journal of Geographical Information Science* 15(7): 591–612.

Smith, E. and D. Medin. 1981. *Categories and concepts.* Cambridge, MA: Harvard University Press.

Smith, E. E. 1988. *Concepts and thought: The psychology of human thought.* Cambridge, UK: Cambridge University Press.

Smith, E. E., D. N. Osherson, L. J. Rips, and M. Keane. 1988. Combining prototypes: A selective modification model. *Cognitive Science* 12(4): 485–527.

Sowa, J. F., 2000. *Knowledge representation: Logical, philosophical, and computational foundations.* Pacific Grove, CA: Brooks Cole Publishing.

Storey, V. C. 1993. Understanding semantic relationships. *VLDB Journal* 2: 455–488.

Strawson, P. F. 1959. *Individuals: An essay in descriptive metaphysics.* Garden City, NY: Doubleday.

Swoyer, C. 2000. Properties. In *Stanford encyclopedia of philosophy.* http://plato.stanford. edu/entries/properties/ (accessed 1 May 2007).

Swoyer, C. 2006. Conceptualism. In *Universals, Concepts and Qualities,* 127–154, ed. P. F. Strawson and A. Chakrabarti. Burlington, VT: Ashgate Publishing.

Turney, P. D. 2006. Similarity of semantic relations. *Computational Linguistics* 32(3): 379–416.

Tversky, B. and K. Hemenway. 1983. Categories of scenes. *Cognitive Psychology* 15: 121–149.

Usery, E. L. 2003. Multidimensional representation of geographic features. Technical Report, U.S. Geological Survey (USGS). http://carto-research.er.usgs.gov/multi-dimension/pdf/usery.996.pdf (accessed 1 May 2007).

USNLM (United States National Library of Medicine). 2007. Unified Medical Language System. http://www.nlm.nih.gov/research/umls/ (accessed 1 May 2007).

Wierzbicka, A. 1996. *Semantics: Primes and universals.* Oxford: Oxford University Press.

Winston, M. E., R. Chaffin, and D. Herrmann. 1987. A taxonomy of part-whole relations. *Cognitive Science* 11: 417–444.

Wittgenstein, L. 1953. *Philosophical investigations.* Englewood Cliffs, NJ: Prentice-Hall.

6 Semantics

6.1 INTRODUCTION

Semantics (also called semasiology) is defined as the study of meaning. The terms originate from derivatives of the Greek verb *semaino* ("to mean" or "to signify"); "semantics" is derived from *semantikos* ("significant"), "semasiology" is derived from *semasia* ("signification") + logos ("account") (*Encyclopædia Britannica*, 2006). Meaning is defined as "the customary significance attached to the use of a word, phrase, or sentence, including both its literal sense and its emotive associations" (*Free On-Line Dictionary of Philosophy*, 2002). Meaning should not be confused with knowledge, although the distinction between them is not always easy to draw. More specifically, dictionaries are supposed to convey the meaning of a term, whereas encyclopedias include scientific and technical knowledge (Wierzbicka, 1996).

Semantics is often contrasted with syntax and pragmatics. *Syntax* (also called syntactics) is the study of the rules and relations between words and other units in a language that determine their arrangement and combination within a sentence. According to certain linguists, syntax and semantics of natural languages are closely interrelated. *Pragmatics* is the study of the rules that determine how different contexts affect the communication of meaning.

Semantics is central to linguistics. Language and meaning are interconnected, because "language is an instrument for conveying meaning" and can only be properly understood and studied under the perspective of this function (Wierzbicka, 1996, 3). However, linguists, and specifically semanticists, face the challenge that they have to use language in order to explain language (Lyons, 1977).

Semantics is further distinguished into the *realistic* and the *cognitive* approach (Gärdenfors, 1999) according to what kinds of entities are the meanings of expressions. The *realistic* approach holds that the meaning of an expression lies in the world, whereas according to the cognitive approach the meaning of an expression coincides with mental entities (see also Section 5.1). Realistic semantics is further categorized into *extensional* and *intensional* semantics (Gärdenfors 1999). *Extensional semantics* associates the constituents of a language L (e.g., terms, predicates) with objects and relations among them in the world. Its purpose is to identify the truth conditions for the sentences in L; thus, meaning is considered to be independent of human perception. *Intensional semantics* associates the constituents of a language L with a set of possible worlds instead of a single one. On the other hand, cognitive semantics focuses on the relation among language, meaning, and cognition. According to the cognitive approach to semantics, the meaning of an expression is mental; thus, semantics provides the association of lexical expressions with mental entities (Jackendoff, 1983; Lakoff, 1987; Talmy, 1988). Gärdenfors (1999) identifies six main tenets of cognitive semantics:

1. Meaning is in the head and not in the world, that is, meaning is a conceptualization in a cognitive model in contrast to the realist view, which considers meaning to be truth conditions in possible worlds.
2. Meaning depends on perception to a certain degree, as perceptual mechanisms determine our cognitive models.
3. Meaning is represented by semantic elements, which are based on spatial or topological constructs.
4. Cognitive models are fundamentally based on image schemata, such as container, source-goal, link, etc., which are transformed by metaphors and metonymies.
5. Semantics is primary to syntax (existed before language was completely developed) and in some degree sets restrictions on it.
6. Concepts exhibit prototypical structure, in contrast to the classical position based on necessary and sufficient conditions.

6.2 MEANING

Aristotle in his work "On Interpretation" was the first to draw the distinction among words, as symbols, experiences (pathēmata), and objects (Sowa, 2000). Ogden and Richards (1923) illustrated the relation among a *concept* (*thought* or *reference*), a *symbol*, and an *object* (*referent*) using a triangle, called the *meaning triangle*, or *semantic triangle*, or even *semiotic triangle* (Figure 6.1). On the lower right there is the object or referent (e.g., a forest), on the top there is the concept of the object and the thoughts it may give rise to, and on the lower left there is the symbol indicating the concept (e.g., the word "forest"). The referent and the reference are connected with a solid line to indicate the direct relation between them; this also holds for the relation between the reference and the symbol, which is represented with a solid line as well. On the contrary, there is no direct relation between the symbol and the referent, as the dashed line indicates. This is the key point denoted by the meaning triangle, that is, symbols do not refer to objects directly, but only through concepts.

Richards (cited in Griffin, 1997) adopted a conception of meaning based on the distinction between signs and symbols. A sign is something we come upon, which, however, stands for something else, for example, a kiss is a sign of affection. Symbols, like words, are a special kind of sign, because there is no natural correlation between a symbol and the object it represents, for example, there is neither visual nor sound correlation between the word "cry" and a burst of tears. Because words bear no inherent meaning, they acquire meaning in the context they are used; therefore, different contexts entail different meanings. The notion of context is used broadly to refer not only to the linguistic but also to the semantic sense (see also Section 7.2).

The study of meaning, called semantic theory, seeks answers to two fundamental questions:

1. What is meaning?
2. How can it be described and represented?

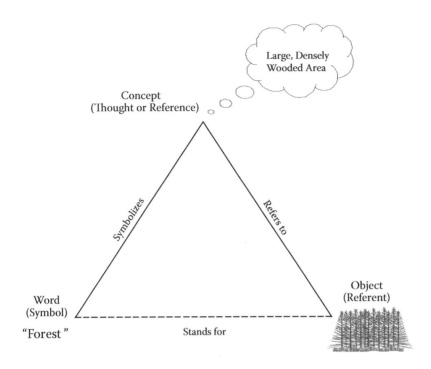

FIGURE 6.1 The meaning triangle. (Modified from Ogden and Richards, 1923).

The first question constitutes a research subject in different disciplines such as philosophy, linguistics, psychology, semiotics, and communication theory, which leads to the formulation of various theories of meaning. The second question is mainly a subject of linguistics and mostly concerns methods of semantic description and representation.

Zlatev (2002) presents an analysis of the notion of meaning as treated by different disciplines. In philosophy and linguistics, meaning is a fundamental notion arousing controversy over whether its ontological status is abstract, physical, mental, or social. More specifically, in philosophy, theories of meaning study "the conditions under which an expression comes to have internal significance and external reference" (*Free On-Line Dictionary of Philosophy*, 2002). The study of meaning in linguistics is closely interrelated to the problem of semantic complexity (Wierzbicka, 1996). Most concepts are considered to be complex since they can be decomposed into simpler ones. In that sense, the meaning of a word can be identified in terms of the arrangement of the simple concepts that comprise it (Wierzbicka, 1996). A contrasting approach towards the notion of meaning considers that meaning is by nature indefinable due to its fuzziness and to the difficulty in defining it (Fodor et al., 1980; Lakoff and Johnson, 1980; Chomsky, 1987). Psychology deals with meaning as being a property of the mind, in contrast to social sciences, where meaning is a result of the "collective consciousness of a social group" (Zlatev, 2002, 256). For semiotics, the subject under debate is whether the production of meaning is a matter of the universe, of life, or of human societies. Artificial intelligence on the other hand, adopts a more superficial attitude towards meaning, considering it superfluous or a subject of external convention and specification (Wierzbicka, 1996).

These disciplines study the notion of meaning from different perspectives (e.g., meaning as use, as concepts, as truth conditions, etc.) leading to corresponding meaning theories. *Denotative or referential theories* explain the meaning of an expression in terms of the objects it denotes. *Use theories* explain the meaning of an expression in terms of its use. *Ideational* or *conceptualist theories* hold that meaning is a mental representation corresponding to ideas, thoughts, or concepts. *Pragmatic theories* explain the meaning of an expression in terms of the result of its implementation. *Truth-conditional theories* interpret the meaning of an expression in terms of the conditions under which it is true.

Lyons (1977) identifies three kinds of meaning conveyed by language: (a) *descriptive*, (b) *social*, and (c) *expressive*. *Descriptive* (also referential, cognitive, propositional, ideational, and designative) meaning is factual, because "it can be explicitly asserted or denied" and in certain cases "objectively verified" (50). *Expressive* meaning depends on the distinctive features of the speaker. *Social* meaning supports the establishment and preservation of social relations. He draws attention to the sense of meaning related to the notions of intention and significance, which are closely associated with the notion of communication, that is, words and sentences acquire meaning with regard to what the speakers of the language mean by using the specific words or sentences (Lyons, 1977).

Zlatev (2002) proposes a theory of meaning based on a biological and sociocultural notion of value. There are two basic tenets of this theory. The first one holds that "all living systems and only living systems are capable of meaning" (257). The second assumes that meaning systems are organized into a hierarchy, which is both evolutionary and epigenetic, that is, subsequent levels presuppose and integrate the previous ones. "Meaning (M) is the relation between an organism (O) and its physical and cultural environment (E), determined by the value (V) of E for O" (258). According to this theory, meaning systems are classified into: (a) cue-based, (b) associational, (c) mimetic, and (d) symbolic. *Cue-based meaning systems* are those possessed by simpler living creatures without central nervous systems (e.g., bacteria, yeast, plants, etc.). For those creatures, the perception of and the reaction to the environment is performed through a set of cues, for example, the need for water and sunrays for plants. *Associational-based meaning systems* concern animals with ability to act, assess their actions, and learn through the development of associations between the environment and their actions. *Mimetic meaning systems* possessed by chimpanzees concern their ability of sign use and communication with humans and among themselves. *Symbolic meaning systems* refer to the human ability for "language, reflection, and cultural variation" (Zlatev, 2002, 269).

6.3   WIERZBICKA'S SEMANTIC AND LEXICAL UNIVERSALS

In order to provide a rigorous account of meaning, Wierzbicka (1996) introduced the notion of *semantic primitives* (or *semantic primes*) to describe the set of fundamental elements that are used to define the meaning of words or any other meanings. These elements cannot be defined themselves; thus, they are considered to be "indefinibilia" (indefinables). This semantic theory is based on the assumption (traced back to Aristotle) that there is an "absolute order of understanding" (10) in meaning acquisition. Some words are semantically more fundamental than others and are the first to be understood in order to capture the meaning of more complex words. For example, one cannot grasp the concept of "information" without first perceiving the concept of "knowing." The definition of a set of primitives, which are used to define all other meanings, contributes to the avoidance of circular definitions. Research focuses on the optimal set of such primitives from the point of view of providing better understanding since the task of these primitives is to disaggregate "complex and obscure meanings in terms of simple and self explanatory ones." Wierzbicka (1996, 11) is based on two hypotheses. The first one states that a set of semantic primitives can be identified on the basis of in-depth analysis of natural languages. The second one states that such a set exhibits universality among different natural languages, since semantic primitives are demonstrations of basic, innate concepts. This hypothesis was confirmed by empirical studies of different languages (Goddard and Wierzbicka, 1994).

Wierzbicka (1996) recognizes that her universalistic position is bound by two restrictions. First, semantic systems from different languages are "unique and culture-specific" in spite of the existence of universals. Second, the premise that universals are "embodied," that is, lexicalized, does not signify their equal use among different languages. Every language includes words with no counterparts in other languages and makes different semantic distinctions. The set of semantic primitives provides the common ground for the description and comparison of meanings among different languages and cultures. This set of fundamental concepts constitutes a shared core, a language-independent metalanguage, called Natural Semantic Metalanguage (NSM). In contrast to artificial languages for meaning representation, NSM is a self-explanatory system, that is, the building blocks of NSM are the shared core concepts of all languages, thus facilitating its understanding without additional explication.

Wierzbicka (1996) identified fifty-five universal semantic primitives; thirty-seven of them underwent long-lasting empirical studies across a variety of languages (Goddard and Wierzbicka, 1994) and are considered to be better established than recently introduced ones, whose status is considered to be rather unsettled. These universal primitives are organized into main categories such as:

- Substantives (e.g., I, you, something)
- Determiners (e.g., this, other)
- Movement, existence, life (e.g., move, live)
- Actions and events (e.g., do, happen)
- Descriptors (e.g., big, small)
- Time (e.g., when, before, after)
- Space (e.g., where, under, above)
- Partonomy and taxonomy (part-of, kind-of)

Apart from the existence of a set of universal semantic primitives, Wierzbicka (1996) also favors the existence of a universal syntax of meaning, that is, the innate and universal combinations of primitive concepts lead to meaning expression. The universal primitives and the universal syntax of meaning comprise a *"language of thought"* or *"lingua mentalis."*

The theory of universal semantic primitives is related to the investigation of the set of semantic elements (properties and relations) which are used for the rigorous definition of geographic concepts (see Chapter 12). Furthermore, the view that semantic primitives exhibit universality is highly relevant to issues such as ontology integration. In case the definition of concepts is based on a universally conceived set of semantic primitives, these may facilitate the "translation" of semantic information even among different languages, thus contributing to the preservation and communication of meaning.

6.4 MODES OF MEANING

An important distinction drawn in semantics is that between the two modes of meaning: *extension* and *intension*, or *reference* and *sense*, or *denotation* and *connotation*.

The *extension* of a term refers to the class of things to which the term correctly applies. For example, the extension of the term "ship" refers to all the ships that exist or existed or will exist in the future. It includes the things, past, present, and future, if any, which belong, or are described by the term. For example, the term "dinosaur" does have an extension consisting of the dinosaurs which existed in the past. In contrast to dinosaurs, there are other terms such as those referring to legendary or fictional creatures (e.g., sirens, dragons, goblins, etc.) which have zero extension since these creatures never existed in the past, neither exist in the present nor will exist in the future. Extension has two synonyms: *denotation* and *reference*.

The *intension* of a term refers to the set of essential properties (sometimes called defining features or characteristics) common to the individual things to which the term applies. Therefore, the intension of the term ship is "a large boat that carries people or goods." Although there may be two expressions with the same extension but different intensions, the opposite is not possible. This is because intension determines extension: the intension of an expression outlines the properties that are used as a criterion to decide which objects belong to its extension. Furthermore, extension and intension are inversely proportional (Swartz, 1997): as the intension increases, the extension decreases. For example, the addition of the property "red" to the intension of the term "roses" leads to the reduction of the extension to only red roses, thus, excluding pink, white, yellow, etc. roses, which belong to the extension of the more general term "roses."

Reference pertains to the relation between an expression and the things (objects or states-of-affairs) which are represented by the expression. Ogden and Richards (1923) embraced a different thesis; the term *referent* describes the object or state-of-affairs in the world specified by a word or expression, whereas the term *reference* relates to the concept, which acts as mediator between the word or expression and the referent (Section 6.2). Reference is distinguished from *sense*, which usually refers to the meaning of an expression. This distinction is usually exemplified by Frege's famous example about the Morning Star and the Evening Star. Frege underlined that the two expressions "the Morning Star" and "the Evening Star" have the same reference, since they refer to the same planet, but different senses.

Connotation on the other hand, as opposed to *denotation* and as originally introduced by Mill in 1843 (cited in Lyons, 1977), signifies not only a class of individuals but also a property ascribed to these individuals and on account of which they all belong to the specific class. Mill gives the example of the word "white" which indicates white things such as snow, paper, etc. and also constitutes an implication (or connotation) of the property "whiteness." *Denotation* refers to the relation that holds between a vocabulary word and "persons, things, places, properties, processes and activities external to the language-system" (Lyons, 1977, 207). The term *denotata* refers to the members of a class of persons, things, etc., to which the expression correctly applies.

6.5   MEANING AND DEFINITIONS

6.5.1   BASIC NOTIONS OF DEFINITIONS

Definitions have been the center of research and of various debates for centuries leading to different theories about their nature, representation, and conceptual relevance. Locke (cited in Wierzbicka, 1996, 212) stated that "a definition is nothing else but showing the meaning of one word by several other not synonymous terms," or else a definition is the decomposition of the meaning of a word into its constituents, that is, the conveyance of the meaning of a complex word based on indefinable simpler ones. Richards (cited in Griffin, 1997) considered definitions to be symbol substitutions providing a linguistic way to communicate meaning. He indicated some definitional directions to guide the communication of meaning: (a) symbolization, (b) similarity, (c) spatial relations, (d) temporal relations, (e) causation, (f) object of mental state, and (g) legal relations.

Since Plato and Aristotle, the feasibility and usefulness of definitions have been dignified by some and condemned by others. Defenders of definitions claim that they are useful in a variety of tasks, such as understanding and knowledge acquisition, and their development is a difficult but attainable task. On the other hand, there are those who question the usefulness of definitions and their ability to represent meaning. According to this position, meanings cannot always be analyzed into smaller parts, and furthermore, most concepts are not definable due to their vague and indeterminate boundaries.

These speculations have led to the formulation of different theories of definitions, which are closely related to theories of concepts (see also Section 5.2 for a more detailed analysis). The classical theory on definitions holds that a definition signifies the necessary and sufficient conditions for the application of the term being defined. Probabilistic theories of concepts have introduced the view that categories exhibit family resemblances and vague boundaries, and therefore, cannot be represented by classical definitions, that is, formulation of necessary and sufficient conditions. Conceptual Atomism (Fodor et al., 1980) expresses a more extreme view towards definitions, according to which, most concepts are innate, primitive, and therefore, indefinable, drawing evidence from the inadequacy of existent definitions to completely and explicitly describe the meaning of words.

Wierzbicka (1996) states that the aim of the definition is to "capture the invariant aspects of a word's use" (240), called the *semantic invariant*; however, this is a difficult task, which is further complicated by the use of quasi-synonyms and circular definitions. This task presupposes the acceptance of the premise that meaning is determinate. This premise, which is also relevant to the discussions about the feasibility of definitions, has undergone serious criticism reinforced by other considerations regarding the deficiency in the specification of a set of indefinables, by the theory of polysemy, and the confusion of lexical meaning with literary or rhetorical instruments such as metaphor, irony, and hyperbole (Wierzbicka, 1996).

Despite the debates and criticism on definitions, arising primarily from philosophy, cognition, and linguistics, there are many other scientific fields where the preservation and communication of meaning through the use of language can be achieved and reinforced by the agreement on the definition of terms. Definitions are expressions that offer "explicit accounts of the meaning of a word or phrase" as used in different contexts and for different purposes (Kemerling, 2001). They consist of two parts: (a) the *definiendum* and (b) the *definiens*. The *definiendum* is the term or expression that is being defined, whereas the *definiens* is the expression serving to define another word or expression.

Regarding the representation of concepts through definitions, theories of lexical semantics are distinguished into *constructive* and *differential* (Miller et al., 1990). In a *constructive* theory, the representation includes adequate information in order to enable a person or a computer to accurately create the concept. In a *differential* theory, meaning is represented with sufficient information in order to distinguish one concept from another. Modern lexical databases such as WordNet, VerbNet, and FrameNet, are based on the differential theory, allowing the user to distinguish between different words, or between different senses of the same word.

The basic rules for the formulation and evaluation of definitions (Wierzbicka, 1996; Kemerling, 2001; Wilson, 2003) are:

1. Definitions should tell the truth.
2. Definitions should focus on the essential properties of the objects belonging to the extension of the definiendum, that is, they should specify those properties that determine the identity and true nature of the objects.
3. Definitions should be neither too broad nor too narrow in order to include the same things (no more, no less) as the term being defined.
4. Definitions should not be superfluous.
5. Definitions should be affirmative rather than negative.
6. Definitions should be literal.
7. Definitions should not be circular, that is, the term to be defined should not be used in the definiens.
8. Definitions should be mutually consistent with other previously introduced definitions.
9. Definitions should not be ambiguous or vague.
10. Definitions included in standard dictionaries should not contain technical or scientific knowledge.

6.5.2   TYPES OF DEFINITIONS

Definitions may be classified into different types, according to different criteria (Swartz, 1997; Kemerling, 2001; Wilson, 2003) (Figure 6.2). According to the *purpose* or the *function* served, definitions are classified into: (a) lexical, (b) stipulative, (c) précising, (d) theoretical, and (e) persuasive.

Lexical or dictionary definitions explicate the conventional meaning or the common usage of a term within a language. Their goal is to inform about the common usage (or usages) of a term, and also (in cases when specific conditions are satisfied) to regulate common usage. Representative examples of lexical definitions are dictionary definitions.

Stipulative definitions specify the meaning of a novel term, thus, generating a usage that did not exist until then. The goal of this type of definition is to introduce the use of a novel term, which will be subsequently accepted and adopted.

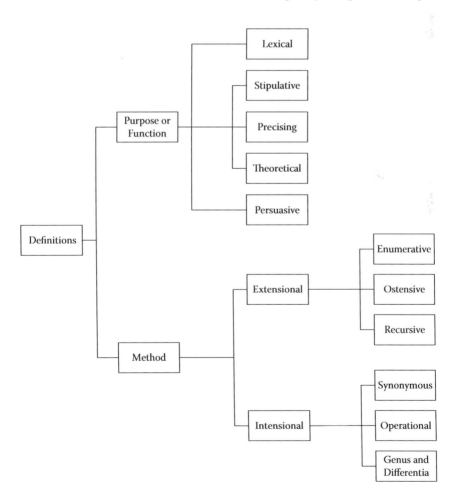

FIGURE 6.2   Classification of definitions according to purpose or function and method of defining.

Stipulative definitions are also used to constrain the meaning in a specific context or for a specific goal.

Précising definitions refine the meaning of a term and eliminate vagueness in its use. They are a combination of lexical and stipulative definitions; they indicate the common usage of a term through lexical definitions and then stipulate a more restricted use, probably in a specific context.

Theoretical definitions explicate the meaning of a term under the framework of a specific theory (e.g., scientific, philosophical, etc.); thus, adopting a theoretical definition implies the acceptance of the underlying theory as well. For example, Newton's definition of "inertia" bears the engagement to Newton's laws of motion (Kemerling, 2001).

Persuasive definitions attempt to perplex lexical meaning and manipulate convictions, so they are not considered to be legitimate.

According to the *method of defining*, definitions are classified into: (a) *extensional (denotative)*, and (b) *intensional (connotative)* (Wilson, 2003). *Extensional definitions* specify the extension of a term or expression, that is, indicate the objects, to which the term or expression pertains. For the majority of terms and expressions extensional definitions are not easy to formulate. First of all, there are numerous objects that belong to the extension of most terms or expressions. The opposite case also holds, that is, some terms have no objects in their extension (e.g., unicorn). Because two terms may have the same extensions but different intensions, extensional definitions may not allow the classification of new instances, that is, it may not be possible to decide whether a new instance does or does not belong to the extension of a term or expression. *Enumerative, ostensive*, and *recursive* definitions are special types of extensional definitions.

Enumerative definitions list the complete extension of a term. This is done by indicating either the objects or the subclasses (also called *definition by subclass*) which belong to the extension of the term. An example of the first case is to define the expression "states of the USA" by enumerating all fifty states of the USA. An example of the second case is to define "citrus" by indicating its subclasses: orange, mandarin, lemon, lime, grapefruit, kumquat, pomelo, citron, and citrange. In some cases, the requirement for a complete listing of the extension is loosened and a partial listing of the extension is provided. For example, the former definition of citrus could be transformed so as to include a partial listing of the most popular citrus fruits, that is, citrus: orange, mandarin, lemon, lime, grapefruit, and the like. *Ostensive definitions* specify the meaning of a term or expression by pointing to an object that belongs to its extension. For example, an ostensive definition of "river" consists of pointing to a river. *Recursive (or inductive) definitions* specify the definiendum in terms of itself, and are very common in mathematics, logic, and computer science.

In contrast to extensional definitions, *intensional definitions* express the meaning of the definiendum by determining its intension, that is, the properties of the objects by virtue of which the term or expression pertains to those objects. There are different types of intensional definitions (Wilson, 2003). *Synonymous definitions* define a term with a synonym, that is, a word or phrase that has the same or nearly the same meaning with the one that is being defined. *Operational definitions* describe an operation or a series of operations that can be performed in order to decide whether the definiendum pertains to a specific instance. Using operational definitions, fundamental scientific concepts (e.g., length and time) are defined on the basis of techniques or operations for measuring them, instead of essential properties, which imply a commitment to a specific theory. However, operational definitions do not provide an objective basis for the comparison and evaluation of measurement techniques and results and hence they are abandoned (Swartz, 1997). *Definitions by genus and differentia* explain the meaning of a term using a more general category or kind of objects (genus) pursued by the specification of those essential properties (*differentia*) that differentiate the members of the definiendum from other objects of the same genus. For example, the WordNet definition "apple: fruit with red or yellow or green skin and sweet with tart crisp whitish flesh," "fruit" is the genus, whereas the other properties (e.g., red or yellow or green skin, crisp, whitish flesh, etc.), which differentiate apples from other fruits, are the differentia.

6.6   GEOSPATIAL SEMANTICS

Geospatial semantics is the focus of interdisciplinary research, that is, philosophy, linguistics, cognition, computer science, GIScience, etc. The semantics of geographic objects have distinct characteristics and can be analyzed into different dimensions, as outlined in Sections 1.2 to 1.4.

Spatial meaning is a focal point of cognitive linguistic research. According to Zlatev (2007), there are two reasons why space has attracted such attention. The first reason concerns the universality of space relative to human experience. Spatial cognition is considered to be central to our thinking and functioning in our physical and social environment. The second reason is that space is a concrete, directly perceptible domain, which may be used in metaphorical mappings for the interpretation of more abstract domains. Metaphor is a cognitive phenomenon, which serves to conceptualize and understand an abstract domain in terms of a more tangible one (Engberg-Pedersen, 1999). Engberg-Pedersen (1999) focuses on the distinction between space and time and examines the claim that there is metaphorical mapping between the two domains. The conclusion drawn from this study is that space and time are not cognitively separate domains but are conceptualized at different levels giving rise to space-time as well as time-space metaphors.

Levinson (2003) studies the relation between language and spatial cognition, and especially *frames of reference*, that is, spatial coordinate systems in language and cognition. In contrast to the view that spatial concepts and relations are innate, Levinson (2003) holds that spatial thinking is largely dependent on culture and especially language; thus, critical language differences induce differences in the conceptualization of spatial relations. He embraces the notion of frames of reference to show that languages impose different semantic parameters in the categorization of spatial relations. Levinson's studies revealed both one universal as well as many cultural characteristics. The universality lies in the existence of three main types of frames of reference used by languages: (a) intrinsic frames of reference use an object-centered coordinate system, (b) relative frames of reference use a coordinate system based on the viewer–speaker, and (c) absolute frames of reference are formed independently of the viewer–speaker but rely on fixed "cardinal directions." Cultural variation is exhibited by the selection and instantiation of the type of reference used in different occasions.

Zlatev (2007) identifies seven spatial categories, which seem to be universally accepted, although they may be described with different terms and definitions. These are:

1. *Trajector* (also called *figure* or *referent*): an entity whose location is relevant, for example, "the *train* is at the station."
2. *Landmark* (also called *ground* or *relatum*): an entity with reference to which the location or trajectory of the trajector is determined, for example, "go to the church."
3. *Frame of reference and viewpoint*: a reference point or coordinate systems, for example, "the store is behind the station."
4. *Region* (also called *conformation*): "a configuration of space" relative to a landmark, a region that relates the trajector and the landmark, for example, "the schoolyard."
5. *Path*: a trajectory of motion (actual or imaginary) relative to a landmark, for example, "the path along the river."
6. *Direction*: a vector along one of the axes of a frame of reference, for example, "the boat sailed to the north."
7. *Motion*: change in position in relation to time, for example, "the men run through the door."

The discussions about geospatial semantics revolve around issues such as the definition, representation, and formalization of the meaning of geospatial concepts, and the probable differences among geospatial and other concepts (which translates to the popular expression of "whether there is anything special about spatial"). Overall, research on geospatial semantics is still primitive relative to the development of a profound and rigorous theory. Research is primarily focused on ontologies to define the semantics of geospatial data, as explicated in Section 4.6.

With the aim of providing a formalization of the semantics of geographic information, Kuhn (2002) applies the notion of conceptual integration, used in cognitive linguistics, to study semantic mapping between conceptual spaces. This notion is shown in a case study involving the association of the conceptual spaces "house" and "boat" (which comprise the source domains) to form the categories "houseboat" and "boathouse" (which comprise the target domains). The approach attempts to enrich an existing ontology (WordNet) using functions (affordances such as sheltering and transferring) and image schemata (i.e., the container, the surface, the path, and the contact schemata) at the higher levels of the WordNet hierarchy, as well as allowing multiple instead of single inheritance. Semantic mappings are formalized as *morphisms* among theories of conceptual spaces (i.e., from theories of house and boat to those of houseboat and boathouse).

The notion of semantic reference systems is also applied to the formalization of thematic data contained in Geographical Information Systems analogously to spatial and temporal reference systems, which describe spatial and temporal data respectively (Kuhn, 2003, 2005). Semantic reference systems may account for the meaning of thematic data and allow information translation and integration among information communities. Although ontologies comprise a core constituent of semantic reference systems, the latter also specify mappings between ontologies and a formalism to support their computation (Kuhn 2003). A semantic reference system consists of two parts, a semantic datum and a reference frame, and supports three processes: semantic referencing, projection, and transformation (Kuhn and Raubal, 2003). A semantic datum is defined as the mechanism for grounding (interpreting) the fundamental terms outside the reference system, that is, by pointing to other ontologies or to the formal semantics of a programming language. A semantic reference frame is a top-level ontology, which acts as a formal, application-independent framework for semantic referencing across application domains. Semantic referencing is the process of associating the terms of a data model to those of the semantic reference frame. Semantic projection is the process of providing mappings from a more complex to a simpler semantic space. Semantic transformation is the change from one reference system (datum, reference frame, and projection) to another.

These notions are implemented in a characteristic example, involving the attachment of semantics to a graph data model for vehicle navigation (Kuhn and Raubal, 2003). For the purposes of this example, the semantics of Hugs (i.e., a dialect of Haskell functional language) provides the semantic datum. The reference frame is implemented by the use of the basic concepts "naming" and "location," and the generic concepts "links," "paths," and "surfaces." Semantic referencing is established by the specification of axioms relating the terms of the data model to those of the reference frame. Semantic projection is exemplified as the simplification of the data model into one that defines only nodes and edges, in contrast to the original one, which also defines road elements and ferry connections. By specifying the necessary axioms, semantic transformation is implemented in order to transform the original system, which supports the distinction between cars and ferries, into another one that does not.

Kokla and Kavouras (2005) developed a method for the extraction and formalization of semantic information from definitions of geographic category terms. Based on specific rules, definitions are decomposed into a set of semantic elements (properties and relations) and values (see also Chapters 12 and 16). Examples of semantic properties are: PURPOSE, AGENT, COVER, LOCATION, SIZE, TIME, FREQUENCY, etc. Examples of semantic relations are: IS-A, IS-PART-OF, SURROUNDNESS, INTERSECTION, CONNECTIVITY, etc. The extraction and formalization of semantic information constitute the basis for the comparison and heterogeneity reconciliation processes.

Karalopoulos, Kokla, and Kavouras (2005) developed an algorithm for the analysis and comparison of geographic concepts' definitions. The method is based on the representation of definitions of geographic concepts using conceptual graphs and the measurement of their semantic similarity; this measurement is based on a modification of the Dice coefficient, which accommodates the special structure and content of conceptual graphs.

Tomai and Kavouras (2004) explore the semantics of geospatial concepts from a linguistic perspective, that is, they identify linguistic terms used by humans to describe their location or motion around space and to identify objects located in space.

Moreover, they investigate the way objects located in space are represented and the descriptions used to communicate information about these spatially located objects.

Cai (2007) follows a context-oriented approach to spatial data semantics. A semantic model consisting of three components is defined: (a) context knowledge, (b) contextualized ontology base, and (c) context-sensitive interpretation. Semantic knowledge is represented using a contextualized geo-ontology, which is defined as "an unambiguous and coherent theory about a piece of geographical reality within a prescribed context" (219). Contexts may refer to domains (e.g., geography) and applications (e.g., travel) at varying levels of generality and are described by schemas, called contextual schemas. Contextual schemas include knowledge, such as: (i) features of the context, (ii) context-specific meanings of concepts, (iii) goals of a context, (iv) actions, and (v) interpretation of new events.

O'Brien (2004) and O'Brien and Gahegan (2005) developed a programmatic framework for representing, manipulating, and reasoning with geographic semantics. The resources of the framework, which are specified by corresponding domain ontologies, are data, methods, and human experts. The framework also incorporates task ontologies, which specify the synthesis of specific information products, and application ontologies. Information products are the results of the interaction of data, methods, and human experts. According to this conceptualization, semantic information is considered as the transformations of geospatial information during the creation of the information product, that is, transformations due to the interactions among data, methods, and human experts. The authors adopt a "light" conception of semantics driven by its purpose to assist the user in inferring meaning for specific tasks. In this framework, geographic semantics play two roles: (1) they identify the appropriate resources (method, data, or human expert) in order to solve a problem, and (2) they measure changes in meaning during data manipulation by methods and human experts.

Sotnykova et al. (2005) propose a method for the integration of spatiotemporal conceptual schemas based on conceptual models and Description Logics. For the description of the semantics of spatiotemporal schemas, a conceptual model, called modeling of application data with spatio-temporal features MADS (Modeling of Application Data with Spatio-temporal features), is used. It represents complex objects and relationships between them described both by attributes and methods and associated with IS-A links. Predefined spatial and temporal abstract data types are used to encompass the spatial and temporal dimensions of objects, relations, and attributes. For example, the attribute "season" adds a temporal characterization to the object "theatre." Relationships are enriched with aggregation, topological, synchronization, and inter-representation semantics. MADS also supports multiple representations in order to incorporate varied requirements in the database design phase as well as multiple levels of detail.

Computer science studies semantic issues relative to integration and interoperability, search and discovery, location-based applications, the Semantic Web, etc. Arpinar et al. (2006) model the thematic, geospatial, and temporal aspects of geographic information in order to provide semantic analytics and discovery capabilities. The Geospatial Semantic Analytics (GSA) framework incorporates geospatial semantics for executing spatial queries using imprecise spatial and temporal information expressed in natural language and other nonmetric descriptions (e.g., near, far, around 5 o'clock, etc.).

Location-based applications are another application of geospatial semantics. They require the interpretation of the trajectory of the moving person. Schlieder (2005) proposes a representational formalism and a reasoning mechanism, called *spatially grounded intentional systems*, in order to express the meaning of the user's behavior in terms of his or her changing intentions. Tryfona and Pfoser (2005) categorize data relevant to Location-Based Services (LBS) according to its semantics and use. They distinguish the following data categories, which are formalized using three corresponding ontologies, that is, the Domain Ontology, the Content Ontology and the Application Ontology (modeled with Protégé ontology editor), and the UML model:

- *Domain data* describe spatial and temporal concepts, which are commonly represented in all LBS applications, that is, position, location, movement, and time. The authors propose a distinction between location and position in order to describe the cases when the actual position is not known (e.g., when the GPS device is shadowed) or insufficient but must be supplemented with information of a surrounding area (e.g., a visit to an archaeological site).
- *Content data* describe information specific to a particular LBS application (e.g., hotels, museums, and historical and transportation information for a tourist LBS) and are further categorized into: (a) descriptive, (b) spatially referenced, and (c) temporally referenced.
- *Application data* describe the user (e.g., interests, preferences, etc.), the device profile (e.g., memory, screen size, etc.), and the services (e.g., for tourists or drivers).

Pike and Gahegan (2007) deal with scientific knowledge representation and communication in the framework of designing a web-based application for capturing and exchanging knowledge representations, called Codex. The goal is to enable the user to evaluate the usefulness and reliability of knowledge representations. Knowledge is considered as the result of a fluid process of interaction between research teams, which involves information capture, manipulation, revision, and reuse. Three components are used to define knowledge or the *perspective* of a particular thinker (or a community): (a) concepts, (b) metadata, and (c) situations. A concept is defined as an abstract category that maintains the same meaning, although it may be described with different terms in different contexts. The authors propose a representation of concepts based on their relationships with other concepts. More specifically, concepts are defined as *values* or a range of values along continuous *dimensions*; a dimension stands for a concept and a value specifies the nature of the relationship. This view is closely related to the notions of affordances (Gibson, 1977) and concept spaces. Metadata describe information independent of a concept's use, such as by whom, where, when, and how the concept was created. A situation describes the process of conceptual manipulation and analysis in order to reach a specific goal.

The development of the Semantic Geospatial Web is another field where the formalization and representation of geospatial semantics is of central importance. This importance is recognized by numerous efforts towards efficient information retrieval and processing, based on its semantics. The World Wide Web Consortium (W3C) Semantic Web Activity (Hendler, Berners-Lee, and Miller, 2002; Shadbolt, Berners-Lee, and Hall, 2006) aims at the provision of a framework for data sharing and reuse across applications and user communities. OGC's Geospatial Semantic Web Interoperability Experiment aims at sharing the meaning of geographic queries among different systems and online services using ontologies and semantic query capabilities. Egenhofer (2002) presents an analysis and formalization of the semantics of a *geospatial request* as the starting point for successful information retrieval from the Semantic Geospatial Web. Roman, Klein, and Skogan (2006) present an overview of the SWING (Semantic Web Services INteroperability for Geospatial Decision Making) project, which explores semantic technologies in order to facilitate geospatial decision making.

The success of such activities may be evaluated according to the degree to which the semantics of the information expressed by the user is matched to that of the available information sources. Egenhofer (2002) identifies two research issues pertaining to the advance of the Semantic Geospatial Web: (a) the specification of a "canonical form" for expressing geospatial queries, and (b) the development of methods for assessing the adequacy of available information sources in executing a specific geospatial query. Fonseca and Sheth (2002) identify three prominent research questions concerning the advance of the Semantic Geospatial Web:

1. Creation and management of geographic ontologies with broad acceptance in order to ensure their adoption
2. Matching geographic concepts in web pages to geographic ontologies using geospatial characteristics for the text interpretation
3. Ontology integration

Geospatial semantics for the Web can be represented using the following (Egenhofer, 2002):

- *Natural language with minimum markup* (e.g., HTML or XHTML) applied by people and documents
- *Simple metadata* (e.g., tags expressed in XML languages) used in interfaces, documents, and search systems
- *Data models* (e.g., RDF) used in interfaces, documents, and search systems
- *Logical (model-theoretic) semantics* (e.g., tags expressed in DAML+OIL) used in interfaces, documents, and search systems

REFERENCES

Arpinar, I. B., A. Sheth, C. Ramakrishnan, E. L. Usery, M. Azami, and M. P. Kwan. 2006. Geospatial ontology development and semantic analytics. *Transactions in GIS* 10(4): 551–575.
Cai, G. 2007. Contextualization of geospatial database semantics for mediating human-GIS dialogues. *Geoinformatica* 11(2): 217–237.
Chomsky, N. 1987. Language in a psychological setting. *Sophia Linguistica* 22: 1–73.

Egenhofer, M. J. 2002. Toward the semantic geospatial web. In *Proc. 10th ACM International Symposium on Advances in Geographic Information Systems*, McLean, VA, 1–4. New York: ACM Press.

Encyclopædia Britannica. 2006. Semantics. http://www.britannica.com/eb/article-9110293 (accessed 8 March 2007).

Engberg-Pedersen, E. 1999. Space and time. In *Cognitive semantics: Meaning and cognition*, ed. J. Allwood and P. Gärdenfors, 131–152. Amsterdam: John Benjamins Publishing.

Fodor, J., M. Garrett, E. Walker, and C. Parkes. 1980. Against definitions. *Cognition* 8: 263–367.

Fonseca, F. and A. Sheth. 2002. The Geospatial Semantic Web. UCGIS White Paper. http://www.ucgis4.org/priorities/research/2002researchagenda.htm (accessed 25 October 2006).

Free On-Line Dictionary of Philosophy (Foldop3.0). 2002. Meaning, edited by the SWIF (Sito Web Italiano per la Filosofia), Terravecchia, G.P. (Chief Editor and Project Coordinator). http://www.swif.it/foldop/dizionario.php?find=meaning (accessed 2 May 2007).

Gärdenfors, P. 1999. Some tenets of cognitive semantics. In *Cognitive semantics: Meaning and cognition*, ed. J. Allwood and P. Gärdenfors, 19–36. Amsterdam: John Benjamins Publishing.

Gibson, J. J. 1977. The theory of affordances. In *Perceiving, acting, and knowing*, ed. R. Shaw and J. Bransford, 67-82. New York: Wiley.

Goddard, C. and A. Wierzbicka. 1994. *Semantic and lexical universals: Theory and empirical findings*. Amsterdam: John Benjamins Publishing.

Griffin, E. 1997. *A first look at communication theory*, 3rd ed. New York: McGraw-Hill.

Hendler, J., T. Berners-Lee, and E. Miller. 2002. Integrating applications on the semantic web. *Journal of the Institute of Electrical Engineers of Japan* 122(10): 676–680.

Jackendoff, R. 1983. *Semantics and cognition*. Cambridge, MA: MIT Press.

Karalopoulos, A., M. Kokla, and M. Kavouras. 2005. Comparing representations of geographic knowledge expressed as conceptual graphs. In *Proc. First International Conference on GeoSpatial Semantics (GeoS 2005)*, ed. M. A. Rodriguez, I. F. Cruz, M. J. Egenhofer, and S. Levashkin, 1–14. Lecture Notes in Computer Science 3799. Berlin: Springer-Verlag.

Kemerling, G. 2001. Definition and meaning. *Philosophy pages*. http://www.philosophypages.com/lg/e05.htm (accessed 5 March 2007).

Kokla, M. and M. Kavouras. 2005. Semantic information in geo-ontologies: Extraction, comparison, and reconciliation. *Journal on Data Semantics* 3: 125–142.

Kuhn, W. 2002. Modeling the semantics of geographic categories through conceptual integration. In *Proc. Second International Conference on Geographic Information Science (GIScience 2002)*, Boulder, CO, September 25–28, 108–118. Lecture Notes in Computer Science 2478. Berlin: Springer-Verlag.

Kuhn, W. 2003. Semantic reference systems. *International Journal of Geographical Information Science* 17(5): 405–409.

Kuhn, W. 2005. Geospatial semantics: why, of what, and how? *Journal on Data Semantics* 3: 1–24.

Kuhn, W. and M. Raubal. 2003. Implementing semantic reference systems. In *Proc. 6th AGILE Conference on Geographic Information Science*, Lyon, France, 63–72. Presses Polytechniques et Universitaires Romandes. http://musil.uni-muenster.de/documents/AGILE_ImplementSemRefSyst.pdf (accessed 2 May 2007).

Lakoff, G. 1987. *Women, fire, and dangerous things: What categories reveal about the mind*. Chicago: Chicago University Press.

Lakoff, G. and M. Johnson. 1980. *Metaphors we live by*. Chicago: Chicago University Press.

Levinson, S. C. 2003. *Space in language and cognition: Explorations in cognitive diversity*. Cambridge, UK: Cambridge University Press.

Lyons, J. 1977. *Semantics*. Cambridge, UK: Cambridge University Press.

Miller, G., R. Beckwith,, C. Fellbaum, D. Gross, and K. Miller. 1990. Introduction to WordNet: An on-line lexical database. *International Journal of Lexicography* 3(4): 235–244.

O'Brien, J. A. 2004. GISmate: A framework for representing, manipulating and reasoning with geographic semantics. Ph.D. thesis, The Pennsylvania State University, The Graduate School, Geography Department.

O'Brien, J. and M. Gahegan. 2005. A knowledge framework for representing, manipulating and reasoning with geographic semantics. In *Developments in spatial data handling, 11th International Symposium on Spatial Data Handling*, ed. P. F. Fisher, 585–598. Berlin: Springer.

Ogden, C. K. and I. A. Richards. 1923. *The meaning of meaning*. London: Routledge & Kegan Paul.

Pike, W. and M. Gahegan. 2007. Beyond ontologies: Toward situated representations of scientific knowledge. *International Journal of Human-Computer Studies*. 65(7): 674–688.

Roman, D., E. Klien, and D. Skogan. 2006. SWING: A semantic web services framework for the geospatial domain. Paper presented at Terra Cognita 2006—Directions to the Geospatial Semantic Web, in conjunction with the Fifth International Semantic Web Conference (ISWC 2006), Athens, GA. http://www.ordnancesurvey.co.uk/oswebsite/partnerships/research/research/TerraCognita_Papers_Presentations/RomanKleinSkogan.pdf (accessed 2 May 2007).

Schlieder C. 2005. Representing the meaning of spatial behavior by spatially grounded intentional systems. In *Proc. First International Conference on GeoSpatial Semantics (GeoS 2005)*, ed. M. A. Rodriguez, I. F. Cruz, M. J. Egenhofer, and S. Levashkin, 30-44. Lecture Notes in Computer Science 3799. Berlin: Springer-Verlag.

Shadbolt, N., T. Berners-Lee, and W. Hall. 2006. The semantic web revisited. *IEEE Intelligent Systems* 21(3): 96–101.

Sotnykova, A., C. Vangenot, N. Cullot, N. Bennacer, and M. A. Aufaure. 2005. Semantic mappings in description logics for spatio-temporal database schema integration. *Journal on Data Semantics* 3: 143–167.

Sowa, J. F. 2000. *Knowledge Representation: Logical, Philosophical and Computational Foundations*, Brooks Cole Publishing Co, Pacific Grove, CA, USA.

Swartz, N., 1997, *Definitions, dictionaries, and meanings.* Department of Philosophy, Simon Fraser University. http://www.sfu.ca/philosophy/swartz/definitn.htm (accessed 2 May 2007).

Talmy, L. 1988. Force dynamics in language and cognition. *Cognitive Science* 12: 49–100.

Tomai, E. and M. Kavouras. 2004. "Where the city sits?" Revealing geospatial semantics in text descriptions. In *Proc. 7th AGILE Conference on Geographic Information Science*, Heraclion, Greece, April 28–May 1. http://www.agile-secretariat.org/Conference/greece2004/papers/3-1-4_Tomai.pdf (accessed 2 May 2007).

Tryfona, N. and D. Pfoser. 2005. Data semantics in location-based services. *Journal on Data Semantics* 3: 168–195.

Wierzbicka, A. 1996. *Semantics: Primes and universals.* Oxford: Oxford University Press.

Wilson, W. K. 2003. *The essentials of logic.* Piscataway, NJ: Research and Education Association.

Zlatev, J. 2002. Meaning = Life (+ Culture): An outline of a biocultural theory of meaning. *Evolution of Communication* 4(2): 255–299.

Zlatev, J. 2007. Spatial semantics. In *Handbook of Cognitive Linguistics*, ed. H. Cuyckens and D. Geeraerts. Oxford: Oxford University Press.

Part 3

Formal Approaches

7 Knowledge Representation Instruments

7.1 KNOWLEDGE REPRESENTATION

Once the problem domain has been identified (see Part 1) and the necessary theoretical framework has been established (see Part 2), it is necessary to employ a sufficient arsenal of formal tools in order to develop practical approaches to semantic integration (see Part 4). The term *knowledge representation instruments* is employed to collectively refer to various representation notions known as tools, technologies, mechanisms, formalisms, structures, approaches, or theories. This is the subject of the present and the subsequent chapters of Part 3. So far in this book, the issues of *Knowledge (K), Representation (R),* and subsequently *Knowledge Representation (KR)* as such have not been rigorously defined but only implicitly touched upon. This is not strange for there are at least different, not to say conflicting, perspectives of K/R/KR. A thorough theoretical discussion on the matter is stimulating and interesting but of questionable practical importance and outside the scope of this book. At the practical level, however, it is important to provide a minimal framework of the most advanced, common, and interesting approaches to KR. The present chapter attempts just that.

KR is an important subject that appears in various fields, such as philosophy, epistemology, cognitive science, linguistics (especially computational linguistics), and artificial intelligence (AI). These different and partial perspectives are not necessarily contradictory but not exactly complementary either. This is because of fundamental unsettled questions about two subissues, the one of *knowledge* and the other of *representation. Knowledge,* introduced in Chapter 2, has received numerous definitions in the literature and has always been a main issue either in philosophical debates or in discussions on the more practical aspects of everyday life. Aristotle's introductory statement in his book "Metaphysics"—*All men by nature desire to know*—sounds clear and profound. Nevertheless, it is the nature of *knowledge,* which can be simple or complex, declarative or procedural, explicit or fuzzy, deterministic or probabilistic, complete or incomplete, generally sound or not, monotonic or nonmonotonic, etc. (Panayiotopoulos, 2001), that makes it hard to define it uniquely.

For most practical purposes, one thing seems to be agreed upon: that *knowledge* involves an important yet complex assortment of various cognitive processes, which, among others, entail *understanding* and the ability of *reasoning.* As such, *knowledge* is often a partial and subjective process, especially at its development. Scientific knowing and institutional knowing may be slightly less subjective than individual knowing. *Representation*, on the other hand, depends not only on the *knowledge* to be represented, but also on the subjective views and interests of the *representor* (Hjørland and Nicolaisen, 2006). Some important aspects in representing knowledge are: (a) which things constitute the knowledge to be represented, (b) the way this representation is formalized (using terms, symbols, or other notation), (c) how inference is supported, and (d) how this representation is implementable in an AI environment. Along these lines, Sowa (2002) deals with KR drawing theories and techniques as regards aspect (a) from the Ontology point of view, as for (b and c) from the Logic point of view, and as for (d) from the Computation point of view.

Loosely speaking, KR is the problem of encoding all various forms of (human) knowledge in such a way that supports reasoning. As Davis, Shrobe, and Szolovits (1993, 29) state, "Representation and reasoning are inextricably intertwined: we cannot talk about one without also, unavoidably, discussing the other." What is more, reasoning does not simply refer to any form of rightful inference but more significantly to *intelligent reasoning* (Davis, Shrobe, and Szolovits, 1993). This goal of KR is also described in B. C. Smith's statement, known as the *KR hypothesis*:

> Any mechanically embodied intelligent process will be comprised of structural ingredients that (a) we as external observers naturally take to represent a propositional account of the knowledge that the overall process exhibits, and (b) independent of such external semantical attribution, play a formal but causal and essential role in engendering the behavior that manifests that knowledge (1982,33).

This KR hypothesis leads to the realization that the only intelligent activity justifying possession of knowledge is reasoning, or reversely put, "intelligence presupposes knowledge" (Kramer and Mylopoulos, 1992, 743). Without getting into the epistemics of KR and the associated alternatives (Hjørland and Nicolaisen, 2006), the following question rises naturally: what can be considered *intelligent reasoning*?

Since Aristotle, philosophers, mathematicians, logicians, psychologists, biologists, economists, AI researchers, and others have proposed various types or variants of reasoning which prove, on the one hand, the tremendous complexity of human intelligence and, on the other hand, the difficulty of simulating that intelligence in machines. For instance, reasoning is used in *proving* or *answering questions* (as in mathematics and mathematical logic) by producing right propositions. It is implemented by using various inference rules. Two prominent examples are (a) the *Modus Ponens** rule, and (b) the *inheritance (generalization†/specialization‡)* rule in concept hierarchies.

Because it has been rather impossible for researchers, at least up to today, to agree upon a widely acceptable formal definition for Knowledge or Knowledge Representation, some have focused their efforts on describing the properties that

---

* *Modus Ponens*: a simple and very common affirmative argument form—if the proposition A → B (i.e., "if A then B") is TRUE, and the proposition A is TRUE, then the proposition B is also TRUE. For instance, "if it is Sunday, the office is closed." "It is Sunday," therefore, "the office is closed."
† The *generalization* rule: if an instance of a class has a property, then every instance of that class, or the class as a whole, has that property.
‡ The *specialization* rule: it is exactly the inverse to generalization. If a class has a property, then all its subclasses or instances have this property.

either a KR or equivalently KR languages should possess. Welty (1995) examines programming languages and their limited satisfaction of the criteria set by the KR hypothesis. Brachman and Levesque (1987) look at reasoning as KR languages computationally implement it, and state that "even simple reasoning capabilities can be computationally hard and lead to serious tractability problems." Efficiency and concurrent updating are implementation concerns at the database level; while expressivity and effective reasoning are concerns at the KR level (Fikes, 1998). Levesque and Brachman (1985), however, have shown that KR expressiveness is intertwined with its computational cost in terms of reasoning; thus, one can trade expressiveness with computational cost and vice versa.

KR languages are formal objects that provide the necessary and appropriate "structural ingredients" mentioned in the KR hypothesis. Woods (1975) describes the properties that a KR language should have as follows: "A KR language must unambiguously represent any interpretation of a sentence (logical adequacy), have a method for translating from natural language to that representation, and must be usable for reasoning" (cited in Welty, 1995, 30) As Welty (1995) pinpoints, this is a simplification of the KR hypothesis; nevertheless, it is included here for it makes a direct reference to natural language as a valuable knowledge source, or a special KR itself.

One of the most interesting and thoughtful descriptions about the properties a KR should have is given by the MIT researchers Davis, Shrobe, and Szolovits (1993). From an AI perspective, their answer lies in five different roles that a representation plays—roles with possibly different demands. These are:

1. A KR is a *surrogate*, namely, an imperfect substitute of a portion of reality and not reality itself.
2. A KR is a *set of ontological commitments*, or in other words, a KR defines the concepts of the portion of the reality being represented and how this model is built.
3. A KR is a *fragmentary theory of intelligent reasoning.* This refers to the important role of KR as an intelligent, yet partial, reasoning mechanism and the set of inferences that could or should be made. Davis, Shrobe, and Szolovits (1993) distinguish two senses as being "fragmentary" and both of them refer to the natural incompleteness of the two processes along the common modeling sequence: *phenomenon* (intelligent reasoning) → *belief* → *representation*. Namely, the two processes are that a belief covers a fraction of the phenomenon and a representation covers a fraction of the belief about the phenomenon.
4. A KR is a *medium for pragmatically efficient computation.* This goes beyond epistemological issues and focuses on the fact that a KR must serve practical purposes and be implemented in a machine; thus, issues of computational complexity and efficiency are involved.
5. A KR is a *medium of human expression* or, in other words, a medium for human communication. Thus, unavoidably, one should not only be concerned about its expressive adequacy but also about its lucidness and neatness.

As Davis, Shrobe, and Szolovits (1993) rightly state, it is important to consider all these roles when implementing a KR language.

Having provided some basic definitions about the nature of KR, a number of important issues are introduced in the remaining sections of this chapter. The first addresses the issue of contextuality of knowledge and how this can be formalized. The second discusses the role of natural language in knowledge representation. The last section refers to advanced formal instruments, which are employed in the representation of knowledge.

7.2 KNOWLEDGE REPRESENTATION AND CONTEXT

7.2.1 CONTEXT

It is widely accepted that Knowledge and Knowledge Representation—as many other notions—are *context* dependent (Bouquet et al., 2003). From this perspective, *context* is conceived as a framework, which to an extent determines how knowledge is acquired or reasoned about. Alfred Korzybski's famous line: "A person does what he does because he sees the world as he sees it"* states more or less the same thing. From a slightly different but not contradicting point of view, *knowledge* and *context* can be examined as two notions with similar natures or characteristics (Pomerol and Brézillon, 2001). It has been widely recognized that context as a point of view is really important, and, as Saffo (1994) forecasted, "It is this avalanche of content that will make context the scarce resource." It is, therefore, worthwhile examining the notion and importance of *context* in the present context.

Various distinctions of the notion of context can be found in the literature. Context is characterized either as "objective" and "metaphysical" or "subjective" and "cognitive" (Penco, 1999), as well as "linguistic" or "semantic" (Sowa, 2000). Dictionaries usually follow the latter distinction and give two major senses for the word context. For example, in the lexical database *WordNet* (Fellbaum, 1998), context is defined as "a discourse that surrounds a language unit and helps to determine its interpretation" or as "the sets of facts or circumstances that surround a situation or event." Both senses suggest that context considerations could help to derive additional information either from a situation or a verbal/written linguistic unit. The ability to disambiguate the meaning of words and clarify some aspects of semantics makes context an important issue in many scientific areas, such as philosophy, cognitive science, linguistics, and artificial intelligence. Furthermore, many applications such as Natural Language Processing (NLP), information retrieval, and information integration use the notion of context to increase their effectiveness. As far as NLP is concerned, context is taken into account to avoid confusion about the meaning of words and supply explanatory information. In information retrieval applications, context is used in order to provide the opportunity of expressing pertinent queries and retrieving the most relevant information. In information integration, context is also used to facilitate interoperability and help the information comparison among different systems.

7.2.2 CONTEXT FORMALIZATION

Context formalization mainly concerns scientists who deal with knowledge representation and reasoning, such as logicians and computer scientists interested in the field

---

* http://www.thisisnotthat.com/sampler/3.html

of artificial intelligence and other related areas. Logicians are concerned about context in terms of its integration to predicate calculus or other forms of logic, its manipulation in combination with logical features, its role in logical deductions, and its contribution to knowledge representation. Computer scientists are mostly interested in how the notion of context can be incorporated and handled in the framework of all the previously mentioned computer applications (NLP, information retrieval, integration, etc). In the following, some approaches to formalizing context are presented.

Peirce laid the foundation for the formalization of the notion of context towards the close of the nineteenth century. He proposed the assertion of contextual information in propositions, in the form of explanatory sentences included in an oval schema (Sowa, 2000; 2003). Moreover, he developed Existential Graphs as a knowledge representation theory that uses and combines contexts in logical propositions.

Sowa's formalization of context was made in the scope of Conceptual Graphs (Chapter 9). Conceptual Graphs are a system of logic based on Peirce's Existential Graphs and semantic networks used in Artificial Intelligence (AI) and computational linguistics (Sowa, 2000). Sowa defined context as "a concept whose referent field contains nested conceptual graphs" (2000, 276–277). Boxes are used to enclose contexts' structured content. Both Peirce's and Sowa's approaches to context can be translated in predicate calculus form.

McCarthy's main motivation in trying to formalize context was to resolve the "problem of generality" (McCarthy, 1987), that is, the lack of generality, from which he considers that the AI domain suffers. In his "Notes on Formalizing Context" (1993), he introduced the expression:

c': ist(c,p),

which stands for: "p is true in the context c, itself asserted in an outer context c'." Upon this basic formula, he proposed the formalization of contexts, relations between contexts, operations like entering and leaving a context, rules for lifting information to outer contexts, etc. According to McCarthy, although some contexts are "poor" objects and can be completely described, some others are "rich," like situations or information associated with a conversation, and, as a result, cannot be completely described.

Guha (1991, 41–42) introduced the notion of "microtheories" to describe contexts corresponding to assumptions and simplifications about the world made in the scope of a specific topic or domain. Each microtheory has its own axioms and vocabulary and therefore, defines appropriate predicates and functions to be used in a specific context. He also defined some specific contexts like the "Problem Solving Contexts," which are related to a particular problem-solving task, and "Discourse Contexts," which correspond to a "very slightly decontextualized representation of the utterances made in a discourse." With regard to logical manipulation, Guha's conception of context formalization is similar to McCarthy's.

Lenat (1998) examined the integration of contexts on the common sense knowledge base Cyc. In Cyc, although each context used to be considered a first-class object constituted of two parts, that is, its assumptions and its content, a more adequate structure for formalizing contexts was proposed later. Contexts were defined as regions in a twelve-dimensional space, the space of assumptions. The dimensions of this space correspond to the most independent parameters along which context may vary, that is, absolute time, type of time, absolute place, type of place, culture, sophistication/security, granularity, epistemology, argument-preference, topic, justification, and anthropacity. Lenat also described a terminology, functions, and predicates to manipulate contexts in a formal way.

7.2.3   CONTEXT IN GEOGRAPHIC APPLICATIONS

In the geospatial domain, context is also of great concern mainly for its semantic, nonlinguistic function. A context can be defined as "the setting, environment, domain in which an entity or topic of interest exists or occurs" (Kokla and Kavouras, 2001, 680). Geographic concepts present a strong context-dependent character since, by nature, they are closely related to parameters like location, time, size, and scale, influencing the context in the scope of which they may be considered (Smith and Mark, 1998). Furthermore, the conceptualization of a geographic reality is often made subjectively, depending on human cognition or on the particular requirements of the application in which it will be used. Therefore, geographic ontologies are also context-dependent because they specify conceptualizations.

Rodríguez and Egenhofer (2004) use the notion of context (i.e., a particular application domain defined by a set of concepts, the actions or operations applying to these concepts, and the semantic relations between concepts) to reinforce the similarity measure applied for concept comparison. The actions are linguistically represented by verbs associated with nouns, which depict the concepts of the similarity model (e.g., "find a hotel," or "buy a land parcel").

To accommodate the possibility of some distinguishing features being more relevant than others in an application domain, two different approaches are used. The first approach is based on the feature's variability, that is, the degree of its informativeness. Features shared by many concepts in the application domain are less relevant than others shared by fewer concepts. For example, in a transportation application where the objective is to find the appropriate transportation means for a passenger with specific demands (e.g., direct access to an airport, ability to carry bulky luggage, a bicycle, or a pet, etc.), the relevant features are those that distinguish those transportation means that meet the passenger's needs (e.g., specific regulations for carriage, etc.). Different weights are assigned to parts, functions, and attributes, according to their variability, such that the type of distinguishing feature more able to differentiate concepts is assigned a greater weight. In that case, the variability of a type of feature is a function of the inverse frequency of the feature's occurrence in the concept's definition.

The second approach is based on the feature's commonality, that is, the degree to which the feature describes the application domain. Features shared by many concepts in the domain are more relevant and are assigned greater weights. For example, in a transportation application where the objective is to find a transportation means to go to a specific destination point, features such as "destination," shared by many transportation means, are more relevant. Commonality is a function of the frequency of the feature's occurrence in the concept's definition.

The two approaches mentioned above boil down to the well-known problem of lumping and splitting in categorization. The lumping approach focuses on similarities, considering them more important to categorization than differences, whereas the splitting approach takes the opposite direction, putting emphasis on discriminating differences rather than similarities.

Stuckenschmidt and Wache (2000) pointed out the need for context modeling in order to achieve information interoperability at the semantic level. In this approach, it is argued that contexts may be regarded as "one possibility to describe the semantic model" (115) and should contain information about data interpretation and data properties (such as its source, quality, and precision). To illustrate this position, an example was presented in which geological data defined in the context of the CORINE database was transformed into the context of ATKIS, the National Topographic Cartographic Information System of Germany. This procedure, called "context-transformation," was performed using the Ontology Inference Layer (OIL) and Description Logics (DL) to model the content of CORINE and ATKIS and integrate the semantics of the information sources in an overall model.

In Kokla and Kavouras (2001), a methodology was proposed for the fusion of different geographic domain ontologies based on context formation and complementarity. According to this approach, geographic contexts include information about geographic types, characteristics, relations, and operations and are formed as cross-tables using Formal Concept Analysis (Chapter 8). Contexts associate the semantic factors derived from geographic categories of different ontologies with their own properties. Thus, this incorporation of different but complementary conceptualizations of geographic space, each of which is suitable for some context and level of detail, facilitates information exchange and reuse.

Tomai and Kavouras (2005) examined how the notion of formalized context can be incorporated into a geographic ontology. They suggested the application of a geographic context formalism using dimensions that are determined as variables that once were combined, providing bounds under which ontological information acquires specific meaning (Stavrakas, 2003). Stavrakas (2003) proposed a formalization of context with respect to a set of dimensions and the relevant dimension values in corresponding domains. According to this approach, a context may be assigned to more than one set of dimensions, depending on the circumstances under which it is considered. This approach leads to the conclusion that incorporating the notion of context increases the robustness and the expressivity of a geographic ontology.

Cai (2007) specified a semantic model of geographical information based on context; this model supports human–GIS interaction and overcomes geosemantic interpretability difficulties. Cai described contexts as first-class objects, and, based on Turner (1998; 1999), he proposed the formation of contexts using a frame-like knowledge structure characterized by a unique identifier and containing both descriptive and prescriptive knowledge. Descriptive knowledge includes information about the necessary features of a situation in order to be considered an instance of a context. Prescriptive knowledge includes information about context-specific meaning of concepts, goals of the context, actions to achieve these goals appropriately in the situation, as well as information for the interpretation of new events." This approach demonstrates that context considerations help to clarify spatial data semantics and facilitate the interpretation of vague spatial concepts.

7.3   KNOWLEDGE REPRESENTATION AND NATURAL LANGUAGE

Historically, KR methods have somehow developed independently of Natural Language (NL). This is strange given that NL is also concerned with semantics and the same tasks of representation and reasoning. One possible explanation is that although NL is very expressive, it also presents ambiguities that are not easy to process systematically. Of course, there have been arguments for the equivalence of NL to any formal language traditionally employed in KR. Such a prominent position is that by the logician and philosopher Montague (1970).

The progress in developing Natural Language understanding/processing (NLU/NLP) techniques has brought KR and NL close again (Ali and Iwanska, 1997; Iwanska and Shapiro, 2000). This is also reinforced by the fact that a great amount of knowledge is already expressed in NL. Bridging this gap is vital, because KR and NLP can contribute not only to each other but more significantly to their common objective. For example, NLP techniques can be used to extract semantic information from a NL text to feed some other KR approach such as Formal Concept Analysis, similarity-based integration and reasoning, automatic metadata collection, etc. Conversely, the efficient but nonexpressive KR reasoning approaches can generate their outcome to NL to gain in expressivity. Of course, there is always the possibility to consider NL neither as a source nor as an outcome, but as a novel KR itself (Sukkarieh, 2003). Another direction exploiting the NL expressivity in knowledge representation is the use of controlled NL-like languages (Sukkarieh, 2003) as a compromise that eases use (transparency), expressivity, and efficiency.

At the practical level, from the broad field of KR and NLP, there is a particular interest in a number of directions. It is most important to develop/employ NLP approaches that exploit the knowledge existing in NL text in order to clarify geographic concepts and build the associated ontologies. A particular kind of NL text that offers this knowledge and needs to be processed is the structured form of concept definitions.

When talking about linguistic issues and NLP, a last remark needs to be made at this point. In accordance with the realist view, the general objective of semantic ontology-based integration of knowledge necessitates that words are just the carriers of the meaning of concepts. Even when linguistic elements are used in comparison and integration, it is not the words that are to be compared and aligned but the implied concepts. As Sowa also states (2006b), this is a common misunderstanding in ontology alignment.

7.4   FORMAL INSTRUMENTS

In searching for appropriate instruments to deal with knowledge representation, some terminology is often loosely or almost synonymously employed, such as KR formalism, conceptual structure, KR notation, KR method, KR technique, KR language, etc. Hypothetically speaking, once these instruments are not associated with high-level knowledge issues but only with their representation in a formal computing application environment, they could be considered equivalent. However, the fact that these instruments have been developed (and are thus, suitable) for different types of knowledge, complicates their association. They present differences in their theories, the assumptions made, the knowledge elements that are being represented, their source, the objective of a KR, etc. For instance, it is different to represent knowledge expressed by some natural language passage, by the categories of a hierarchical classification system, by a scientific corpus, by cognitive or common semantics, by a dictionary or thesaurus, by a formal ontology, by a geospatial database with instances, etc. Therefore, it is important to realize that certain representation instruments/mechanisms are better suited to some forms than others and that converting such an instrument to another is not always a straightforward process.

However, maintaining the realist's ontological perspective (Chapters 1 and 4), it could be quite safely claimed that some key notions in KR are:

1. *Semantics*, that is, concepts and other ontological elements describing what kind (or sort) of knowledge is represented.
2. *Formalism* (terms, symbols, notation, conventions) and *Structure* (rules of use and inference support).
3. *Language*, that is, implementation in a computation-AI environment.

Given that *semantics* refers to knowledge and its association to reality, already addressed in Parts 1 and 2, and *languages* refer to the "low-level" implementation perspective (Chapter 1), not directly addressed in this book, a main focus of the present chapter is on *KR formalisms* and *structures*.

It is obvious that *KR formalisms and structures* play the role of mediator between *KR* and *implementation languages*; thus, they could be considered a high-level or metalanguage for KR. Moreover, they are closely interrelated to the knowledge as this is expressed by ontologies. Ontologies document and explain knowledge but need mechanisms for examining concepts and their relations. This role is supported by KR structures, which provide the ability to share, compare, and visualize the knowledge expressed by ontologies. Ontologies provide a model of reality that KR structures employ to support reasoning and in their turn to engineer new ontologies. Furthermore, structures are used as tools for extracting ontological information from existing knowledge sources such as a Natural Language passage.

Due to the variety of what is considered as knowledge and what the objective of KR is, there are numerous KR instruments proposed in the literature under different names. These include logic-based formalisms (Brachman and Nardi, 2003), (such as first-order logic, propositional logic, rules, and higher-order logics); semantic networks; mathematical theories of concepts and concept hierarchies (Wille, 2005); situation theory (Aczel, Katagiri, and Peters 1993); discourse representation theory (Guenther, 1987; Hess, 1991); frames and scripts (Minsky, 1975; Schank and Abelson, 1977); visualization of knowledge (Kremer, 1998; Judelman 2004); logic-based mental spaces (Fauconnier, 1985); and geometry-based conceptual spaces (Gärdenfors, 2000). More specifically, the most prominent approaches, which not only find a wider application in KR representation but are also better suited to the objective of semantic integration, are the following: Formal Concept Analysis (FCA) (Ganter and Wille, 1999), Conceptual Graphs (CGs) (Sowa, 1991), Channel theory-Information Flow (IF) (Barwise and Seligman, 1997), Description Logics (DL) (Brachman and Nardi, 2003), Natural Language Processing (NLP) (Bolshakov and Gelbukh, 2004), and also concept similarity and mapping (Kremer, 1998; Hahn and Ramscar, 2001; Sowa, 2006a).

All these KR instruments satisfy the first criterion by Davis, Shrobe, and Szolovits (1993), that is, that of being an alternative surrogate of reality. Depending on the reality being represented or, more accurately, on our limited and biased knowledge of it, and on the objective served by KR, some surrogates are better (or more appropriate) than others. This connection to reality is a two-way one. From the above approaches, FCA, CGs, and partly IF employ some visual abilities, thus allowing for a more elucidatory understanding of reality and the concepts involved. Conversely, Description Logics and Discourse Representation theory are more formalistic than KR approaches.

All KR instruments employ some ontological commitments in order to describe a portion or view of reality by putting emphasis on the basic *structural ingredients*— concepts, properties, and relations. This procedure follows the human cognitive system and hierarchical structuring of knowledge. Nevertheless, in approaches that are more formalistic, *concepts* resemble the *objects* described by some properties, and relations resemble the relations described by some operators in logic. This particularly holds for Description Logics which is mainly a technique for making inferences, limiting itself to logic relations of the type true/false with respect to a given situation. The case of Situation Theory is similar. On the other hand, concepts in FCA or CGs are more abstract, allowing the description and structuring of relations among them; relations that are more flexible than the formal relations in, for example, first-order logic.

After these brief preliminary remarks, the approaches are introduced in the remaining chapters of Part 3.

REFERENCES

Aczel, P., V. Katagiri, and S. Peters. 1993. *Situation theory and its applications.* Cambridge, UK: Cambridge University Press.

Ali, S. S. and L. M. Iwanska. 1997. Knowledge representation for natural language processing in implemented systems. *Natural Language Engineering* 3(2/3): 97–101.

Barwise, J. and J. Seligman. 1997. *Information flow: The logic of distributed systems.* Cambridge Tracts in Theoretical Computer Science. Cambridge, UK: Cambridge University Press.

Bouquet, P., C. Ghidini, F. Giunchiglia, and E. Blanzieri. 2003. Theories and uses of context in knowledge representation and reasoning. *Journal of Pragmatics* 35(3): 455–484.

Bolshakov, I. A. and A. Gelbukh. 2004. *Computational linguistics: Models, resources, applications.* Mexico: National Polytechnic Institute—Center for Computing Research. <http://www.gelbukh.com/clbook/Computational-Linguistics.htm> (accessed 10 September 2007).

Brachman, J. R. and H. Levesque. 1987. Expressiveness and tractability in knowledge representation and reasoning. *Computer Intelligence* 3: 78–93.

Brachman, J. R. and D. Nardi. 2003. An introduction to description logics. In *The description logic handbook: Theory, implementation and applications*, Chapter 1. Cambridge, UK: Cambridge University Press.

Cai, G. 2007. Contextualization of geospatial database semantics for human–GIS interaction. *Geoinformatica* 11(2): 217–237.

Davis, R., H. Shrobe, and P. Szolovits. 1993. What is a knowledge representation? *AI Magazine* 14(1): 17–33.

Fauconnier, G. 1985. *Mental Spaces*. Cambridge, MA: MIT Press.

Fellbaum, C., ed., 1998. *WordNet: An electronic lexical database*. Cambridge, MA: MIT Press.

Fikes, R. 1998. *Knowledge representation overview*. CS222 lecture notes, Computer Science Department, Stanford University. http://www-ksl.stanford.edu/people/fikes/cs222/1998/Introduction/tsld008.htm (accessed 1 May 2007).

Ganter, B. and R. Wille. 1999. *Formal concept analysis: Mathematical foundations*. Heidelberg: Springer.

Gärdenfors, P. 2000. *Conceptual spaces*. Cambridge, MA: MIT Press.

Guenthner, F. 1987. Discourse representation theory and the semantics of natural Languages. Presented at a Panel on "Discourse Theory and Speech Acts," Third Conference on Theoretical Issues in Natural Language Processing TINLAP-3, New Mexico State University, January 7–9. http://acl.ldc.upenn.edu/T/T87/T87-1022.pdf (accessed 1 May 2007).

Guha, R. V. 1991. Contexts: A formalization and some applications. Ph.D. thesis, Computer Science Department, Stanford University, Stanford, CA.

Hahn, U. and M. Ramscar. 2001. *Similarity and categorization*. Oxford: Oxford University Press.

Hess, M. 1991. Recent developments in discourse representation theory. In *Communication with Men and Machines*, ed. M. King. ISSCO, University of Geneva, Switzerland. http://www.ifi.unizh.ch/CL/hess/drt.html (accessed 1 May 2007).

Hjørland, B. and J. Nicolaisen. 2006. *The epistemological lifeboat: Epistemology and philosophy of science for information scientists*. http://www.db.dk/jni/lifeboat/home.htm (accessed 1 May 2007).

Iwanska, L. M. and S. C. Shapiro. 2000. *Natural language processing and knowledge representation: Language for knowledge and knowledge for language*. Cambridge, MA: MIT Press.

Judelman, G. B. 2004. Knowledge visualization: Problems and principles for mapping the knowledge space. M.Sc. thesis, University of Lübeck, Germany.

Kokla, M. and M. Kavouras. 2001. Fusion of top-level and geographical domain ontologies based on context formation and complementarity. *International Journal of Geographical Information Science* 15 (7): 679–687.

Kramer, B. and J. Mylopoulos. 1992. Knowledge representation. In *Encyclopedia of artificial intelligence*, Vol 1, ed. S. C. Shapiro, 743–759. New York: John Wiley & Sons.

Kremer, R. 1998. Visual languages for knowledge representation. Paper presented at KAW'98: Eleventh Workshop on Knowledge Acquisition, Modeling and Management, Voyager Inn, Banff, Alberta, Canada, April 18–23. http://pages.cpsc.ucalgary.ca/~kremer/papers/KAW98/visual/kremer-visual.html (accessed 1 May 2007).

Lenat, D. 1998. *The dimensions of context-space*. Austin, TX: Cycorp.

Levesque, H. and R. Brachman. 1985. A fundamental tradeoff in knowledge representation and reasoning. In *Readings in knowledge representation*, ed. R. Brachman and H. Levesque, 42–70.

McCarthy, J. 1987, Generality in artificial intelligence. *Communications of the ACM* 30(12): 1030–1035.

McCarthy, J. 1993. Notes on formalizing context. In *Proc. 13th International Joint Conference on Artificial Intelligence IJCAI '93*, Chambery, France, August 28–September 3, 1993. Los Altos, CA: Morgan Kaufmann, 555–562.

Minsky, M. 1975. A framework for representing knowledge. In *The psychology of computer vision*, ed. P. H. Winston. New York: McGraw-Hill. http://web.media.mit.edu/~minsky/papers/Frames/frames.html (accessed 1 May 2007).

Montague, R. 1970. Universal grammar. In *Theoria* Vol. 26, 373–398. Reprinted in *Formal philosophy: Selected papers of Richard Montague*, ed. R. H. Thomason. New Haven, CT: Yale University Press.

Panayiotopoulos, T. 2001. Logic programming, Course Notes (in Greek), Department of Informatics, University of Peiraeus, Peiraeus.

Penco, C. 1999. Objective and Cognitive Context. In *Proc. Second International and interdisciplinary Conference, CONTEXT'99: Modeling and Using Context*, ed. P. Bouquet, L. Serafini, P. Brezillon, M. Benerecetti, and F. Castellani, 270–283. Berlin/Heidelberg: Springer-Verlag.

Pomerol, J. C. and P. Brézillon. 2001. About some relationships between knowledge and context. In *Proc. 5th International and Interdisciplinary Conference on Modeling and Using Context*, ed. P. Bouquet, 461–464. Berlin: Springer-Verlag.

Rodríguez, A. and M. Egenhofer. 2004. Comparing geospatial entity classes: An asymmetric and context-dependent similarity measure. *International Journal of Geographic Information Science* 18(3): 229–256.

Saffo, P. 1994. It's the context, stupid. *Wired Magazine*, Issue 2.03. http://www.wired.com/wired/archive/2.03/context.html (accessed 1 May 2007).

Schank, R. C. and R. P. Abelson. 1977. *Scripts, plans, goals, and understanding: An inquiry into human knowledge*. Hillsdale, NJ: Lawrence Erlbaum Associates.

Smith, B. 1982. Prologue to 'Reflection and semantics in a procedural language'. In *Readings in Knowledge Representation*, ed. R. J. Brachman and H. J. Levesque, 31–39. Los Altos, CA: Morgan Kaufmann.

Smith, B. and D. M. Mark. 1998. Ontology and geographic kinds. Paper presented at 8th International Symposium on Spatial Data Handling, Vancouver, Canada, July 12–15.

Sowa, J. F. 1991. *Principles of semantic networks: Explorations in the representation of knowledge*. San Mateo, CA: Morgan Kaufmann Publishers.

Sowa, J. F. 2000. *Knowledge representation: Logical, philosophical, and computational foundations*. Pacific Grove, CA: Brooks Cole Publishing.

Sowa, J. F. 2002. Conceptual graphs, Working draft of the proposed ISO standard. http://www.jfsowa.com/cg/cgstand.htm (accessed 1 May 2007).

Sowa, J. F. 2003. Laws, facts and contexts: Foundations for multimodal reasoning. In *knowledge contributors*, ed. V. F. Hendricks, K. F. Jørgensen, and S. A. Pedersen, 145-184. Dordrecht: Kluwer Academic Publishers. http://www.jfsowa.com/pubs/laws.htm (accessed 1 May 2007).

Sowa, J. F. 2006a. Concept mapping. Paper presented at the track on Technology, Instruction, Cognition and Learning (TICL), AERA Conference, San Francisco, April 10. http://www.jfsowa.com/talks/cmapping.pdf (accessed 1 May 2007).

Sowa, J. F. 2006b. Problems of ontology—and terminology, ontac-forum. http://colab.cim3.net/forum/ontac-forum/2006-05/msg00052.html (accessed 1 May 2007).

Stavrakas, Y. 2003. Multidimensional semistructured data. Ph.D. thesis, National Technical University of Athens, Athens, Greece.

Stuckenschmidt, H. and H. Wache. 2000. Context modeling and transformation for semantic interoperability. Paper presented at Workshop "Knowledge Representation meets Databases KRDB 2000," ECAI'2000, Berlin, Germany.

Sukkarieh, J. Z. 2003. An expressive efficient representation: Bridging a gap between NLP and KR. In *KES 2003*, LNAI 2773, ed. V. Palade, R. J. Howlett, and L. C. Jain, 800–815. Berlin: Springer-Verlag.

Tomai, E. and M. Kavouras. 2005. Context in geographic knowledge representation. In *Proc. 2nd International Conference and Exhibition on Geographic information*, Estoril, Portugal, May 30–June 2, 2005.

Turner, R. M. 1998. Context-mediated behavior for intelligent agent. *International Journal of Human-Computer Studies, Special issue on Using Context in Applications* 48: 307–330.

Turner, R. M. 1999. A model of explicit context representation and use for intelligent agents. *Lecture Notes In Computer Science* 1688: 375–388.

Welty, C. A. 1995. An integrated representation for software development and discovery. Ph.D. thesis, Rensselaer Polytechnic Institute. http://www.cs.vassar.edu/faculty/welty/papers/phd/ (accessed 1 May 2007).

Wille, R. 2005. Formal concept analysis as mathematical theory of concepts and concept hierarchies. In *Formal Concept Analysis,* 1–33. Lecture Notes in Computer Science. Berlin: Springer.

Woods, B. 1975. What's in a link: Foundations for semantic networks. In *Representation and understanding: Studies in cognitive science*, ed. D. Bobrow and A. Collins. New York: Academic Press. Reprinted 1985 in *Readings in knowledge representation*, ed. R. Brachman and H. Levesque, 35–82. San Mateo, CA: Morgan Kaufmann. Also reprinted 1988 in *Readings in cognitive science*, ed. A. Collins and E. Smith. San Mateo, CA: Morgan Kaufmann.

8 Formal Concept Analysis

8.1 BASIC NOTIONS OF FORMAL CONCEPT ANALYSIS

Formal Concept Analysis (FCA) (Wille 1992, Ganter and Wille, 1999) is a theory of concept formation and conceptual classification founded in the 1980s by a research group in Darmstadt, Germany. FCA provides a model for analysis and formal representation of conceptual knowledge (Wille, 1992). It is based on mathematical order theory and particularly the theory of complete lattices. Therefore, before introducing FCA, we first present some basic notions of these theories.

Definition
A binary relation on a set P is called an *order relation* ($\leq$) if for all elements, x, y, $z \in P$ the following conditions are satisfied:

1. $x \leq x$ (reflexivity)
2. $x \leq y$ and $y \leq x$ implies that $x = y$ (antisymmetry)
3. $x \leq y$ and $y \leq z$ implies that $x \leq z$ (transitivity)

The inverse relation $\geq$ of an order relation $\leq$ is also an order relation and is called the *dual order* of $\leq$. An *ordered set* (or *partially ordered set* or *poset*) $(P, \leq)$ is a set P with an order relation $\leq$ defined on that set. An illustration of a partially ordered set is given in Figure 8.1. As shown in the figure, a partially ordered set allows an element to have multiple parents, in contrast to a tree, where an element has only one parent. Hence, the structure of a partially ordered set is a generalization of that of a tree structure.

Definition
Let $(P, \leq)$ be a partially ordered set and $S \subseteq P$. An element $x \in P$ is an *upper bound* of S if $s \leq x$ for all $s \in S$. A *lower bound* of S is an element $x \in P$ with $x \leq s$ for all $s \in S$. If the set of all upper bounds of S has a least element, then that element is called the *least upper bound* (*supremum*) of S and is denoted by $\vee S$. If the set of all lower bounds of S has a greatest element, then that element is called the *greatest lower bound* (*infimum*) of S and is denoted by $\wedge S$. The join of two elements $x, y \in S$ when it exists is the least upper bound of $\{x, y\}$ and is denoted by $x \vee y$. Similarly the meet of two elements is the greatest lower bound and is denoted by $x \wedge y$.

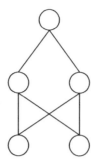

FIGURE 8.1   A partially ordered set.

Definition
Let *P* be a partially ordered set. Then:

1. If for any two elements *x*, *y* ∈ *P* the least upper bound *x* ∨ *y* and greatest lower bound *x* ∨ *y* always exist, then *P* is called a *lattice*.
2. If the greatest lower bound ∧ *S* and least upper bound ∨ *S* exist for all *S* ⊆ *P*, then *P* is called a *complete lattice*.

A lattice is a specialization of a partially ordered set in which each pair of elements always has a least upper bound and a greatest lower bound. Figure 8.2 shows the comparative representation of a tree, a partially ordered set, and a lattice.

Generally, FCA is based on the notion of concept defined as "a unit of thoughts consisting of two parts: the extension and the intension" (Wille, 1992, 493). The extension comprises the objects that pertain to the concept, whereas the intension describes the properties of the concept (see Section 6.4, Modes of Meaning). An object as used by FCA is a general notion, which may refer either to a particular object (instance) or to a kind of object. Formal concepts are defined as subsets of objects and properties (attributes). These concepts and the relationships between them form a lattice in which the meet and join of any combination of elements are

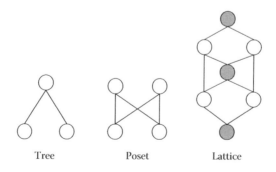

Tree   Poset   Lattice

FIGURE 8.2   Comparative representation of a tree, a partially ordered set, and a lattice.

given by definition. Thus, a concept lattice contains not only concepts corresponding to each object, but also concepts corresponding to the meet and join of other concepts. The remainder of this section presents only the basic notions of FCA, as they apply to the objective of semantic integration. These notions and the definitions provided below are drawn from the work of Wille (1992) and Ganter and Wille (1999). Proofs of theorems or further details can be found in these works.

Definition

A *formal context* K is defined as a triple (G, M, I) where G is a set of objects, M is a set of attributes, and I is a binary relation between G and M. The notation gIm denotes that "the object g has the attribute m."

A formal context may be represented by a cross table (Table 8.1). The rows of the table correspond to objects, and the columns to attributes. A cross in a cell of the table denotes when an object has an attribute. For example, Great Britain is a member of the European Union (EU) but does not have Euro as a currency.

Definition

For a set $A \subseteq G$ of objects and a set $B \subseteq M$ of attributes:

$A' = \{m \in M \,|\, gIm \text{ for all } g \in A\}$ is the set of attributes shared by the objects in A, whereas
$B' = \{g \in G \,|\, gIm \text{ for all } m \in B\}$ is the set of objects possessing all attributes in B.

The root of FCA is the *formal concept* or *conceptual class* or *category*, which is logically defined by two parts: the *extent* and the *intent*; the extent includes all objects (or entities) belonging to the concept, while the intent comprises all attributes (or properties) valid for all those objects.

Definition

A *formal concept* of the context (G, M, I) is a pair (A, B) with $A \subseteq G, B \subseteq M, A' = B$ and $B' = A$. A is the *extent* and B is the *intent* of the concept (A, B), respectively. By definition, the extent A and the intent B of a formal concept determine each other and hence, the formal concept. The set of all formal concepts of (G, M, I) is denoted $\mathfrak{B}(G, M, I)$.

TABLE 8.1

Cross Table of a Context

	EU Member	Currency: Euro	Official Language: French	Maritime
Austria	x	x		
France	x	x	x	x
Great Britain	x			x
Switzerland			x	

Proposition

If (G, M, I) is a formal context, $A, A_1, A_2 \subseteq G$ are sets of objects, and $B, B_1, B_2 \subseteq M$ are sets of attributes, then it holds that:

$$A_1 \subseteq A_2 \Rightarrow A'_2 \subseteq A'_1 \tag{1}$$

$$B_1 \subseteq B_2 \Rightarrow B'_2 \subseteq B'_1 \tag{1'}$$

$$A \subseteq A'' \text{ and } A' = A''' \tag{2}$$

$$B \subseteq B'' \text{ and } B' = B''' \tag{2'}$$

Each concept of a context (G, M, I) has the form (A'', A') for a subset $A \subseteq G$ and the form (B', B'') for a subset $B \subseteq M$. (A'', A') and (B', B'') are derived by (2) and (2') correspondingly. For every set $A \subseteq G$, A' is an intent of a concept, since (A'', A') is always a concept. A'' is the smallest extent containing A. Thus, a set $A \subseteq G$ is an extent if and only if $A = A''$. The same holds for intents.

Proposition

If T is an index set and for every $t \in T$, $A_t \subseteq G$ is a set of objects, and $B_t \subseteq M$ is a set of attributes, then:

$$\left(\bigcup_{t \in T} A_t \right)' = \bigcap_{t \in T} A'_t \tag{3}$$

$$\left(\bigcup_{t \in T} B'_t \right)' = \bigcap_{t \in T} B'_t \tag{3'}$$

From the above proposition, it is evident that the intersection of any number of extents is always an extent and the intersection of any number of intents is always an intent.

Definition

For concepts (A_1, B_1) and (A_2, B_2) of a context, *the subconcept-superconcept relation* (or hierarchical order) $(\leq)$ is defined as follows: the concept (A_1, B_1) is a *subconcept* of the concept (A_2, B_2) (namely, $(A_1, B_1) \leq (A_2, B_2)$), if $A_1 \subseteq A_2$ (which is equivalent to $B_2 \subseteq B_1$). Then, (A_2, B_2) is a *superconcept* of (A_1, B_1). The set of all concepts of (G, M, I) ordered by the subconcept-superconcept relation is called the *concept lattice* of the context (G, M, I) and is denoted by $\mathfrak{B}(G, M, I)$.

Theorem

Let (G, M, I) be a context. Then, $\mathfrak{B}(G, M, I)$ is a complete lattice in which the greatest lower bound (infimum) and the least upper bound (supremum) are given by:

$$\bigwedge_{t \in T}(A_t, B_t) = \left(\bigcap_{t \in T} A_t, \left(\bigcup_{t \in T} B_t \right)'' \right) \tag{4}$$

$$\bigvee_{t \in T}(A_t, B_t) = \left(\left(\bigcup_{t \in T} A_t \right)'', \bigcap_{t \in T} B_t \right) \tag{4'}$$

According to the above theorem, which is considered as the *basic theorem on concept lattices*, the set of concepts produced by a context is a complete lattice; the greatest lower bound and the least upper bound of a group of concepts are defined by the above formulas. The greatest lower bound is the largest common subconcept of the concepts (A_t, B_t). The extent of this concept is the intersection of the extents of (A_t, B_t).

Since $At = Bt'$ for each $t \in T$, according to (3) the formula for the greatest lower bound (infimum) (4):

$$\left(\bigcap_{t \in T} A_t, \left(\bigcup_{t \in T} B_t \right)'' \right)$$

can be transformed to:

$$\left(\left(\bigcup_{t \in T} B_t \right)', \left(\bigcup_{t \in T} B_t \right)'' \right)$$

that is, it takes the form (B', B'') and is, therefore, always a concept. Correspondingly, the formula for the least upper bound (supremum) (4'):

$$\left(\left(\bigcup_{t \in T} A_t \right)'', \bigcap_{t \in T} B_t \right)$$

can be transformed to

$$\left(\left(\bigcup_{t \in T} A_t \right)'', \left(\bigcup_{t \in T} A_t \right)' \right)$$

that is, it takes the form (A'', A') and is, therefore, certainly a concept.

According to Wille (1992), the concept lattice of a given context (G, M, I) is formed by the concepts (A'', A') (for each $A \subseteq G$) and (B', B'') (for each $B \subseteq M$), which are derived using (2) and (2') correspondingly. However, in order to determine all

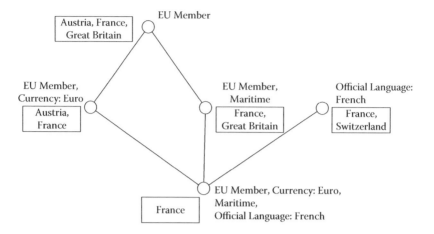

FIGURE 8.3 Concept lattice of the context of Table 8.1.

concepts of a given context, it is preferable to draw a list of object intents or attribute extents and use the following formulas, which are special cases of (3) and (3'):

$$A'_t = \bigcap_{t \in T} \{g\}' \tag{5}$$

or

$$B'_t = \bigcap_{t \in T} \{m\}' \tag{5'}$$

This way, it is possible to form (A'', A') or (B', B'') correspondingly. $\{g\}'$ (or simply g') is the *object intent* $\{m \in M \mid gIm\}$ of an object g, whereas $\{m'\}$ (or m') is the *attribute extent* $\{g \in G \mid gIm\}$ of the attribute m. According to the formulas (3) and (3'), every intent is the intersection of object intents and every extent is the intersection of attribute extents. Figure 8.3 shows the concept of lattice of the context of Table 8.1.

8.2 MANY-VALUED CONTEXTS

Besides one-valued attributes, that is, attributes that an object either has or has not, in reality the majority of attributes takes different values, for example, color, height, temperature, etc.

Definition
A *many-valued context* is a quadruple (G, M, W, I) and is comprised by the sets G, M, and W, as well as the ternary relation I among G, M, and W $(I \subseteq G \times M \times W)$, for which it holds that:

$$(g, m, w) \in I \text{ and } (g, m, v) \in I \text{ imply that } w = v.$$

The set G consists of objects, the set M of many-valued attributes, and the set W of attribute values. If the set W has n elements, then (G, M, W, I) is an n-valued context. The notations $(g, m, w) \in I$ or $m(g) = w$ denote that "the attribute m has the value w for the object g" (Ganter and Wille, 1999, 36).

Many-valued contexts are also represented by a table whose rows correspond to objects and its columns to many-valued attributes. The cell at the intersection of a row g and a column m includes the attribute value $m(g)$. In case the object g has no value for the attribute m, then the cell is empty. Table 8.2 shows three vegetation categories (forest, cereals, and grassland) and their many-valued attributes. For example, the attribute "use" takes the values "logging," "flour production," and "grazing," respectively.

In order to determine the formal concepts of a many-valued context, it is necessary to transform it to a one-valued context following certain rules. The formal concepts of the derived one-valued context constitute the formal concepts of the original many-valued context. This process, which is not always uniquely defined, is called *conceptual scaling*. Every many-valued attribute m is interpreted using a context called *conceptual scale*. The selection of a conceptual scale is not the result of a mathematical process but an issue of interpretation.

Definition

A conceptual scale for the attribute m of a many-valued context is a one-valued context $S_m = (G_m, M_m, I_m)$ with $m(G) \subseteq G_m$. The objects of a scale are called *scale values*, whereas the attributes are called *scale attributes*. Every context may be used as a scale. Formally there is no difference between a context and a scale, but the term "scale" is only used for contexts that are conceptually explicit and meaningful.

After selecting a scale for every attribute, the process of conceptual scaling proceeds with the unification of scales in order to form a one-valued context. *Plain scaling* is the simplest case; scales are placed together without any association. From the many-valued context and the scale S_m, a one-valued context is derived in the following way: the set of objects G is not altered, whereas every many-valued attribute m is replaced with the values of the scale S_m. In the table of a many-valued context, plain scaling is performed with the replacement of every attribute value $m(g)$ with the row of the scale S_m that corresponds to the value $m(g)$. For example, the many-valued context shown in Table 8.2 is transformed to the one-valued context shown in Table 8.3 and the concept lattice shown in Figure 8.4.

TABLE 8.2

Example of a Many-Valued Context

	Vegetation Type	Use	Nature
Forest	woody	logging	natural
Cereal crops	cereals	flour production	cultivated
Grassland	grass	grazing	natural

TABLE 8.3

Transformation of a Many-Valued Context to a One-Valued Context

	Vegetation Type			Use			Nature	
	Woody	**Cereals**	**Grass**	**Logging**	**Flour Production**	**Grazing**	**Natural**	**Cultivated**
Forest	x			x			x	
Cereal crops		x			x			x
Grassland			x			x	x	

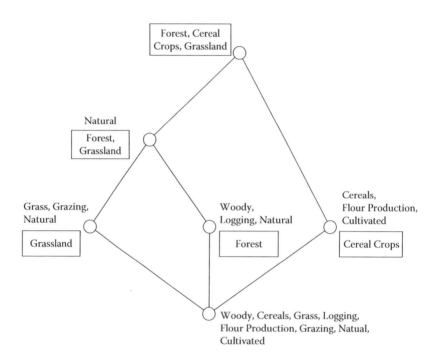

FIGURE 8.4 Concept lattice of the many-valued context of Table 8.3.

8.3 APPLICATIONS OF LATTICES AND FORMAL CONCEPT ANALYSIS

Several researchers have applied FCA in different domains, such as knowledge representation, processing, and discovery (Cole, Eklund, and Walker, 1998; Stumme, Wille, and Wille, 1998; Stumme, 2002; Valtchev, Missaoui, and Godin, 2004; Wille, 2006); ontology construction (Bain, 2003); information retrieval (Carpineto and Romano, 2005); software engineering (Hesse and Tilley, 2005; Snelting, 2005; Tilley et al., 2005); sociology (Freeman and White, 1993); linguistics (Priss, 2005); economics (Wille, 2005); and others.

The application of Formal Concept Analysis has not been widely explored in geographic information science. The advantages of concept lattices are mainly exploited for ontology development and integration. Kokla and Kavouras (2001) and Kavouras and Kokla (2002) introduced a methodology for ontology integration that uses concept lattices as a tool for the formalization and integration of geographic concepts and relations encoded in different ontologies, in order to reveal their association and interaction. Yi, Huang, and Chan (2004) developed a method for geographic information retrieval based on the design of an upper ontology using FCA and on concept brokering rules.

Lattices have been used to model relationships between spatial regions, since they are considered to be a better mathematical structure than hierarchies. Saalfeld (1985) presented lattice structures in geographic files of regional areal subdivisions in order to examine "additional levels of geography" (482) (addition/deletion of areal references, addition of special elements, etc.). Kainz, Egenhofer, and Greasley (1993) used order relations in partially ordered sets, as well as lattice operations to answer spatial queries (determine areas with specific characteristics, model spatial overlay, calculate topological operations of intersection, etc.).

Hirtle (1995) considered lattices an alternative representational scheme for the cognitive representation of spatial knowledge. He compared trees, ordered trees, and semilattices for the representation of multiple, overlapping relationships in geographic space and concluded that semilattices are richer structures and appear to be consistent with many aspects of cognitive space. The structure of a lattice permits the emergence of overlapping relationships, allowing geographic concepts to be associated with multiple superordinate concepts (e.g., "canal" may be considered a subordinate concept of both "transportation" and "hydrography"). The multidimensionality of geographic concepts is a property that reflects human cognition of geographic phenomena and must be preserved in GISs (Buttenfield and DeLotto 1989; Usery 1993). This possibility is highly useful in the integration of different ontologies, especially when the original ontologies correspond to different views, which must be preserved in the integrated ontology.

8.4   SEMANTIC FACTORING

Semantic Factoring is a conceptual analysis process that decomposes a complex concept into its defining, primitive concepts, called *semantic factors* (Ganzmann, 1990; Soergel, 1997; Sowa, 2000). For example, Semantic Factoring decomposes the concept "father" into the primitive concepts "male and parent" (Hjorland, 2007) and the concept "ship" into "vehicle and water transport" (Ganzmann, 1990). Correspondingly, the concept "mountainous terrain" may be decomposed into the concepts "high elevation" and "steep slope," and the concept "pond" into the concepts "small" and "lake."

Semantic Factoring is used in multilingual thesauri construction, development of Knowledge Organization Systems, and ontology development and integration. During multilingual thesauri construction, Semantic Factoring deals with cases where terms from different natural languages are similar but not exactly equivalent (Ganzmann, 1990); this process results in a conceptually solid vocabulary and the formation of a well-structured hierarchy, which contribute to "conceptual economy" (Soergel, 1997). In ontology development and integration, Semantic Factoring may be implemented at all hierarchical levels, from the most general to the most specialized (Sowa, 2000) and consists of the analysis-decomposition of the concepts of the original ontologies into a set of fundamental concepts, that is, semantic factors. The semantic factors correspond to the most clear, unambiguous concepts, which constitute the building blocks of the integrated ontology. They may further be associated to each other through multiple, hierarchical relations using Formal Concept Analysis to form a concept lattice. In Section 16.3, Semantic Factoring is applied to disambiguate concepts during ontology integration, and then Formal Concept Analysis is used to produce a concept lattice, which represents the final integrated ontology.

REFERENCES

Bain, M. 2003. Inductive construction of ontologies from formal concept analysis. In *Advances in artificial intelligence*, ed. T. D. Gedeon and L. C. C. Fung, 88–99. LNAI 2903. Berlin: Springer-Verlag.

Buttenfield, B. P. and J. S. Delotto. 1989. Multiple representations. Scientific report for the specialist meeting, Technical Paper 89-3, National Center for Geographic Information and Analysis, State University of New York.

Carpineto, C. and G. Romano. 2005. Using concept lattices for text retrieval and mining. In *Formal concept analysis*, ed. B. Ganter, G. Stumme, and R. Wille, 161–179. Foundations and Applications, Lecture Notes in Computer Science 3626. Berlin: Springer.

Cole, R., P. Eklund, and D. Walker. 1998. Using conceptual scaling in formal concept analysis for knowledge and data discovery in medical texts. In *Proc. Second Pacific Asian Conference on Knowledge Discovery and Data Mining*. Melbourne, Australia, April 15–17, 1998, ed. X. Wu, K. Ramamchanarao, K. B. Korb. Research and Development in Knowledge Discovery and Data Mining, Lecture Notes in Computer Science 1394. Springer, 378–379.

Freeman, L. C. and D. R. White. 1993. Using Galois lattices to represent network data. *Sociological Methodology* 23: 127–146.

Ganter, B. and R. Wille. 1999. *Formal concept analysis, mathematical foundations*. Berlin: Springer-Verlag.

Ganzmann, J. 1990. Criteria for the evaluation of thesaurus software. *International Classification* 17(3/4): 148–157. http://www.willpower.demon.co.uk/ganzmann.htm (accessed 22 March 2007).

Hesse, W. and T. Tilley. 2005. Formal concept analysis used for software analysis and modeling. In *Formal concept analysis*, ed. B. Ganter, G. Stumme, and R. Wille, 288–303. Formal Concept Analysis, Foundations and Applications, Lecture Notes in Computer Science 3626. Berlin: Springer.

Hirtle, S. C. 1995. Representational structures for cognitive space: Trees, ordered trees and semi-lattices. In *Spatial information theory: A theoretical basis for GIS*, ed. A. U. Frank and W. Kuhn, 327–340. Berlin: Springer.

Hjorland, B. 2007. Semantic primitives. http://www.db.dk/bh/Lifeboat_KO/CONCEPTS/semantic_primitives.htm (accessed 21 March 2007).

Kainz, W., M. J. Egenhofer, and I. Greasley. 1993. Modeling spatial relations and operations with partially ordered sets. *International Journal of Geographical Information Systems* 7(3): 215–229.

Kavouras, M. and M. Kokla. 2002. A method for the formalization and integration of geographical categorizations. *International Journal of Geographical Information Science* 16(5): 439–453.

Kokla, M. and M. Kavouras. 2001. Fusion of top-level and geographical domain ontologies based on context formation and complementarity. *International Journal of Geographical Information Science* 15(7): 679–687.

Priss, U. 2005. Linguistic applications of formal concept analysis. In *Formal concept analysis*, ed. B. Ganter, G. Stumme, and R. Wille, 149–160. Formal Concept Analysis, Foundations and Applications, Lecture Notes in Computer Science 3626. Berlin: Springer.

Saalfeld, A. 1985. Lattice structures in geography. Digital representations of spatial knowledge. In *Proc. AUTO-CARTO 7*, 482–489. Washington, DC: American Society of Photogrammetry and American Congress on Surveying and Mapping.

Snelting, G. 2005. Concept lattices in software analysis. In *Formal concept analysis*, ed. B. Ganter, G. Stumme, and R. Wille, 272–287. Formal Concept Analysis, Foundations and Applications, Lecture Notes in Computer Science 3626. Berlin: Springer.

Soergel, D. 1997. Multilingual thesauri in cross-language text and speech retrieval. In *AAAI Symposium on Cross-Language Text and Speech Retrieval*, ed. D. Hull and D. Oard. American Association for Artificial Intelligence. http://www.ee.umd.edu/medlab/filter/sss/papers/soergel.ps (accessed 22 March 2007).

Sowa, J. F. 2000. *Knowledge representation: Logical, philosophical and computational foundations*. Pacific Grove, CA: Brooks Cole Publishing.

Stumme, G. 2002. Formal concept analysis on its way from mathematics to computer science. In *Conceptual structures: Integration and interfaces*, ed. U. Priss, D. Corbett, and G. Angelova, 2–19. 10th International Conference on Conceptual Structures, Lecture Notes in Computer Sciences, 2393. Berlin: Springer.

Stumme, G., R. Wille, and U. Wille. 1998. Conceptual knowledge discovery in databases using formal concept analysis methods. In *Proc. 2nd European Symposium*, Principles of Data Mining and Knowledge Discovery, PKDD '98, LNAI 1510, ed. J. M. Zytkow and M. Quafofou, 450–458. Heidelberg: Springer.

Tilley, T., R. Cole, P. Becker, and P. W. Eklund. 2005. A survey of formal concept analysis support for software engineering activities. In *Formal concept analysis*, ed. B. Ganter, G. Stumme, and R. Wille, 250–271. Formal Concept Analysis, Foundations and Applications, Lecture Notes in Computer Science 3626. Berlin: Springer.

Usery, E. L. 1993. Category theory and the structure of features in geographic information systems. *Cartography and Geographic Information Systems* 20: 5–12.

Valtchev, P., R. Missaoui, and R. Godin. 2004. Formal concept analysis for knowledge discovery and data mining: The new challenges. In *Proc. Concept Lattices: Second International Conference on Formal Concept Analysis*, ed. P. Eklund, 352–371. Lecture Notes in Computer Science, 2961. Berlin: Springer.

Wille, R. 1992. Concept lattices and conceptual knowledge systems. *Computers and Mathematics with Applications* 23(6-9): 493–515.

Wille, R. 2005. Conceptual knowledge processing in the field of economics. In *Formal concept analysis*, ed. B. Ganter, G. Stumme, and R. Wille, 226–249. Formal Concept Analysis, Foundations and Applications, Lecture Notes in Computer Science 3626. Berlin: Springer.

Wille, R. 2006. Methods of conceptual knowledge processing. In *Proc. 4th International Conference on Formal Concept Analysis (ICFCA 2006)*, Dresden, Germany, Ed. R. Missaoui and J. Schmid, 1–29. Lecture Notes in Computer Science, 3874. Berlin: Springer.

Yi, S., B. Huang, and W. T. Chan. 2004. Retrieval of heterogeneous geographical information using concept brokering and upper level ontology. In *ER Workshops 2004*, ed. S. Wang et al. 172–183. Lecture Notes in Computer Science, 3289. Berlin: Springer-Verlag.

9 Conceptual Graphs

9.1 DEFINING CONCEPTUAL GRAPHS

Conceptual graphs (CGs) were introduced by Sowa in 1984. They constitute a visual, advanced knowledge-based representation formalism grounded on philosophical, linguistic, and object-oriented principles (Sowa, 1984, 2000) and are considered the most general and flexible logic notation that can be transformed to any form of logic (Sowa, 2004). One of the major advantages of CGs in relation to other knowledge representation approaches is that of being close to humans' perception and reasoning. Initially, the purpose of CGs was to facilitate the transition from natural languages to structured logical forms and vice versa (Sowa, 2000), but in time they found use in many other fields, such as text mining (Cyre, 1997; Montes-y-Gómez, Gelbukh, and López-López, 2002; Hensman, 2004) and information retrieval (Huibers, Ounis, and Chevallet, 1996; Zhong et al., 2002). More recently, CGs have started to find application in the geospatial domain, as will be shown in the next sections.

A formal definition of CG is given by the ISO standard working draft on CGs (Sowa, 2002):

> A conceptual graph g is a bipartite graph, which consists of two kinds of nodes called concepts and conceptual relations.

Nodes are linked together by arcs. The term bipartite, which characterizes a CG in the definition above, refers to the fact that arcs always link one concept node to one conceptual relation node. Concept nodes represent entities, attributes, or actions. Conceptual relation nodes are used to identify the kind of relationship between concepts. Different notations exist to represent a CG. These are discussed in the following.

9.2 REPRESENTING CONCEPTUAL GRAPHS

The notations used to represent CGs may be *normative* or *informative*. *Normative* representations, like Conceptual Graph Interchange Form (CGIF), intend to facilitate the transmission of CGs' information through computer systems/networks and are not of concern here. On the other hand, *informative* notations intend to represent knowledge in order to help human readability and reasoning. Two informative notations exist: a *graphical* and a *linear* one. The former is called *Display Form* (DF) and the latter *Linear Form* (LF).

In the *Display Form* (DF), CGs are graphically represented. Concepts are drawn as rectangles; conceptual relations are represented by circles or ovals; concepts and relations are linked through arcs upon which a direction is indicated by an arrow

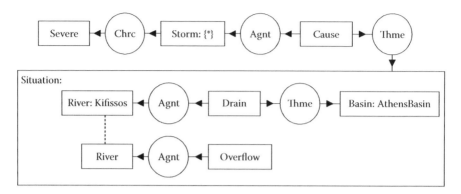

FIGURE 9.1 Conceptual graph representing the sentence: "Severe storms cause the Kifissos River, which drains the Athens Basin to overflow," using Display Form.

head. For example, let's suppose that a CG is to be constructed to represent the sentence: "Severe storms cause the Kifissos River, which drains the Athens Basin, to overflow." The concepts and the conceptual relations of this sentence are identified and associated with each other in a way that forms the DF CG illustrated in Figure 9.1.

Like the DF, the *Linear Form* (LF) is also an easily readable notation. Its main advantages consist in taking less space than DF and being suitable for keyboard entry. In LF, concepts are represented as words in brackets, while conceptual relations are represented as words in parentheses. The CG of the example presented in Figure 9.1 is transformed into the following LF:

[Sever] ← (Chrc) ← [Storm: {*}] ← (Agnt) ← [cause] → (Thme) –
  [Situation:
    [River: Kifissos *x] ← (Agnt) ← [Drain] → (Thme) → [Basin:
      AthensBasin]
    [River: ?x] ← (Agnt) ← [Overflow]
    ]

9.3 BASICS OF CONCEPTUAL GRAPHS

As CGs have been well documented in the literature (Sowa, 2000; 2002), only the basics will be presented below.

A *concept* consists of a *concept type* and a *referent*. A *concept type* may be considered as a category characterizing concepts with similar properties. Moreover, concept types may be organized in hierarchies called *type hierarchies*, each one including two primitive types, *entity* (at the top of the hierarchy) and *absurdity* (at the bottom), as well as defined types partially ordered in subtypes and supertypes. The instances or individuals of concept types are called *referents*. For example, the referent of the concept [River: Kifissos] is Kifissos, whereas its concept type is River.

The referent of a concept is characterized by a *quantifier* and a *designator*. A *quantifier* may be *existential* or *defined*. An *existential quantifier*, represented by the symbol ∃ or by the absence of any symbol, reveals the existence of at least one referent. A *defined quantifier*, represented by the symbols ∀, @, {*}, or by a collection, e.g., {a,b,c}, refers to the quantity of the referent. For example, the referent of the concept [Storm: {*}] has the defined quantifier {*} which indicates plural. A designator shows the form of a referent and may be expressed by: (a) a literal, (b) a locator, which can be thought as a pointer specifying how the referent can be found, or (c) a descriptor, which is a CG describing the referent.

A *relation* has a *type*, a *valence*, and a *signature*. The *relation type* determines: (a) the *valence*, which corresponds to the number of arcs belonging to the relation and (b) the *signature*, which specifies what concept types the relation can be related to. Like concept types, relation types can be organized in hierarchies called *relation hierarchies*, which define a partial ordering of relation types. In Figure 9.1, the *valence* of the relation (Agnt) is equal to 2 since there are always two arcs going to and from the relation. A valid *signature* of this relation can be <Action, Actor> meaning that the concepts pointing *to* the relation must be of type Action or of any subtype of Action and the concepts pointed *by* the relation must be of type Actor or of any subtype of Actor. For example, in the CG: [River: Kifissos] ← (Agnt) ← [Overflow], Overflow has to be a subtype of Action and River a subtype of Actor in the concept type hierarchy.

A *context* corresponds to a concept whose referent's designator is a nonblank CG. In Figure 9.1, a context exists and can be identified by the rectangle enclosing its content. Its type is named Situation and its referent is specified by a designator corresponding to the CG enclosed in the rectangle.

A *coreference set* of a CG is a set of concepts, belonging to this CG or to the nested CG's contexts, which refer to the same individual. The concepts belonging to the same coreference set are linked through *coreference links*. A *coreference link* is represented by a dotted line. In Figure 9.1, the coreference link shows that the concepts [River: Kifissos] and [River] refer to the same individual.

9.4   OPERATIONS OVER CONCEPTUAL GRAPHS

Logical manipulation of CGs may be performed through operations based on canonical formation rules called *equivalence, specialization,* and *generalization rules* (Sowa, 2002). These rules consist of more specific ones named, respectively: *copy/ simplify rules, restrict/join rules,* and *unrestrict/detach rules.* Rules define operations that transform a CG into another CG either logically equivalent, or more specialized or more generalized. Rules can be combined in pairs in such a way that the corresponding operations have the inverse result upon a CG. These pairs of rules are presented subsequently.

Copy and *simplify* rules are equivalence rules. They define operations that result in the transformation of a CG into a logically equivalent one. The *copy* rule consists of copying a subgraph of a CG and inserting it into the initial one. The *simplify* rule deletes the redundant features of a CG without logically modifying it. Figure 9.2 shows how the conceptual relation (Agnt) and the subgraph (Thme) -> [Basin: AthensBasin] are copied during two consecutive steps. Since the latter formed CG includes redundant features, it can be simplified in order to form the initial one.

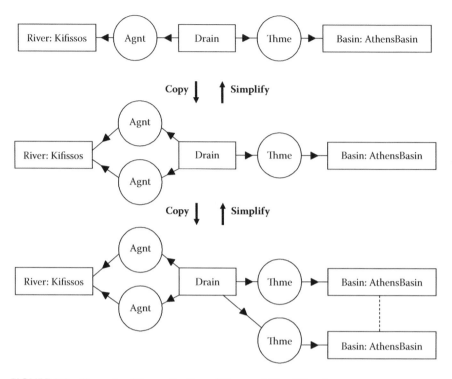

FIGURE 9.2   Example of the application of Copy and Simplify rules.

The *restrict* rule is a specification rule while *unrestrict* is a generalization rule. These rules define inverse operations that result in the specialization or the generalization of a CG, respectively. The *restrict* rule determines how a concept can be replaced by another whose type is a subtype of the previous one or by a referent of the previous one. The *unrestrict* rule performs the exactly inverse operation. Figure 9.3 illustrates how the concepts [Watercourse] and [Basin] are restricted and replaced by the more specialized concepts [River] and [Basin: AthensBasin] and how, inversely, the concepts [River] and [Basin: AthensBasin] are unrestricted and replaced by the more generalized concepts [Watercourse] and [Basin].

The *join* rule is a specification rule, whereas *detach* is a generalization rule. These rules define inverse operations. The *detach* rule performs an operation that splits a CG at a certain concept node into two unrelated CGs derived from the first one. The *join* rule defines the exactly inverse operation and attaches two unrelated CGs over the same concept. Figure 9.4 illustrates how two CGs are joined over the concept [River] and detached at the same concept.

9.5   CONCEPTUAL GRAPHS IN GEOGRAPHIC APPLICATIONS

Recently, the need to explore the potential of CGs in geospatial applications has been recognized (Roddick, Hornsby, and de Vries, 2003; MacEachren, Gahegan, and Pike, 2004; Karalopoulos, Kokla, and Kavouras, 2005; Kontaxaki and Kavouras,

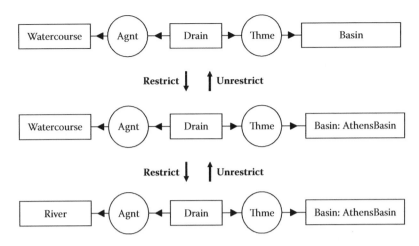

FIGURE 9.3 Example of the application of Restrict and Unrestrict rules.

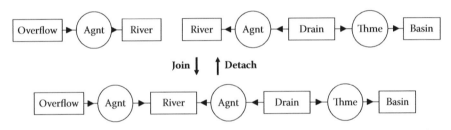

FIGURE 9.4 Example of the application of Join and Detach rules.

2005). Karalopoulos, Kokla, and Kavouras (2005) propose a methodology for the representation of geographic knowledge using CGs. The purpose of this approach is to take advantage of the potential of CGs to represent semantic information and make the semantics of geographic concepts explicit. More specifically, an algorithm that takes a geographic concept definition as input and produces the corresponding CG representation was introduced. This algorithm starts with a tagging step, which consists in recognizing the different parts or tokens of the definition and attributing tags to each of them depending on its grammar functionality. Tags correspond to articles, nouns, verbs, adjectives, prepositions, or conjunctions. The tagging process results in a set of associated tokens and tags. This set is parsed during the next and last step of the algorithm in order to create the CG representation of the definition.

To accomplish the retrieval of geospatial information in both a human-friendly and an effective way, Kontaxaki and Kavouras (2005) proposed an approach based on conceptual graphs and a Controlled English Query Language enriched with geo-spatial concepts and relations in a way that supports geospatial reasoning. More specifically, the controlled query language, called Geo-Q, was introduced for the formulation of questions submitted to geographical database systems in order to extract spatial knowledge. According to this methodology, Geo-Q question's parts are mapped into the components of a Conceptual Query Graph, which is successively transformed into an SQL query. Spatial knowledge is therefore extracted from the execution of the SQL query by the database system.

REFERENCES

Cyre, W. R. 1997. Knowledge extractor: A tool for extracting knowledge from text. In *Proc. Fifth International Conference on Conceptual Structures (ICCS'97)*, 607–610. Berlin: Springer-Verlag.

Hensman, S. 2004. Construction of conceptual graph representation of text. In *Proc. The Student Research Workshop at HLT-NAACL*, Boston, 49–54.

Huibers, T., I. Ounis, and J. P. Chevallet. 1996. Conceptual graph aboutness. In *Conceptual structures: Knowledge representation as interlingua. 4th International Conference on Conceptual Structures (ICCS'96)*, ed. P. W. Eklund, 130-144. Lecture Notes in Artificial Intelligence 1115, ed. P. W. Eklund, G. Ellis, and G. Mann, Berlin: Springer-Verlag.

Karalopoulos, A., M. Kokla, and M. Kavouras. 2005. Comparing representations of geographic knowledge expressed as conceptual graphs. In *Proc. of First International Conference on GeoSpatial Semantics (GeoS 2005)*, ed. M. A. Rodríguez, I. F. Cruz, M.J, Egenhofer, and S. Levashkin, 1–14. Lecture Notes in Computer Science 3799. Berlin: Springer-Verlag.

Kontaxaki S. and M. Kavouras. 2005. Geo-Q: Spatial knowledge extraction based on a Controlled English Query Language and conceptual graphs. In *Proc. International Conference and Exhibition on Geographic Information*, May 29–June 2, Estoril, Portugal.

MacEachren, A. M., M. Gahegan, and W. Pike. 2004. Visualization for constructing and sharing geo-scientific concepts. In *Proc. National Academy of Sciences of the United States of America (PNAS)* 101: 5279–5286. http://www.pnas.org/cgi/reprint/101/suppl_1/5279. pdf (accessed 2 May 2007).

Montes-y-Gómez, M., A. Gelbukh, and A. López-López. 2002. Text mining at detail level using conceptual graphs. In *Proc. 10th International Conference on Conceptual Structures (ICCS 2002)*, Borovets, Bulgaria. Lecture Notes in Artificial Intelligence, 2393. Berlin: Springer-Verlag.

Roddick, J. F., K. Hornsby, and D. de Vries. 2003. A unifying semantic distance model for determining the similarity of attribute values. In *Proc. Twenty-Sixth Australasian Computer Science Conference (ACSC2003)*, Vol. 16. Adelaide, Australia: Australian Computer Society.

Sowa, J. F. 1984. *Conceptual structures: Information processing in mind and machine.* Boston, MA: Addison-Wesley, Longman Publishing Co.

Sowa, J. F. 2000. *Knowledge representation: Logical, philosophical and computational foundations.* Pacific Grove, CA: Brooks Cole Publishing.

Sowa, J. F. 2002. Conceptual graphs, Working Draft of the Proposed ISO Standard. http://www.jfsowa.com/cg/cgstand.htm (accessed 2 May 2007).

Sowa, J. F. 2004. Graphics and languages for the flexible modular framework. In *Conceptual structures at work*, ed. K. E. Wolff, H. D. Pfeiffer, and H. S. Delugach, 31-51. Lecture Notes in Artificial Intelligence 3127. Berlin: Springer-Verlag.

Zhong, J., H. Zhu, J. Li, and Y. Yu. 2002. Conceptual graph matching for semantic search. In *Proc. 10th International Conference on Conceptual Structures (ICCS 2002)*, Borovets, Bulgaria, July 15–19. Lecture Notes in Computer Science 2393. Berlin: Springer.

10 Channel Theory

10.1 INTRODUCTION

Barwise and Seligman (1997) developed *Channel Theory*, a mathematical theory for modelling the flow of information between the components of a system. Channel Theory, also known as *Information Flow Theory* or just *Information Flow* (IF), is inspired by *Category Theory* (CT). *Category Theory* studies mappings/transformations, called *morphisms*, over abstract groups of objects, called *categories*. Categories and morphisms are the two primitive concepts of category theory. Morphisms are considered as the processes preserving categories' structures. Access to the internal structure of categories is prevented and their properties have to be specified by morphisms' properties (Asperti and Longo, 1991).

Comparing CT and IF, Schorlemmer et al. (2002) argue that the manipulation of mathematical structures through structure-preserving relationships in CT can be parallelized with the information management in Channel theory, which focuses on the flow of information rather than on the information itself. Channel theory can also be considered the evolution of *Situation Theory* (ST). ST started as a discipline concentrated on natural language's semantics, but it provides techniques that can be used in many areas of logic theory (Restall, 2005). ST lacked the capacity of making inferences from information existing in situations. Channel theory overcomes this deficiency by permitting the determination of channels that connect situations and support information flow between them (Restall, 2005).

The basic idea behind the notion of information flow reflects the notion of *containment,* which can be explained as the information an object contains about another. Information flow is better understood within the context of a *distributed system*. The expression "distributed system" does not necessarily refer to a hard system as such but to a conceptual one. In Channel theory, distributed systems are regarded as wholes whose parts are related. Formulated by Barwise and Seligman (1997, 8), the first principle, "Information flow results from regularities in a distributed system", reflects the role of regularities in ensuring the flow of information between the parts of such systems. Consequently, the more random a system is, the less information can flow (Bremer and Cohnitz, 2004).

10.2 BASICS OF CHANNEL THEORY

10.2.1 CLASSIFICATION

One of the basic concepts of Channel theory is the concept of *classification. Classifications* can be thought as structures used to model a domain of knowledge. In

(Goguen, 2006), Channel theory classifications are described as a type of primitive ontology. Goguen (2006) also found them comparable to FCA formal contexts (Chapter 8) by corresponding instances to objects, types to attributes, and classification relations to incidence relations. The remaining of this section presents only the basic notions of Channel theory, as they apply to the objective of semantic integration. These notions and the definitions provided below are drawn from the work of Barwise and Seligman (1997), unless otherwise stated.

FIGURE 10.1   Classification A.

In Channel theory, a *classification* $A = \langle A, \Sigma_A, \vDash_A \rangle$, consists of a set A of objects to be classified, called the *tokens* of A, a set Σ_A of objects for the classification of the tokens, called the *types* of A, and a binary relation $\vDash_A$ between A and Σ_A which determines which tokens are classified of being of which types. A diagrammatic representation of a classification is shown in Figure 10.1.

For example, in order to model the Greek river Acheloos and its branches, the classification $A = \langle A, \Sigma_A, \vDash_A \rangle$ may be defined as a set of types $\Sigma_A = \{$WATERCOURSE, RIVER, BRANCH, RIVULET$\}$ containing geographic concepts and a set of tokens $A = \{$Acheloos, Agraphiotis, Tavropos, Trikeriotis, Aspros, Platanias$\}$ containing the instances to be classified according to the classification relation $\vDash_A$ specified in Table 10.1.

A special kind of classification exploited by Barwise and Seligman (1997) is that of a *state space*. *State spaces* are used to model the state of objects or more generally of abstract structures that change under certain circumstances, e.g., with the course of time. They formally define a state space as the classification in which each token is exactly of one type; the types of a state space are called *states*.

For example, let R be a sequence of Greek rivers and T the ordered set of time T = \{January, February, ..., December\}. In order to model the state of Greek rivers' seasonal depth, a state space S can be defined consisting of a set of tokens $S = \{\langle R, t \rangle, t \in T\}$, a set of types $\Sigma_S = \{$LOW, NORMAL, HIGH, OVERFLOW$\}$ and a classification relation $\vDash_S$ that relates each token to an appropriate state. At the

TABLE 10.1

Specification of Relation $\vDash_A$ of Classification A

$\vDash_A$	WATERCOURSE	RIVER	BRANCH	RIVULET
Acheloos	1	1	0	0
Agraphiotis	1	1	1	0
Tavropos	1	1	1	0
Trikeriotis	1	1	1	0
Aspros	1	0	1	1
Platanias	1	0	1	1

TABLE 10.2

Specification of Relation $\models_S$ of State Space S

$\models_S$	LOW	NORMAL	HIGH	OVERFLOW
Acheloos	0	1	0	0
Kifissos	0	0	0	1
Axios	0	0	0	1
Eurotas	0	0	1	0
Nedontas	1	0	0	0

specific time t = February, the rivers (tokens) stand in relation ($\models_S$) to the states (types) as it is shown in Table 10.2.

Furthermore, Barwise and Seligman (1997) introduce the notion of *theory* to model the *constraints* supported by classifications. A *theory* consists of a set of types and an entailment relation $\vdash$ between the sets of types. A pair of such sets $\langle \Gamma, \Delta \rangle$ is called a *sequent*.

Given a classification A and a sequent $\langle \Gamma, \Delta \rangle$ of classification A. A token α of A *satisfies* the sequent provided that if α is of type γ for every $\gamma \in \Gamma$, then α is of type γ for some $\gamma \in \Delta$. If every token α of A satisfies $\langle \Gamma, \Delta \rangle$, then it is said that Γ *entails* Δ in A, and is written $\Gamma \vdash_A \Delta$.

Barwise and Seligman define five types of constraints: (a) entailment, (b) necessity, (c) exhaustive cases, (d) incompatible types, and (e) incoherent types. For instance, regarding the classification A presented before, the constraint RIVULET $\vdash$ BRANCH indicates that all rivulets are branches. This is an entailment constraint. $\vdash$ WATERCOURSE is a necessity constraint, which means that it is necessary for all the tokens to be of type watercourse. $\vdash$ RIVER, RIVULET corresponds to an exhaustive case of constraint. It indicates that every token has to be inevitably either a river or a rivulet. RIVER, RIVULET $\vdash$ indicates incompatible types, that is, it is impossible for a token to be both a river and a rivulet. The constraint CANAL $\vdash$ corresponds to an incoherent type, since, according to the above example (Table 10.1), there is no token of type "canal."

The *complete* theory of a classification, denoted by Th(A), consists of the set of all constraints and represents all the regularities of the system modeled by the classification.

10.2.2 Sum of Classifications

In order to combine classifications, Barwise and Seligman (1997) define the *sum* of classifications. The sum of classifications results in a new classification whose parts consist of the initial ones. Given two classifications A and B, their sum $A + B$ is a classification whose components are defined as in the following. The tokens of $A + B$ are the set of pairs $\langle \alpha, b \rangle$, where $\alpha \in A$ and $b \in B$; thus, the set $A+B$ is the Cartesian product of A and B. The set Σ_{A+B} is the disjoint union of Σ_A and Σ_B. The types of

$A + B$ are pairs $<i, \gamma>$, where $i = 0$ and $\gamma \in \Sigma_B$ or $i = 1$ and $\gamma \in \Sigma_B$. The tokens $<\alpha, b>$ stand in relation $\vDash_{A+B}$ to the types of $A + B$, so that for any type A of A and B of B:

$$<\alpha, b> \vDash_{A+B} <0, \gamma> A_1 \text{ iff } \alpha \vDash_A \gamma \text{ and } < \alpha, b > \vDash_{A+B} <1, \delta> \text{ iff } b \vDash_B \delta.$$

For example, the geographic concept of a river considered as a watercourse, which empties into a target body of water, can be modeled as a sum of classifications. Let A be the classification corresponding to "watercourse" and B the classification corresponding to "target body of water," it can be said that the classification $A = A + B$ models the concept "river." The tokens of the sum are pairs $<\alpha, b>$ where $\alpha \in A$ and $b \in B$, e.g., <Kifissos, Mediteranean Sea>, <Agraphiotis, Acheloos>, etc. Σ_{A+B} results from the disjoint union of Σ_A and Σ_B. Thus, in case the target body of water is also a river, two distinct types of river have to be defined in the sum, one corresponding to watercourses and one to target bodies of water. Barwise and Seligman (1997) generalize the definition of the sum of two classifications for i classifications, where $i = 1, 2, 3, \ldots$.

10.2.3 Local Logic

According to Barwise and Seligman (1997), a *local logic* $\mathcal{L} = <A, \vdash_\mathcal{L}, N_\mathcal{L}>$ consists of three parts, a classification A, a set $\vdash_\mathcal{L}$ of sequents that involve the types of A, and are called the *constraints* of $\mathcal{L}$, and a subset $N_\mathcal{L} \subseteq A$, that satisfy all the constraints of $\vdash_\mathcal{L}$, and are called the *normal tokens* of $\mathcal{L}$. A local logic is said to be *sound* if every token is normal, and *complete* if every sequent satisfied by every normal token is defined as a logic constraint.

Furthermore, each classification has a local logic that governs its types (Bremer and Cohnitz, 2004). The logic generated by a classification is called *natural*. A natural logic models the regularities of a classification and is always sound and complete. For example, the classification A presented before, modeling the Greek river Acheloos and its branches, generates a natural logic whose theory's constraints are:

RIVULET $\vdash$ BRANCH, $\vdash$ WATERCOURSE, $\vdash$ RIVER, RIVULET,   RIVER, RIVULET $\vdash$.

10.2.4 Infomorphism

For relating classifications with each other, Barwise and Seligman (1997) introduce the notion of *infomorphism*. Let $A = < A, \Sigma_A, \vDash_A>$, and $B = < B, \Sigma_B, \vDash_B>$, be two classifications. An infomorphism f that relates A and B consists of a pair of two contravariant functions: $f^\wedge$ from the types of A to the types of B, and $f^\vee$ from the tokens of B to the tokens of A (Figure 10.2). f is such that:

$$f^\vee(b) \vDash_A \alpha \text{ iff } b \vDash_B f^\wedge(\alpha) \text{ for all types } \alpha \text{ of } A \text{ and for all tokens } b \text{ of } B.$$

For example, using the classification A presented above, which models the Greek river Acheloos and its branches, an infomorphism f can be defined to relate the geographic concepts contained in classification A with definitions provided by

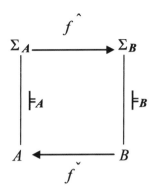

FIGURE 10.2   Infomorphism f from A to B.

the Wordnet ontology. Let W be the classification whose set of tokens, W, contains the definitions of the nouns "watercourse," "river," "branch," and "rivulet," whose set of types, Σ_W, contains the nouns "watercourse," "river," "branch," and "rivulet," and classification relation, $\vDash_W$, maps each noun with its definition as shown in Table 10.3.

The infomorphism f allows information to flow from A to W and vice-versa. Furthermore, the condition $f^\vee(w) \vDash_A \alpha$ iff $w \vDash_W f^\wedge(\alpha)$ is satisfied for all the types α of A and for all the tokens w of W. For instance, the following relations hold:

$$f^\vee \text{ ("a small stre}\vDash\text{m")} \vDash_A \text{RIVULET and "a smal}\vDash \text{ stream" } \vDash_W f^\wedge(\text{RIVULET}),$$

where "a small stream" is a token of W and RIVULET is a type of A.

In essence, infomorphisms describe how the components of a system are related to the whole of a system, since they model the part-whole relations among instances of a whole (Barwise and Seligman, 1997). Infomorphisms can also be seen as the constituents of channels, which express the structure of distributed systems (Goguen, 2006). Channels are examined in the following sections.

TABLE 10.3

Specification of Relation $\vDash_W$ of Classification W

$\vDash_S$	NOUN WATERCOURSE	NOUN RIVER	NOUN BRANCH	NOUN RIVULET
A natural body of running water flowing on or under the earth	1	0	0	0
A large natural stream of water (larger than a creek)	0	1	0	0
A stream or river connected to a larger one	0	0	1	0
A small stream	0	0	0	1
A long and narrow strip of water made for boats or for irrigation	0	0	0	0

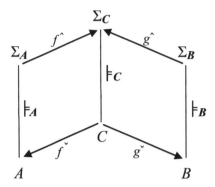

FIGURE 10.3   Channel C connecting classifications A and B.

10.2.5 CHANNEL

Channels constitute a central notion of Barwise and Seligman's Channel theory. Channels allow the formalization of the context in which the flow of information takes place. In other words, a channel is the medium for information flow between classifications.

Formally, given two classifications A and B, a channel consists of a core classification C that connects them via two infomorphisms f and g (Schorlemmer, 2003) (Figure 10.3). This is the case of a binary channel.

In Figure 10.3, the tokens α of type A are connected to the tokens b of type B through the tokens c of type C as following: $f\check{}(c) = \alpha$ and $g\check{}(c) = b$. The token c is considered a connection between α and b. Furthermore, it is said that the token α of type A carries the information that b is of type B. In general, it is said that tokens carry information and types represent the information that is carried. Thus, both tokens and types are involved in the flow of information as the second principle of Barwise and Seligman's (1997, p. 27) theory points out: "Information flow crucially involves both types and its particulars".

For example, in the framework of a GIS tool, geospatial features may be represented according to the vector or the raster model. Let V be the classification of these features under the vector model and R the classification under the raster model. Thus, a linear feature may correspond to the token $((x_1, y_1), (x_2, y_2))$ of the type LINE_V in V and to the token (pixel1, pixel2, pixel3) of the type LINE_R in R. A channel C connecting these classifications has to model the mappings between them through specific infomorphisms f and g (Figure 10.4).

As a result, in the case of the linear feature presented previously, there must be a token $c \in C$ such that $f\check{}(c) = ((x_1, y_1), (x_2, y_2))$ and $g\check{}(c) = $ (pixel1, pixel2, pixel3).

However, information channels are not always binary. A more general definition of an information channel is that of an indexed family C of infomorphisms such that: $C = \{f_i : A_i \rightleftarrows C\}$, where C is a common codomain

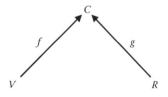

FIGURE 10.4   Channel C connecting classifications V and R.

called the *core* of the channel and A_i are the classifications/components of a system connected through that channel. Barwise and Seligman's third principle, states that "it is by virtue of regularities among connections that information about some components of a distributed system carries information about other components" (Barwise and Seligman, 1997, p. 35). These regularities are represented by the classification C and its corresponding theory with the constraints, and it can be said that, by being parts of a system C, the components A_i carry information about one another.

REFERENCES

Asperti, A. and Longo, G., 1991, *Categories, Types and Structures:An introduction for the working computer scientist*, Foundations of Computing Series, MIT Press, Cambridge, MA, USA.

Barwise, J. and Seligman, J., 1997, *Information flow: The logic of distributed systems.* Cambridge Tracts in Theoretical Computer Science. Cambridge, UK: Cambridge University Press.

Bremer, M. and Cohnitz, D., 2004, *Information and Information Flow: An Introduction*, Ontos-Verlag.

Goguen, J., 2006, Information integration in institutions. To appear in Jon Barwise Memorial Volume edited by Larry Moss, Indiana University Press. Draft available from http://www.cs.ucsd.edu/users/goguen/pps/ifi04.pdf (accessed 2 May 2007).

Restall, G., 2005, Logics, situations and channels. *Journal of Cognitive Science,* 6: 125-150.

Schorlemmer, M. 1993. Channel theory. The Information Flow Framework, IEEE P1600.1, Standard Upper Ontology Working Group (SUO WG), http://suo.ieee.org/IFF/work-in-progress/IF/chth.pdf (accessed 10 September 2007).

Schorlemmer, M., Potter, S., and Robertson, D., 2002, Automated Support for Composition of Transformational Components in Knowledge Engineering, Informatics Research Report EDI-INF-RR-0137, Division of Informatics, University of Edinburgh.

11 Description Logics

11.1 INTRODUCTION

Description logics (DL) are logic-based knowledge representation languages used for modeling and reasoning about knowledge. DL were developed by artificial intelligence (AI) researchers as an evolution of earlier knowledge representation formalisms, such as *semantic networks* (Quillian, 1967; Lehmann, 1992) and *frames* (Minsky, 1975), supplied with formal logic-based semantics and particularly *first-order logic* (FOL). In fact, DL can be seen as a mathematical restriction on or a structured fragment of FOL (Nardi and Brachman, 2003).

Nardi and Brachman (2003) provide a historical analysis of the development of DL. The first DL systems were called *terminological systems* since they were used to describe the basic terminology adopted in specific domains. Later, the term *concept languages* was adopted to stress the constructs for the development of concepts provided by the language. Finally, the term Description Logics prevailed in order to place emphasis on the "properties of the underlying logical systems" (Nardi and Brachman, 2003, 7). An issue of concern in DL research is the trade-off between expressiveness and computational complexity of reasoning (Brachman and Levesque, 1984).

In the following, the basic concepts and formalisms of DL are presented.

11.2 DESCRIPTION LOGICS SYNTACTIC ELEMENTS

DL, like most contemporary conceptual models adhere to an object-centered view of the world; according to this view, the world consists of individual objects, which are related to each other and which form classes of objects with common properties (Borgida and Brachman, 2003). The syntactic elements used in DL are:

- Atomic concepts
- Atomic roles
- Individuals
- Operators (constructors)

More specifically, *atomic concepts* represent sets of individuals, and, in terms of mathematical logic, they are regarded as unary predicates. *Atomic roles* describe relations among individuals, and in a similar way, in terms of mathematical logic, they are regarded as binary predicates. *Individuals* correspond to particular instances of concepts and are used in the way constants are used in FOL. In standard notation, the letters A and B denote atomic concepts, the letter R atomic roles, and the letters

C and D concept descriptions, that is new concepts derived from other concepts. Finally, *operators*, such as intersection ⊓, union ⊔, and complement ¬ of concepts, define the operations used by the language in order to construct new concepts from already defined ones.

The intersection of concepts (concept conjunction), denoted as C ⊓ D, returns the set of individuals that belong to both C and D. The union of concepts (concept disjunction), denoted as C ⊔ D, returns the set of individuals that belong to either C or D. The complement of concepts (concept negation), denoted as ¬C returns the individuals which do not belong to the concept C.

Relations are also used to generate new concepts (Roos, 2006). *Value restrictions* are constructors used for generating relations between concepts. For example, the value restriction denoted as ∀R.C returns all individuals that are in relation R with the individuals that belong to the concept C. The *universal concept*, denoted as ⊤, includes every individual, while the *bottom concept,* denoted as ⊥, does not include any individual.

For example, if the concepts Watercourse and Natural are atomic concepts and hasTributary is a defined role, then the expression:

$$\text{Watercourse} \sqcap \text{Natural} \sqcap \forall \text{ hasTributary.River}$$

defines the concept of a natural watercourse that has tributary rivers flowing into it, that is, the concept "River."

Constructors such as union, intersection and complement, also apply to roles. For example, the case of an object that both overlaps and is situated to the north of another object can be expressed as the intersection of the roles: Overlaps ⊓ OnTheNorthOf.

11.3 DESCRIPTION LOGICS LANGUAGES

DL languages are considered the central part of knowledge representation systems dealing with the structure of the underlying knowledge bases and their relative reasoning services (Nardi and Brachman, 2003). They are distinguished according to the constructors they include.

Brachman and Levesque (1984) studied the simple DL language called *Frame Language* (*FL*). The *Attributive Language* (*AL*), introduced in 1991 by Schmidt-Schauß and Smolka, is a minimal practical language, which consists the basis of the family of attributive languages (cited in Baader and Nutt, 2003). The syntax of *AL* includes:

- A set of atomic concepts
- The universal concept (⊤)
- The bottom concept (⊥)
- Atomic negation (¬A)
- Intersection (C⊓D)
- Value restriction (∀R.C)
- Limited existential quantification (∃R.⊤)

Many extended languages have been constructed from the $\mathcal{AL}$ language (Baader and Nutt, 2003). Intuitively, any extension in the expressiveness of a DL language makes the reasoning harder from the aspect of computability. The $\mathcal{AL}$ enriched with the capability of expressing the negation of arbitrary concepts (and not only of atomic concepts), the union of concepts, and the full existential quantification is called *Attributive Language with Complements* ($\mathcal{ALC}$) (Baader and Nutt, 2003). $\mathcal{ALC}$ is considered to be the smallest DL that is propositionally closed (Franconi, 2002; Lutz and Sattler, 2005). In the geospatial domain, a family of DL called $\mathcal{ALCI}_{\mathcal{RCC}}$ is introduced for qualitative spatial representation and reasoning at various levels of granularity (Wessel, 2002, 2003).

11.4 DESCRIPTION LOGICS KNOWLEDGE ORGANIZATION AND MODELING

DL formalism in knowledge representation distinguishes between *intensional* and *assertional* (*extensional*) knowledge. According to this formalism, a Knowledge Base consists of two components: the *TBox* and the *ABox* (Donini et al., 1997; Baader and Nutt, 2003; Nardi and Brachman, 2003; Breitman, Casanova and Truszkowski, 2007). The *TBox* (T stands for terminology) contains the *intensional* knowledge or, in other words, the general knowledge of the domain of interest in the form of a *terminology*. A *terminology* defines the vocabulary of the domain and consists of concepts, which correspond to sets of individuals and roles, which correspond to binary relations among individuals (Baader and Nutt, 2003). Apart from atomic concepts and roles, the *Tbox* is used to define new and more complex concepts and roles using different description languages.

In general, definitions of concepts and roles have the form (Baader and Nutt, 2003):

$$C \sqsubseteq D \text{ or } C \equiv D$$

$$R \sqsubseteq S \text{ or } R \equiv S$$

where C and D are concepts, R and S are roles, and $\sqsubseteq$ and $\equiv$ stand, respectively, for the inclusion and the equality axioms. In case $C \sqsubseteq D$, it is said that C constitutes a specialization of D, while $C \equiv D$ means that C and D are equivalent concepts. Similar remarks hold for the roles R and S. For example, the concept "Canal" is defined using the equality axiom as follows:

$$\text{Canal} \equiv \text{Watercourse} \sqcap \text{Artificial}$$

The *ABox* set (A stands for assertional) contains the *assertional* knowledge of the domain, that is the knowledge regarding the individuals called *membership assertions*. Membership assertions are further classified into *concept assertions* and *role assertions* (Nardi and Brachman, 2003) denoted respectively as C(α) and R(α,b), where C stands for a concept, R stands for a role, and α, b stand for individuals (Baader and Nutt, 2003). For example, the concept assertion:

Canal(Panama)

states that the individual "Panama" is an instance of the concept "Canal," that is, an artificial watercourse. Similarly the role membership assertion:

hasTributary(Euphrates, Tigris)

states that the individual "Tigris" is tributary to "Euphrates."

Generally, intensional knowledge is diachronic, in other words, it does not change through time, wheres assertional knowledge is not diachronic and is possible to change under certain circumstances, e.g., through time (Nardi and Brachman, 2003). For example, the definition of the Concept "River" as well as the relation between a "River" and its "Tributaries" are specific and not subject to changes. However, human intervention may cause a particular river to be desiccated and cease to exist or the natural course of a particular tributary to be diverted and flow into another river or lake.

11.5 REASONING IN DESCRIPTION LOGICS

Reasoning is an important capability of DL-systems, since it "allows one to infer implicitly represented knowledge from the knowledge that is explicitly contained in the knowledge base" (Baader and Nutt, 2003, 47). Reasoning in DL-systems is based both on the *TBox* which deals with axioms that express concept properties and the *ABox* which deals with individuals (Donini et al., 1997). Donini et al. (1997) analyze reasoning services into four categories: (1) reasoning with plain concept expressions, (2) reasoning with individuals, (3) reasoning with axioms that express concept properties, and (4) reasoning with both individuals and axioms. More specifically, the basic kinds of reasoning services provided by DL languages are:

- *Subsumption*: for given concepts C and D, checking whether C is subsumed by D, that is if C constitutes a specialization of D.
- *Concept satisfiability*: checking whether a concept is satisfiable in a specific knowledge base, that is checking if a concept expression under a specific interpretation does not correspond to an empty concept.
- *Consistency*: checking whether all the concepts of a knowledge base have at least one instance.
- *Instance checking*: checking whether the assertion $C(\alpha)$ is satisfied, that is checking if α is indeed an instance of C.
- *Retrieval*: given a concept C, retrieving its instances.
- *Realization*: if α is an individual, finding whose concept C, α is the instance.

However, it is possible for DL languages to additionally provide other kinds of reasoning tasks such as finding the least common subsumer of a set of concepts, finding the most specific concept of an *ABox* individual, etc.

REFERENCES

Baader, F. and W. Nutt. 2003. Basic description logics. In *The description logic handbook: Theory, implementation and applications*, Eds. F. Baader, D. Calvaneze, D.L. McGuinness, D. Nardi, and P.F. Patel-Schneider, New York: Cambridge University Press, 43–95.

Borgida, A., and R.J. Brachman. 2003. Conceptual modeling with description logics. In The description logic handbook: Theory, implementation and applications, Eds. F. Baader, D. Calvaneze, D.L. McGuinness, D. Nardi, and P.F. Patel-Schneider, New York, NY, USA: Cambridge University Press, 349–372.

Brachman, R. and H. Levesque. 1984. "The Tractability of Subsumption in Frame-Based Description Languages", R.J. Brachman (Ed.): Proceedings of the National Conference on Artificial Intelligence. Austin, TX, August 6–10, 1984. AAAI Press, 34–37

Breitman, K.K., M.A. Casanova, and W. Truszkowski. 2007. Knowledge Representation in Description Logic. In Semantic Web: Concepts, Technologies and Applications, NASA Monographs in Systems and Software Engineering, Springer London, 35–55.

Donini, F.M., M. Lenzerini, D. Nardi, and A. Schaerf. 1997. Reasoning in Description Logics. In CSLI Studies In Logic, Language And Information Series: Principles of knowledge representation, Center for the Study of Language and Information, Stanford, CA, USA, 191–236.

Franconi E. 2002. Description Logics: Propositional Description Logics, Tutorial course, Master of Science in Computer Science, Free University of Bozen-Bolzano, Italy, http://www.inf.unibz.it/~franconi/dl/course/slides/prop-DL/propositional-dl.pdf (accessed 10 September 2007).

Lehmann F. 1992. Semantic Networks. In *Semantic networks in artificial intelligence*, Fritz Lehmann (ed.) Elsevier Science Inc. New York, NY, 1–50.

Lutz, C. and U. Sattler. 2005, Course: Description logics, ICCL Summer School 2005: Logic-based Knowledge Representation, 2–17 July 2005 Department of Computer Science, Technical University of Dresden, Germany, , http://www.computational-logic.org/content/events/iccl-ss-2005/lectures/lutz/index.php?id=24 (accessed 1 September 2007).

Minsky, M. 1975. A framework for representing knowledge. In *The psychology of computer vision*, ed. P. H. Winston. New York: McGraw-Hill. http://web.media.mit.edu/~minsky/papers/Frames/frames.html (accessed 1 May 2007).

Nardi D. and R.J. Brachman, 2003. An introduction to description logics. In The description logic handbook: Theory, implementation and applications, Eds. F. Baader, D. Calvaneze, D.L. McGuinness, D. Nardi, and P.F. Patel-Schneider, New York: Cambridge University Press, 1–40.

Quillian, M.R. 1967. Word concepts: a theory and simulation of some basic semantic capabilities. *Behavioral Science*, Vol. 12, No. 5. (September 1967), 410–430.

Roos, N. 2006. Description logics. Course on Knowledge representation and reasoning, Master on Computer Science, Artificial Intelligence and Knowledge Engineering, University of Maastricht, The Netherlands. http://www.fdaw.unimaas.nl/education/4.2FKRR/description%20logic.pdf (accessed 1 September 2007).

Wessel, M., 2002, On spatial reasoning with description logics - position paper. In *Proc. Int. Workshop in Description Logics 2002 (DL2002)*, Toulouse, France, 19–21.

Wessel, M., 2003, Qualitative spatial reasoning with the $ALCI_{RCC}$ family — first results and unanswered questions. Tech. Rep. FBI-HH-M-324/03, FB Informatik, University Hamburg.

12 Natural Language and Semantic Information Extraction

12.1 SEMANTIC INFORMATION EXTRACTION FROM TEXT

Natural Language Processing (NLP) is based on Computational Linguistics (Bolshakov and Gelbukh, 2004). It focuses on problems regarding human-computer interaction, such as natural language generation and understanding, information retrieval and information extraction, machine translation, speech recognition, etc. Generally NLP has to face the problem of the representation of world knowledge (commonsense, general, or domain) and of the association of such knowledge representations with linguistic structures such as vocabulary and grammar (Bateman, 1992). Ontologies, as classifications and representations of the things that exist in the world in general or in a particular domain, are highly relevant to the objectives of NLP. Ontologies are called on to fulfill different functions of NLP; they contribute to the organization of world knowledge (or the world itself) and the formalization of semantics of natural language expressions and constitute the "interface between system external components, domain models, etc. and NLP linguistic components" (Bateman, 1992).

For the purposes of this book we focus on NLP tasks relative to semantic information extraction. Semantic information identified in natural language text is represented by the notion of semantic relations. In the field of NLP, *semantic relations* or *semantic roles* or *thematic roles* refer to the relation of a constituent to the main verb in a clause. Semantic relations may be either general, application-independent constructs or specific, domain/application dependent ones. The main semantic relations are: AGENT, EXPERIENCER, PATIENT, INSTRUMENT, CAUSE, LOCATION, PURPOSE, SOURCE, and PATH. For example, in the sentence, "The hunter fired at the jackrabbit," the "hunter" is the AGENT and the "jackrabbit" is the PATIENT of the action.

The extraction of semantic information is based on the mapping of linguistic expressions with syntactic relations (e.g., subject-verb-object triples) to semantic relations. Pattern matching is the primary method for the automatic identification of semantic relations in text (Khoo and Myaeng, 2002). However, other approaches, such as probabilistic parsing for trees augmented with annotations for entities and relations or clustering of semantically similar syntactic dependencies, are also used (Specia and Motta, 2006).

Patterns are distinguished into two categories (Khoo and Myaeng, 2002): (a) linear patterns of words and syntactic categories in text and (b) graphical patterns in the syntactic structure, that is, the parse tree of the text. In case a text segment or a part of a parse tree matches a pattern, then this entails that the text segment or the part of the parse tree includes the semantic relation. For example, the pattern "[effect] is the result of [cause]" expresses the cause-effect relation. Both categories of patterns have advantages and disadvantages (Khoo and Myaeng, 2002). Linear patterns offer simplicity and do not require the construction of a complete syntactic parse tree; thus, they obviate the problems caused due to the parsing process. On the other hand, a large number of linear patterns may be needed to cover the demands of a specific application, therefore creating problems during their generation, management, and in-between interaction.

The identification of semantic relations may be performed either based on inferences from general or domain knowledge or without such a requirement. The first case is appropriate for applications relevant to specific domains that require high accuracy and is based on manually codified knowledge. The second case is appropriate for general approaches relevant to large subject areas that do not require high precision.

The automatic identification of semantic relations in text can be useful for different tasks, such as information retrieval (e.g., relation matching), information extraction, question answering, semantic web annotation, construction of lexical resources and ontologies, and other natural language processing applications (Khoo and Myaeng, 2002; Specia and Motta, 2006). Furthermore, it is used in different scientific domains, such as bioscience (Rosario, 1995; Roserio and Hearst, 2004), medicine (Bhooshan, 2005), law (Lame, 2003), and geographic information science (Kokla and Kavouras, 2005).

The Berkeley FrameNet (Baker, Fillmore, and Lowe, 1998; Fillmore and Baker, 2001; Fillmore, Wooters, and Baker, 2001) is a project that aims at the development of a lexical database and software tools. The lexical database includes frame-based semantic descriptions for several thousand English words (mainly nouns, adjectives, and verbs) and semantically annotated sentences from contemporary English corpora. Frames are conceptual structures that involve frame elements, that is, semantic relations either general or domain-specific. For example, the ACTION frame involves frame elements ACTOR, MEANS, and MANNER; the TRANSPORTATION frame includes frame elements such as MOVER, MEANS, and PATH. Gildea and Jurafsky (2002) developed a statistical approach for automatically labeling both general (e.g., AGENT or PATIENT) and domain-specific (e.g., SPEAKER, MESSAGE, and TOPIC) semantic roles, based on frames identified by the FrameNet project (Baker, Fillmore, and Lowe, 1998).

Polaris (LCC, 2007) is a semantic parser that uses a hybrid approach in order to extract a set of forty semantic relations (e.g., CAUSE, INSTRUMENT, PURPOSE, AGENT, LOCATION-SPACE, etc.), which present a balance between generality and specificity, and usefulness and manageability. This approach consists of classifiers that extract one specific semantic relation, others that extract all forty semantic relations, and manually encoded rules to extract the most explicit and unambiguous ones.

Specia and Motta (2006) employ an approach using both supervised and unsupervised knowledge-based techniques to extract semantic relations from texts for semantic web annotation, and ontology learning and population. The components of the method are: a parser, a part-of-speech tagger, a named entity recognition system, a pattern-based classification, and word sense disambiguation models. Also, the method exploits resources such as a domain ontology, a knowledge base, and lexical databases. Entities and relations are either known or new. Known entities and relations are provided by the knowledge base; new entities are specified by the named entity recognition system based on the entity types of the ontology, whereas new relations are either predicted (from the ontology) or unpredicted (totally new).

Loper (2000) developed SQUIRE, a Semantic Question-answering Information Retrieval Engine, which converts sentences into semantic relations, which are recorded in a database. A user's question is also converted into semantic relations and the engine searches the database to find the best matches for these semantic relations.

Some efforts focus on the identification of specific relations. A prominent semantic relation is the causal (cause-effect) relation (Khoo et al., 1998; Girju, 2003; Cole et al., 2006). Khoo et al. (1998) extracted the cause-effect relation from newspaper text using a computational method based on linguistic clues and pattern matching that did not require full parsing of the sentences. Linguistic clues that are used to explicitly denote causal relations may be:

1. Causal links between two phrases, clauses, or sentences (e.g., because of, as, hence, etc.).
2. Causative verbs, that is, transitive verbs, which denote "the result of an action, event, or state or the influence of some object" (Szeto, 1988, cited in Khoo et al., 1998, 178) (e.g., cause, result in, force, enable, kill, etc.).
3. Resultative constructions, that is, a phrase that describes the state of an object as a result of an action signified by a verb (e.g., "I painted the car yellow").
4. Conditionals (e.g., "if … then …").
5. Causative adverbs and adjectives (e.g., fatal).

Girju (2003) developed a method for the extraction and disambiguation of causal relations for a question answering system. Cole et al. (2006) developed a tool for the extraction of causal (cause-effect) relations from text and the automatic creation of Bayesian networks.

12.2 SEMANTIC INFORMATION EXTRACTION FROM DEFINITIONS

The extraction of semantic relations may be used for the formalization and comparison of geographic concept definitions during ontology integration. Hence, for the purposes of this book, we focus on information extraction not from text in general but from definitions, that is, structured text. This is due to the fact that ontology integration usually refers to existing ontologies, whose concepts' semantics is mainly described by definitions. Other elements, such as terms, attributes, and hierarchical relations, are not sufficient for the comparison of similar concepts. Furthermore, existing sources of geographic information do not include other elements that might contribute to an adequate description of category semantics; functions and parts are examples of such elements. For example, "riverbed," "estuary," and "rapid" are parts of a "river," whereas a "hotel's" function is to "provide rooms and meals for people." These additional semantic elements may be accurately extracted from concept definitions.

Semantic information extraction from free text is also important but is more useful in the case of ontology engineering. For example, the process of developing a domain ontology may be facilitated by the analysis of free texts that describe the concepts pertaining to the domain, their meaning, and their between relations. However, the extraction of semantic information from free text is difficult to be performed with high accuracy due to a number of reasons depending on the texts used for the processing, the domain, the semantic relations to be identified, the method used (quality of syntactic processing, amount of text training, use of knowledge-based inferencing, etc.) (Khoo and Myaeng, 2002).

Despite some controversy about their nature, relevance, and importance (see Chapters 5 and 6), definitions are considered essential for the recording, systematization, and exchange of general and domain-specific knowledge. Definitions are considered a kind of text with special structure and content. They constitute important sources of knowledge expressed in natural language (Jensen and Binot, 1987; Klavans, Chodorow, and Wacholder, 1993; Swartz, 1997); the language of dictionary definitions should be considered a sublanguage of natural language (Calzolari, 1984). Moreover, definitions retain the meaning of information and reduce vagueness and misinterpretation during information exchange and integration. Research on definitions is seeking ways to exploit the wealth of knowledge immanent in this special kind of text through the automatic extraction of semantic information from definitions (Jensen, Heidorn, and Richardson, 1993; Barriere, 1997).

According to Wilks et al. (1988) three methodological assumptions underlie the extraction of semantic information from dictionaries: (a) sufficiency, (b) extricability, (c) bootstrapping. The first assumption refers to the sufficiency of knowledge included in dictionaries. The second assumption concerns the possibility of extracting this knowledge using a set of computational procedures without human intervention. The third assumption relates to the collection of initial information required by the set of computational procedures. This information can be either extracted from the dictionary itself using the computational procedures (internal bootstrapping methods) or acquired by a method other than the use of these procedures (external bootstrapping methods).

Geographic information sources use natural language definitions to describe the essential features of their concepts. Definitions describe the meaning of concepts and, therefore, contain sufficient information to disambiguate similar concepts. By analyzing geographic concept definitions, our aim is to identify and represent the knowledge contained in an explicit form in order to identify similarities and heterogeneities between similar concepts. Geographic concept definitions contain general and domain knowledge, that is, knowledge related to the field of geographic information (e.g., land cover, land use, transportation, etc.). As far as their generation method is concerned, geographic concept definitions are intensional, that is, describe meaning by specifying the essential properties of geographic concepts, for example:

"well: a hole drilled or dug into the earth or sea bed for the extraction of liquids or gases"

"grassland: area composed of uncultured plants which have little or no woody tissue"

Intensional definitions are composed of two parts: *genus* and *differentiae*. The genus, or hypernym, is the superordinate term of the defined concept. For example, in the definition: "river: large natural stream of water," "stream" is the genus of "river." The differentiae are other elements of the definition apart from the genus, which distinguish terms with the same genus. In the definition: "creek: a small stream, often a shallow or intermittent tributary to a river," "creek" has the same genus as "river," but they are distinguished by the differentiae (e.g., "large," "natural," "small," and "shallow").

Patterns applied to the genus part of the definition extract the hypernym or the IS-A relation. Chodorow, Byrd, and Heidorn (1985) developed a heuristic for locating the genus terms in noun definitions. The mechanism is based on the observation that the genus term for noun definitions is typically the head of the noun phrase. The special structure of definitions simplifies the analysis process. This consists in isolating the substring of the definition that contains the head and then finding the head in the substring. Empty heads, for example, "one," "kind," "class," "any," followed by the preposition "of" also indicate the IS-A relation. The head in this case is the head of the noun phrase following the preposition "of" (Markowitz, Ahlswede, and Evens, 1986). However, empty heads are not common in geographic concept definitions. Patterns applied to the differentiae part extract other semantic information such as PURPOSE, LOCATION, TIME, SIZE, PART-OF, etc.

In the literature, there is no complete list of semantic information that can be extracted from definitions because they may vary according to the dictionary from which they are extracted (Barriere, 1997), or they may be domain-specific. Furthermore, research on automatic acquisition of semantic information from definitions has focused more on identification of hypernyms or IS-A relations (Chodorow, Byrd, and Heidorn, 1985; Markowitz, Ahlswede, and Evens, 1986; Ide and Veronis, 1993) and less on other semantic elements. Therefore, it is necessary to analyze geographic concept definitions in order to specify which semantic elements are used for the identification of geographic concepts.

For this reason, different geographic ontologies, standards, and categorizations (e.g., Cyc Upper Level Ontology, WordNet, CORINE Land Cover, DIGEST, SDTS, etc.) can be analyzed in order to identify patterns that are systematically used to express specific semantic elements and formulate the corresponding rules. The most commonly used are shown in Tables 12.1 and 12.2 (Kokla and Kavouras, 2005). Although, as mentioned in the previous section, *semantic relations* denote all semantic information extracted generally from text or more specifically from definitions, here, for purposes of conceptual clarity, we further distinguish two types of semantic elements: (a) *semantic properties*, which describe characteristics of the concept itself (internal characteristics) and (b) *semantic relations*, which describe characteristics of the concept relative to other concepts (external characteristics).

Besides general semantic elements (e.g., PURPOSE, CAUSE, TIME, etc.), other "geographically oriented" elements were also identified. For example, geographic definitions include wealth of information on location (e.g., on, below, above, etc.), topology (e.g., adjacent-to, surrounded-by, connected-to, etc.), proximity (e.g., near, far, etc.), and orientation (e.g., north, south, towards, etc). Furthermore, other context-specific semantic elements were also identified. For example, concepts relative to hydrography are described by semantic elements such as nature (natural or artificial)

TABLE 12.1

Main Semantic Properties of Geographic Concepts

Semantic Properties

PURPOSE

AGENT

PROPERTY-DEFINED LOCATION

COVER

PROPERTY-DEFINED TIME

POINT IN TIME

DURATION

FREQUENCY

SIZE

SHAPE

TABLE 12.2

Main Semantic Relations of Geographic Concepts

Semantic Relations

IS-A

IS-PART-OF

HAS-PART

UPWARD VERTICAL RELATIVE POSITION

DOWNWARD VERTICAL RELATIVE POSITION

IN FRONT OF HORIZONTAL RELATIVE POSITION

BEHIND HORIZONTAL RELATIVE POSITION

BESIDE HORIZONTAL RELATIVE POSITION

SOURCE-DESTINATION

SEPARATION

ADJACENCY

CONNECTIVITY

OVERLAP

INTERSECTION

CONTAINMENT

EXCLUSION

SURROUNDNESS

EXTENSION

PROXIMITY

DIRECTION

and flow (flowing or stagnant). Concepts relative to agriculture are described by semantic elements such as crop rotation, tillage, irrigation, vegetation type, etc. The next sections analyze the main semantic properties and relations found in geographic concept definitions and the rules for their extraction.

12.3 MAIN SEMANTIC PROPERTIES IN GEOGRAPHIC CONCEPT DEFINITIONS

The PURPOSE semantic property is determined by specific phrases containing the preposition "for" (e.g., for (the) purpose(s) of, for, used for, intended for) followed by a noun phrase, present participle, or infinitival clause. For example, the definition "canal: a manmade or improved natural waterway used for transportation" includes a PURPOSE semantic property with the value "transportation."

The AGENT semantic property is identified by specific phrases including the preposition *"by"* (e.g., by, created by, influenced by, affected by, dominated by, used by, etc.) followed by a noun phrase. For example, the definition "quarry: an excavation created by removal of stone" includes an AGENT semantic property with the value "removal of stone."

The PROPERTY-DEFINED LOCATION semantic property implies that an action or activity (given by the value of the semantic property) takes place in the defined concept. This relation is identified in the following definition: "airfield: a place where planes take off and land." The above definition includes the PROPERTY-DEFINED LOCATION semantic property with the value "take off and land," and the subject "planes."

The COVER semantic property is identified by specific phrases including the preposition "of" or by phrases such as "covered by" or "covered with." For example, both of the following definitions imply a COVER semantic property with the value "water":

"lake: body of water surrounded by land"
"body of water: the part of the earth's surface covered with water"

The PROPERTY-DEFINED TIME semantic property implies that an action or activity takes place during the defined concept. This semantic property is identified by dependent or subordinate clauses including the subordinating conjunctions "when" and "while" or the preposition "during" followed by the relative pronoun "which." For example, the following definition includes a PROPERTY-DEFINED TIME semantic property with the value "viewers remain in their vehicles": "drive-in theatre: a place where motion pictures are shown while viewers remain in their vehicles."

The POINT IN TIME semantic property is expressed by prepositional phrases including the prepositions "in," "at," "on" followed by temporal expressions such as "Sunday," "week," "summer," and "today." For example, a POINT IN TIME semantic property with the value "winter" is identified in the following definition: "inland marshes: low-lying land usually flooded in winter."

Adjectives or adjective phrases expressing size, such as "large," "small," "big," "of a large volume," and "tall," indicate a SIZE semantic property. For example, the following definitions include a SIZE semantic property:

"river: *large*, natural stream of water"
"grain elevator: a *tall* structure, equipped for loading, unloading, processing
   and storing grain"

Adjectives or adjective phrases expressing shape or form, such as "square,"
"round," "rectangular," and "oblong," indicate a SHAPE-FORM semantic property,
as in the following definition: "runway: a defined area, usually *rectangular*, used for
the conventional landing and take-off of aircraft."

12.4 MAIN SEMANTIC RELATIONS IN GEOGRAPHIC CONCEPT DEFINITIONS

Patterns applied to the genus part of the definition extract the hypernym or IS-A
relation. In noun definitions, which are our case, the head of the noun phrase most
frequently indicates the genus, as in the following definitions:

"hotel: a *building* where travelers can pay for lodging and meals and other
   services"
"hospital: a medical *institution* where sick or injured people are given medical
   or surgical care"

However, empty heads, such as "kind of," "any," "any of," etc., followed by a
noun phrase also indicate the IS-A relation but are not common to geographic concept definitions. In these cases, the hypernym is the head of the noun phrase; in the
following definition "drink" is the hypernym of "nectar" (Markowitz, Ahlswede,
and Evens, 1986): "nectar: any delicious drink."

Prepositional phrases such as "part of" followed by a noun phrase indicate an
IS-PART-OF semantic relation, as in the following definitions:

"aerial cableway: cables which are strung between elevated supports as part of a
   conveyor system on which cars, buckets, or other carrier units are suspended"
"foreshore: that part of the shore or beach which lies between the low water
   mark and the coastline/shoreline"

The HAS-PART semantic relation is determined by phrases such as "consist of,"
"comprised of," and "composed of." For example, the following definition contains
a HAS-PART semantic relation with "road or path" as the value: "way: artifact consisting of a road or path affording passage from one place to another."

The RELATIVE POSITION semantic relation together with TOPOLOGY,
PROXIMITY, and DIRECTION constitute spatial relations. In contrast to the
PROPERTY-DEFINED LOCATION semantic property, spatial relations describe
the location of the defined concept relative to other concepts. The RELATIVE POSI-
TION semantic relation is indicated by prepositional phrases introduced by prepositions such as "on," "in," "below," "under," "above," and so on. More specifically
prepositions such as "above," "over," "beneath," "below," "under," "underneath,"

express VERTICAL RELATIVE POSITION; prepositions such as "before," "behind," "beside," "beyond," express HORIZONTAL RELATIVE POSITION. The following definitions express a VERTICAL RELATIVE POSITION semantic relation:

"diffuser: an artificial installation at or below water level, where liquids are spread out"
"stream: natural body of running water flowing on or under the earth"

The SOURCE-DESTINATION semantic relation is expressed by phrases including the prepositions "from" and "to." In this case, the head of the noun phrase after the preposition "from" is the value of the SOURCE semantic relations, whereas the head of the noun phrase after the preposition "to" is the value of the DESTINATION semantic relation, as in the following definition: "conveyor: an apparatus for moving materials *from place to place* on a moving belt or series of rollers."

Topological relations are divided in the following categories:

- The SEPARATION semantic relation is expressed by phrases such as "separated from."
- The ADJACENCY semantic relation is expressed by phrases such as "adjacent to," "next to," or by prepositional phrases introduced by the preposition "against."
- The CONNECTIVITY semantic relation is expressed by phrases such as "connected to."
- The OVERLAP semantic relation is expressed by the verb "overlap," or by phrases such as "partly covered by" and "extend beyond."
- The INTERSECTION semantic relation is expressed by the verb "cross," phrases such as "intersect with," or by prepositional phrases introduced by the prepositions "through" and "via."
- The CONTAINMENT semantic relation is expressed by phrases such as "contained in" or by prepositional phrases introduced by the prepositions "within" and "inside."
- The EXCLUSION semantic relation is expressed by prepositional phrases introduced by the preposition "outside."
- The SURROUNDNESS semantic relation is expressed by phrases such as "surrounded by," "enclosed by," or by prepositional phrases introduced by the prepositions "around," "among," and "between."
- The EXTENSION semantic relation is expressed by verb phrases including the verbs "extend" and "span" or by prepositional phrases introduced by the prepositions "along" and "across."

The following definitions express topological relations:

"lake: body of water *surrounded by land*"
"circular irrigation system: an elevated irrigation system revolving *around a central pivot point*"

"substation/transformer yard: a facility, *along a power line route*, in which electric current is transformed and/or distributed"

"coastal lagoons: stretches of salt or brackish water in coastal areas which are *separated from the sea* by a tongue of land or other similar topography"

The PROXIMITY semantic relation is indicated by phrases such as "at a distance of" or by prepositional phrases including the prepositions "near" and "far," as in the following definition: "coastal water: surface water on the landward side of a line every point of which is *at a distance of one nautical mile* on the seaward side from the nearest point of the baseline from which the breadth of territorial waters is measured, extending where appropriate up to the outer limit of transitional waters."

The DIRECTION semantic relation is expressed by phrases including words such as "north," "south," "east," "west" (as nouns, adjectives, or adverbs), or by prepositional phrases including the preposition "towards." Adverbs ending "–ward" or "–wards," for example, northwards, also express the DIRECTION semantic relation. For example: "navigation line: line generated by the straight line connection between two navigational aids, and which extends *towards the area of navigational interest*."

REFERENCES

Baker, C. F., C. J. Fillmore, and J. B. Lowe. 1998. The Berkeley FrameNet project. In *Proc. of the COLING-ACL*, Montreal, Canada, 86–90.

Barriere, C. 1997. From a children's first dictionary to a lexical knowledge base of conceptual graphs. Ph.D. thesis, School of Computing Science, Simon Fraser University, British Columbia, Canada.

Bateman, J. A. 1992. The theoretical status of ontologies in natural language processing. In *Text representation and domain modeling: Ideas from linguistics and AI,* ed. S. Preuß and B. Schmitz, 50-99. KIT-Report 97. Technische Universitaet, Berlin, http://arxiv.org/PS_cache/cmp-lg/pd/9704/9704010v1.pdf (accessed 10 September 2007).

Bhooshan, N. 2005. Classification of semantic relations in different syntactic structures in medical text using the MeSH hierarchy. B.Sc. and M.Sc. thesis, Department of Electrical Engineering and Computer Science, Massachusetts Institute of Technology.

Bolshakov, I. A. and A. Gelbukh. 2004. *Computational linguistics: Models, resources, applications*, Mexico: National Polytechnic Institute–Center for Computing Research. http://www.gelbukh.com/clbook/Computational-Linguistics.htm (accessed 10 September 2007).

Calzolari, N. 1984. Detecting patterns in a lexical data base. In *Proc. 22nd Conference on Association for Computational Linguistics*, Stanford, California, 170–173.

Chodorow, M. S., R. J. Byrd, and G. E. Heidorn. 1985. Extracting semantic hierarchies from a large on-line dictionary. In *Proc. 23rd Annual Meeting on Association for Computational Linguistics (ACL)*, Chicago, IL, 299–304.

Cole, S. V., M. D. Royal, M. G. Valtorta, M. N. Huhns, and J. B. Bowles. 2006. A lightweight tool for automatically extracting causal relationships from text. In *Proc. IEEE SoutheastCon*, March 31–April 2, Memphis, TN, 125–129.

Fillmore, C. J. and C. F. Baker. 2001. Frame semantics for text understanding. In *Proc. WordNet and Other Lexical Resources Workshop*, NAACL, Pittsburgh. http://framenet.icsi.berkeley.edu/papers/FNcrime.pdf (accessed 21 March 2007).

Fillmore, C. J., C. Wooters, and C. F. Baker. 2001. Building a large lexical databank which provides deep semantics. In *Proc. Pacific Asian Conference on Language, Information and Computation*, Hong Kong. http://framenet.icsi.berkeley.edu/papers/dsemlex16.pdf (accessed 1 May 2007).

Gildea, D. and D. Jurafsky. 2002. Automatic labeling of semantic roles. *Computational Linguistics* 28(3): 245–288.

Girju, R. 2003. Automatic detection of causal relations for question answering. In *Proc. ACL 2003 Workshop on Multilingual Summarization and Question Answering*, July 11, Sapporo, Japan, 76–83.

Ide, N. and J. Veronis. 1993. Refining taxonomies extracted from machine readable dictionaries. In *Research in humanities computing*, ed. S. Hockey and N. Ide, 145–159. Oxford: Oxford University Press.

Jensen, K. and J. L. Binot. 1987. Disambiguating prepositional phrase attachments by using on-line dictionary definitions. *Computational Linguistics* 13(3-4): 251–260.

Jensen, K., G. Heidorn, and S. Richardson, eds. 1993. *Natural language processing: The PLNLP approach*. Norwell, MA: Kluwer Academic Publishers.

Khoo, C. and S. H. Myaeng. 2002. Identifying semantic relations in text for information retrieval and information extraction. In *The semantics of relationships: An interdisciplinary perspective*, ed. R. Green, C. A. Bean, and S. H. Myaeng, 161–180. Dordrecht: Kluwer Academic Publishers.

Khoo, C. S. G., J. Kornfilt, R. N. Oddy, and S. H. Myaeng. 1998. Automatic extraction of cause-effect information from newspaper text without knowledge-based inferencing. *Literary and Linguistic Computing* 13(4): 177–186.

Klavans, J., M. Chodorow, and N. Wacholder. 1993. Building a knowledge base from parsed definitions. In *Natural language processing: The PLNLP approach*, ed. K. Jensen, G. Heidorn, and S. Richardson, 119–133. Norwell, MA: Kluwer Academic Publishers.

Kokla, M. and M. Kavouras. 2005. Semantic information in geo-ontologies: Extraction, comparison, and reconciliation. *Journal on Data Semantics* 3: 125–142.

Lame, G. 2003. Using text analysis techniques to identify legal ontologies' components. In *ICAIL 2003 Workshop on Legal Ontologies and Web Based Legal Information Management*, Edinburgh, UK, June 24–28.

LCC (Language Computer Corporation). 2007. Polaris Semantic Parser. http://www.languagecomputer.com/solutions/Polaris/index.html (accessed 21 March 2007).

Loper, E. 2000. Applying semantic relation extraction to information retrieval. Master's thesis, Massachusetts Institute of Technology, Cambridge, MA.

Markowitz, J., T. Ahlswede, and M. Evens. 1986. Semantically significant patterns in dictionary definitions. In *Proc. 24th Annual ACL Conference*, New York, 112–119.

Rosario, B. 1995. Extraction of semantic relations from bioscience text. PhD Thesis in Information Management and Systems and the Designated Emphasis in Communication, Computation and Statistics, University of California, Berkeley, USA.

Rosario, B. and M. A. Hearst. 2004. Classifying semantic relations in bioscience texts. *Proceedings of the 42nd Annual Meeting on Association for Computational Linguistics*, 21–26 July, 2004, Barcelona, Spain, Association for Computational Linguistics, Morristown, NJ, 430–437.

Specia, L. and E. Motta. 2006. A hybrid approach for extracting semantic relations from texts. In *Proc. 2nd Workshop on Ontology Learning and Population*, Association for Computational Linguistics, Sydney, 57–64.

Swartz, N. 1997. Definitions, dictionaries, and meanings. http://www.sfu.ca/philosophy/swartz/definitions.htm (accessed 1 May 2007).

Szeto, Y. K. 1988. The semantics of causative and agentive verbs. *Cahiers Linguistiques d'Ottawa* 16: 1–51.

Wilks, Y., D. Fass, C. Guo, J. McDonald, T. Plate, and B. M. Slator. 1988. Machine readable dictionaries as tools and resources for natural language processing. In *Proc. 12th International Conference on Computational Linguistics*, Budapest, Hungary, Vol. 2, 750–755.

13 Similarity

13.1 THE NOTION OF SIMILARITY

The notion of *similarity* is of concern to many scientists. A lot of research has been performed by psychologists, mathematicians, and researchers in artificial intelligence and other disciplines. In the cognitive domain, similarity refers to the way humans perceive and classify objects, events, and, more generally, concepts of life into categories. In other scientific domains, similarity is assessed and represented into a specific context. In the geospatial domain, the notion of similarity refers to the degree of resemblance of geographic information. Geographic information is characterized by both geometric and thematic properties, and, as a result of this complexity, similarity is hard to be evaluated. Furthermore, the determination of similarity between geographic features of different geographic systems has to overcome another, more intractable difficulty due to the great diversity of the applied modeling methods. Heterogeneity among information models can refer to the syntactic, schematic, or semantic level. As far as the semantic level is concerned, the determination of a measure of similarity between geographic concepts modeled in different geographic systems can facilitate the exchange and the use of information, namely *interoperability* (see Chapter 3). Therefore, several approaches have been proposed to tackle the task of modeling similarity between geographic concepts based on different similarity measures (Rodriguez, Egenhofer, and Rugg, 1999; Yaolin, Molenaar, and Kraak, 2002; Doan, 2002; Rodriguez and Egenhofer 2003) (see Sections 15.11 and 16.2.2).

Besides interoperability, semantic similarity assessment may also be used to perform tasks such as ontology checking for consistency and coherency, and geographic information retrieval from digital libraries, databases, and the World Wide Web. Recently, the concept of Semantic Geospatial Web has been envisioned as a framework providing an efficient way to retrieve geospatial information, based on the comparison between formal expressions of semantics and the available semantics of information resources (Egenhofer, 2002). In the framework of the Semantic Geospatial Web, Win and Hla (2006) proposed the development of a Geospatial Semantic Query System to achieve interactive geospatial information search and retrieval. This system is designed to assess similarity between geospatial entity classes by combining the similarity function proposed by Rodriguez and Egenhofer (2003) and a new one called Geospatial Similarity function (GSS) that takes into account the spatial relations (coordinates, direction, topology, and proximity) between geospatial entities.

13.2 MEASURES OF SIMILARITY

To perform a semantic comparison between geographic concepts, several approaches of assessing similarity have been proposed. Many of them, combined with each other, have given birth to hybrid semantic similarity measures. In the following, the main approaches are presented.

Feature similarity measure approaches assume that the concepts to be compared are represented by a set of features (synonyms, attributes, parts, functions, etc.). Then, similarity is assessed based on the number of common descriptive features between two geographic concepts. The greater the number of shared features, the greater is the similarity. For example, Tversky's (1977) similarity measure is based on the features' sets of the concepts to be compared. The more common features and the fewer noncommon features two concepts have, the more similar they are. Tversky proposed the following similarity measure:

$$S(a,b) = \frac{|A \cap B|}{|A \cap B| + a(a,b)|A/B| + (1 - a(a,b))|B/A|} \quad \text{for } 0 \leq a \leq 1$$

where A, B correspond to the sets of features of the concepts a and b, respectively, $|A \cap B|$ to the number of common features, $|A/B|$, to the number of features of A that are not features of B, $|B/A|$, to the number of features of B that are not features of A, and $a \in [0, 1]$ defines the relative importance of the distinguishing features. This measure takes values from 0, for semantically different concepts, to 1, for semantically similar concepts. In addition, it is evident that this measure of similarity does not respect the property of symmetry.

The *geometric similarity measure* is a different strategy for feature-based similarity models. In fact, it is a measure of semantic dissimilarity. Using the features of the concepts to be compared, a measure of semantic distance is determined as a geometric distance (euclidean, manhattan, etc.) in a semantic representational space (Rips, Shoben, and Smith, 1973). The shorter the distance between the concepts, the greater the similarity. Euclidean distance is an example of such a dissimilarity measure. More specifically, supposing that concepts a and b are represented into an n-dimensional space, a measure of semantic similarity can be assessed according to the euclidean distance formula:

$$d(a,b) = \sqrt{\sum_{n=1}^{m} (x_{in} - x_{jn})^2}$$

where x_{in} and x_{jn} are the coordinates or values of the concepts' features corresponding to the dimension n ($n = 1, \ldots, m$). Inasmuch as the features' values are often measured in different units, the euclidean distance formula is usually applied after each value has been standardized to mean 0 and 1 standard deviation.

Network or edge-counting similarity measure approaches can be applied to the elements of an is-a taxonomy (hierarchy, network, ontology). Similarity is assessed from the number of edges linking two different nodes of the taxonomy. The shorter the path of edges linking two nodes, the greater is the similarity. Variations of this kind of measure exist taking into account some coefficients attributed to the edges, calculating in parallel the number of times that the path between two nodes changes direction, etc. For example, according to the *shortest path similarity measure*, a simple way to evaluate semantic similarity is to find how closely two concepts are represented in a taxonomy. Rada et al. (1989) suggest that semantic distance between these concepts is given by the minimum number of edges separating them. The shorter the path between them is, the more similar they are.

Information content similarity measure approaches are quite comparable to the previously mentioned one but they have the advantage of not being sensitive to the problem of varying link distances (Resnik, 1999). Measures of this kind can be implemented on concepts represented in an is-a taxonomy and rely on the fact that the more information two concepts have in common, the more similar they are. To be more specific, similarity between two concepts is determined according to the shared information content given by the first superconcept that subsumes them both in the taxonomy. Resnik's (1999) information content similarity measure is an example of such a similarity measure. More specifically, if a and b are concepts participating in an is-a taxonomy, their similarity measure is determined by the formula:

$$S(a,b) = \max_{c \in S(a,b)} \left[-\log p(c) \right]$$

where p is a function that evaluates the probability of encountering an instance of concept c in the taxonomy and $S(a,b)$ is the set of concepts that subsume a and b. P is increasing as the concept is found higher in the taxonomy. As $p(c)$ can take values between 0 and 1, $-\log p(c)$ varies from $+\infty$ to 0. As a result, the similarity measure can vary between infinity, for semantically very similar concepts, and 0, for semantically different concepts.

13.3 PORTRAYING SIMILARITY

During the last decade, in order to deal with the rapidly growing volumes of data, new methods of information visualization have been developed enabling people to search among large amounts of complex data and find the desired information. Information visualization constitutes an excellent way of obtaining a quick and better understanding of information because the human mind processes and assimilates information more efficiently in graphical or optical form than in a raw data form, like alphanumeric, numeric, etc. (Tufte, 1983). Various techniques to visually represent information exist. These techniques have been developed according to two directions: (a) *scientific visualization* deals with data of the physical world, such as human organism, physical phenomena, molecules, etc., whereas (b) *information visualization* deals with abstract data and concepts, such as documents, statistical data, etc. A very interesting research direction in information visualization is towards representing abstract data in representational space using spatial metaphors or different methods of dimension reduction. This kind of visualization is called *spatialization*.

Spatialization defines a mapping between the abstract and usually high-dimensional data space, and the representational space of one, two, or three dimensions that the human mind is able to perceive. Spatialization can be considered as a method of portraying similarity because similar data are projected to objects with similar appearance, similar spatial properties, and contiguous locations in the representational space. Following Kuhn (1996), "spatialization maps physical space to abstract computational domains through spatial metaphors." According to them, a spatialization defines a partial mapping between information and physical space, so that these two spaces are always distinct. Spatial metaphors help human cognition because "most of our fundamental concepts are organized in terms of one or more spatialization metaphors" (Lakoff and Johnson, 1980, 17) and allow the organization of abstract concepts based on familiar experiences (Kuhn and Blumenthal, 1996).

Skupin and Buttenfield (1997) define spatialization as "... a projection of elements of a high-dimensional information space into a low-dimensional, potentially experiential, representational space." This second definition is considered more generic and lays emphasis on the reduction of data dimensions during the mapping between information and representational space. Spatialization leads up to the creation of graphical representations often called *information terrains, information landscapes, information spaces, information world*, etc. These representations facilitate information access, exploration, and retrieval in very large data collections because they allow navigating and exploring information spaces like humans do in the real geographical and familiar space (Fabrikant and Buttenfield, 2001). An example of spatialization that uses the spatial metaphor of a landscape is illustrated in Figure 13.1. Data corresponding to forty-four categories of the CORINE land cover classification schema are mapped according to their semantic similarity: the more semantically similar the entities are, the closer they are mapped in the landscape. The landscape relief corresponds to the density of geographical categories that clusters of similar ones (mountains, hills) are revealed.

Another method to portray similarity is by defining and visualizing *clusters* of similar data previously projected in a representational space. Many clustering algorithms have been developed to classify objects on the basis of similarity. Their goal is to find particular data clusters for further analysis and to extract knowledge from a large data collection. *Spatial clustering* methods can be combined in spatialization

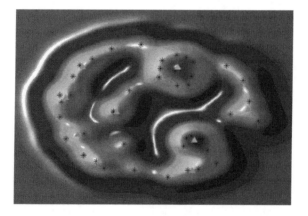

FIGURE 13.1   Similarity landscape of CORINE land cover categories.

to help the interpretation of the resulting visualization. Clustering methods can be broadly classified into two main categories, (a) the *hierarchical* and (b) the *partitioning* approaches, although several other approaches have been proposed, for example, the *density-based* and the *grid-based* clustering methods.

Hierarchical clustering methods made their appearance in 1951 (Florek et al., 1951). Since then, a lot of variants have been proposed but they all rely on the same idea. Hierarchical clustering methods create a hierarchy of clusters, represented by a dendrogram or a nested sequence of partitions, with an all-inclusive cluster at the top of the hierarchy. These methods may be categorized into *divisive* or *agglomerative methods* depending on the assumption made when the clustering algorithm begins. *Divisive methods* assume that there is a single cluster at the top of the hierarchy which splits into nested subclusters, whereas *agglomerative methods* assume that there are as many objects as clusters that are sequentially merged into a larger one.

Partitioning clustering methods divide data into several subsets, creating clusters of similar objects according to some spatial criterion. One of the most popular, k-means (Hartigan and Wong, 1979) regroups data into k clusters represented by their centroids, in order to minimize the mean, usually weighted average of distances between the points of each cluster and its centroid.

The basic idea behind *density-based* clustering methods is the creation of clusters of objects according to their density in a representational space. A cluster is defined as a dense region of objects whose limits are determined by the regions of low-density. These methods, as opposed to the partitioning one, are not sensible to noise or outliers and can detect clusters of arbitrary shapes. For example, the algorithm DENCLUE (DENsity-based CLUstEring) uses a summative density function to model the overall density of objects' distribution in representational space (Hinneburg and Keim, 1998). DENCLUE leads to a density "landscape" in which local peaks, that is, density maxima, identify the clusters of data.

To perform *grid-based* data clustering, the representational space should be divided in order to form a grid mesh consisting of rectangular cells. Then clusters are detected by grouping together similar nearby cells, for example, by discarding low-density grid-cells and combining adjacent high-density cells. In addition, different levels of cells may be defined that correspond to different levels of data granularity. STING (Statistical INformation Grid-based method) (Wang, Yang, and Muntz, 1997) is a representative grid-based clustering algorithm. It creates a hierarchical index structure over a multilevel grid whose cells contain statistical information derived from the input data. The grid is then accessed in a top-down way, from the root to the leaves, by processing the statistical information and following the path given by the relevant cells. Once the last level is reached, contiguous cells whose density is greater than a given threshold are merged to form clusters.

13.4 SIMILARITY AND CONTEXT

Common sense and intuition tell us that similarity has undoubtedly a *context-dependent* nature. In the cognitive domain, Goldstone, Medin, and Halberstadt (1997) showed through adequate experiments that similarity depends on the context of human judgment and argued that "similarity is not just a relation between two objects; rather, it is a relation between two objects and a context" (238). According to Hahn and Ramscar (2001) similarity is a nonstatic construct and it may be altered depending on the context in which it is examined. Swoyer (2006) outlined that similarity is very flexible and strongly influenced by the context in which it is assessed.

Furthermore, context is incorporated in various applications of semantic data modeling, which evaluate measures of context-dependent similarity. Sheth and Kashyap (1992) and Kashyap and Sheth (1995) use the notion of semantic proximity to characterize semantic similarities between database objects. In this approach, one of the components of the four-tuple semantic proximity definition refers to the context in which a particular semantic similarity holds. Moreover, in order to measure similarity within and between ontologies regarding a specific task, Ehrig et al. (2004) present a framework consisting of three layers: data, ontology, and context layers. The latter includes information beyond the ontology, for example, the application context. In each layer, similarity is assessed in different ways as the information taken into account varies. Finally, an overall similarity measure is computed using an amalgamation of the individual ones. As far as similarity between geospatial concepts is concerned, Kuhn (2005) pointed out the importance of capturing the context-dependence of human similarity judgments and observed that approaches exist, which are based on "'factoring' out context through relative similarities" or "modeling the use of entities" (15). Such an approach was proposed by Rodriguez and Egenhofer (2004) for the determination of a context-based semantic similarity measure between entity classes. Context is incorporated in the assessment of a matching-distance similarity measure and affects the relative importance of the entities' distinguishing features.

REFERENCES

Doan, A. 2002. Learning to map between structured representations of data. Dissertation study, University of Washington, Seattle.

Egenhofer, M. J. 2002. Toward the semantic geospatial web. In *Proc. 10th ACM International Symposium on Advances in Geographic Information Systems*, McLean, VA, 1–4. New York: ACM Press.

Ehrig, M., P. Haase, M. Hefke, and N. Stojanovic. 2004. Similarity for ontologies: A comprehensive framework. In *Proc. Workshop Enterprise Modelling and Ontology: Ingredients for Interoperability*, PAKM 2004. New York: Springer.

Fabrikant, S. I. and B. P. Buttenfield. 2001. Formalizing semantic spaces for information access. *Annals of the Association of American Geographers* 91(2): 263–280.

Florek, K., J. Lukaszewicz, J. Perkal, H. Steinhaus, and S. Zubrzycki. 1951. Sur la liaison et la division des points d'un ensemble fini. *Colloquium Mathematicum* 2: 282–285.

Goldstone, R., D. Medin, and J. Halberstadt. 1997. Similarity in context. *Memory and Cognition* 25(2): 237–255.

Hahn, U. and M. Ramscar. 2001. *Similarity and categorization.* Oxford: Oxford University Press.

Hartigan, J. A. and M. A. Wong. 1979. A k-means clustering algorithm. *Applied Statistics* 28, 100–108.

Hinneburg, A. and D. A. Keim. 1998. An efficient approach to clustering in large multimedia databases with noise. In *Proc. The 4th International Conference on Knowledge Discovery and Data Mining, (KDD98)*, New York, 58–65.

Kashyap, V. and A. Sheth. 1995. *Semantic and schematic similarities between databases objects: A context-based approach*. Technical Report, LSDIS Laboratory, University of Georgia.

Kuhn, W. 1996. Handling data spatially: Spatializing user interfaces. In *Proc of 7th International Symposium on Spatial Data Handling*, SDH'96, Advances in GIS Research II, ed. M.-J. Kraak, and M. Molenaar, Delft, the Netherlands (August 12–16, 1996). Published by IGU, Vol. 2, 13B.1–13B.23, ftp://ftp.geoinfo.tuwien.ac.at/kuhn/2318_SDH96_Spatialization.pdf (accessed 12 September 2007).

Kuhn, W. 2005. Geospatial semantics: Why, of what, and how? *Journal on Data Semantics, Special Issue on Semantic-based Geographical Information Systems*. Lecture Notes in Computer Science, 3534 1–24.

Kuhn, W. and B. Blumenthal. 1996. Spatialization: Spatial metaphors for user interfaces. In *Proc.* Conference on Human Factors in Computing Systems, Vancouver, British Columbia, Canada, 346–347.

Lakoff, G. and M. Johnson. 1980. *Metaphors we live by*. Chicago: University of Chicago Press.

Rada, R., H. Mili, E. Bicknell, and M. Blettner. 1989. Development and application of a metric on semantic nets. *IEEE Transactions on Systems, Man, and Cybernetics* 19(1): 17–30.

Resnik, O. 1999. Semantic similarity in taxonomy: An information-based measure and its application to problems of ambiguity and natural language. *Journal of Artificial Intelligence Research* 11: 95–130.

Rips, L., J. Shoben, and E. Smith. 1973. Semantic distance and the verification of semantic relations. *Journal of Verbal Learning and Verbal Behavior* 12: 1–20.

Rodriguez, A. and M. Egenhofer. 2003. Determining semantic similarity among entity classes from different ontologies. *IEEE Transactions on Knowledge and Data Engineering* 15(2): 442–456.

Rodríguez, A. and M. Egenhofer. 2004. Comparing geospatial entity classes: An asymmetric and context-dependent similarity measure. *International Journal of Geographic Information Science* 18(3): 229–256.

Rodriguez, A., M. Egenhofer, and R. Rugg. 1999. *Assessing semantic similarities among geospatial feature class definitions*, ed. A. Vckovski, K. Brassel, and H. J. Schek, 189–202. Interoperating Geographic Information Systems, Second International Conference (Interop '99), Zurich, Switzerland. Lecture Notes in Computer Science, 1580. Dordrecht: Springer-Verlag.

Sheth, A. and V. Kashyap. 1992. So far (schematically) yet so near (semantically). In *Proc. IFIP DS-5 Conference on Semantics of Interoperable Database Systems*, Lorne, Australia, 283–312.

Skupin, A. and B. P. Buttenfield. 1997. Spatial metaphors for display of information spaces. *Proc. AUTO-CARTO 13*, Seattle, WA, 116–125.

Swoyer, C. 2006. Conceptualism. In *Universals, concepts and qualities*, 127–154. ed. P. F. Strawson and A. Chakrabarti. Burlington, VT: Ashgate Publishing.

Tufte, E. R. 1983. *The visual display of quantitative information*. Cheshire, CT: Graphics Press.

Tversky, A. 1977. Features of similarity. *Psychological Review* 84(4): 327–352.

Wang, W., J. Yang, and R. Muntz. 1997. STING: A statistical information grid approach to spatial data mining. *Proc. The 23rd Very Large Databases Conference (VLDB 1997)*, Athens, Greece.

Win, K. K. and H. K. S. Hla. 2006. Measuring geospatial semantic similarity between geospatial entity classes. In *Proc. 2nd International Conference on Information and Communication Technologies*, ICTTA'06. 24–28 April 2006, Vol. 1, 558–563.

Yaolin, L., M. Molenaar, and M. J. Kraak. 2002. *Semantic similarity evaluation model in categorical database generalization*. In Symposium on Geospatial Theory, Processing and Applications, Ottawa.

Part 4

Ontology Integration

14 Integration Framework

14.1 THE ONTOLOGY OF INTEGRATION

Interoperability was presented in Chapter 3, as the ability of heterogeneous information systems to communicate, process, and interpret the information exchanged. Information represents domain knowledge about reality, in the form of concepts structured in representations. Because we are particularly interested in the correct interpretation and reuse of the information conveyed, the emphasis, according to the conceptualization perspective (Section 1.5), is put on semantics (Chapter 6), that is, the intended meaning of concepts in representations. As a result, the aim of semantic interoperability is the ability to associate and compare different concepts of interoperating systems on the basis of their semantics, in order to match them and possibly create integrated knowledge bases. Therefore, integration is a core issue in semantic interoperability.

There are various approaches of matching, detecting, resolving conflicts, and eventually associating heterogeneous information. This problem is poorly understood (Partridge, 2002; Calvanese, De Giacomo, and Lenzerini, 2001a; Uschold and Gruninger, 2002), and there is also confusion in the terms found in the literature. This is expected for there is not a general agreement on what *semantics* is or what *integration* is, and what is more, what *semantic integration* denotes. Uschold and Gruninger (2002), for instance, introduce a "gold standard" of "complete semantic integration," when two foreign agents successfully exchange information on the web. Any such definition, however, serves as a general target notion, needing of course further clarification of what "complete" and "successful" imply or how they are materialized. Semantic differences are often difficult to identify, comprehend, and resolve, not only via a consensual reconciliation but also in a "right" way. Thus, complete semantic integration, even in a broad sense of "complete," appears impossible or unattainable (to many, even a chimera). On the other hand, pursuing complete semantic integration, is very costly, and above all, not always necessary. Users can perform various reasoning tasks with incomplete types of integration; not forgetting, on the other hand, that heterogeneities are also attributable to incomplete, inaccurate, or unknown conceptualizations. It is therefore important that less complete types of integration are also understood and supported.

Integration is used here in a broad sense to refer to a number of related but not quite equivalent terms, such as coordination, alignment, mapping, merging, partial compatibility, unification, etc. (Sowa, 2000; Klein, 2001; Bouquet et al., 2004). This distinction is clarified in Section 14.7. Also, the subject of "integration" is presented differently in the literature: "database integration" (Parent and Spaccapietra 1998), "data integration" (Lenzerini, 2002), "schema integration" (Batini, Lenzerini, and Navathe, 1986), "ontology-based integration" (Wache et al., 2001), "ontology integration" (Calvanese, De Giacomo, and Lenzerini, 2001a), "representation matching" (Doan, 2002), "semantic integration" (Gruninger and Kopena, 2003; Doan, Noy, and Halevy, 2004), etc. Since the explicit semantic knowledge about a domain is usually set and clarified by an ontology (see Chapters 2 and 4), semantic integration is ontology-based (Uschold and Gruninger, 2002), as expected. Ontologies are usually connected to information sources, such as databases (Wache et al., 2001). This is accomplished by: (a) a connection to the database schema (structural or schematic resemblance), and (b) the use of concept definitions to describe terms used in the database. When different databases, which are to be integrated, comply with a standard or reference ontology, the latter is used to assist, guide, or enforce the matching. Despite various efforts aimed at developing information and knowledge systems that use a common shared ontology, such as an upper ontology (e.g., Cyc or SUO), it is unrealistic (and to some even frightening) to imagine that all people and organizations will commit to such an idea. Especially in the milieu of the World Wide Web, it is widely evident that we are living in a multiple-ontology world. In such a heterogeneous setting where different ontologies are used, semantic integration inevitably leads to inter-ontology mapping, also called *ontology integration*. It is thus a common belief (Bouquet et al., 2004) that data integration methods require that the ontologies have to be integrated. Therefore, ontology integration differs from data integration for it does not presuppose the presence of a single model, as is the case in databases (Calvanese, De Giacomo, and Lenzerini, 2001a).

14.2 ONTOLOGY INTEGRATION

Ontology integration is concerned with existing ontologies. It is a crucial issue for the reuse and exchange of knowledge among different domains and also for creating knowledge pools. Furthermore, since knowledge evolves and new knowledge is created, so do the associated ontologies. It is therefore of utmost importance to provide tools for associating present and future ontologies when evolution occurs. This is a common case in the spatiotemporal geographic domain, for instance when national statistical agencies change or expand their classification nomenclatures from one census to the next.

Another distinction that also applies here is the two perspectives presented in Section 1.5, similar to the terminology of Visser et al. (1997), further elaborated by Klein (2001). Conflicts at the higher ontological level—called semantic conflicts— are due to different conceptualizations and models of the domain in an information system. These show up as different concepts and/or relations among concepts. Conflicts at the "lower" explication level are due to differences in the specification of the conceptualization. These vary from trivial encoding differences (formats, units, etc.) to representation language mismatches. Notice that terminological conflicts can be treated at the explication level; nevertheless, they often carry some semantic weight, which should not be ignored.

For the reader who is interested in ontology mappings addressing mainly issues at the explication level, there is a lot of literature to consider. Doan in his Ph.D. thesis (2002) wrote about mappings between structured representations of data. Based on several survey results (Uschold and Jasper, 1999; Wache et al., 2001; Ding and

Foo, 2002; Noy, 2004; Shvaiko and Euzenat, 2005), when concepts from different ontologies are to be associated (mapped, merged, integrated, etc.), most of what many approaches offer today are tools to assist experts perform manual mappings. Semi- or full automation methods have achieved very little. Some (e.g., Hameed, Prreece, and Sleeman, 2003) accept the position that none of the available tools can resolve all types of discrepancies, focusing thereinafter on interoperability between the tools. There have also been several studies on database integration (heterogeneous database systems, federated information systems, etc.). Batini, Lenzerini, and Navathe (1986), Litwin and Abdellatif (1987), E. Kuhn, Puntigam, and Elmagarmid (1991), Arens, Knoblock, and Shen (1996), Goh (1997), Busse, Kutsch, and Leser (1999), Rahm and Bernstein (2001) are just a few. A large part of research in this area focuses on query processing and reformulation (Capezzera, 2003), and the technical issues of answering queries using views (Halevy, 2000; Pottinger and Halevy, 2001), instead of tackling more semantic issues (Firat, 2003).

One can also identify characteristics of integration approaches or tools focusing on lower technical differences of otherwise equivalent and conversable structures, languages, and formats. These can be knowledge representation paradigms (first-order logic, description or frame-based logic), conceptual structures used (conceptual graphs, concept lattices, etc.), ontology markup languages (RDF(S), OIL, DAML+OIL, OWL, etc.), and ontology management tools—commercial (LinkFactory, OntoEdit Professional), open (Protégé, KAON, OilEd, OntoEdit, Apollo, OpenKnoME), or restricted distributions (WebODE, Ontolingua, Ontosaurus, WebOnto). Denny (2004) has presented a detailed survey on ontology editing tools, which also includes merging. These ontology management tools are considered secondary for they do not concentrate on conceptualization or semantic aspects of integration, which is the central theme of Part 4. Some other ontology merge tools include PROMPT, a plugin for Protégé, Chimaera built on top of the Ontolingua server, MoA, an OWL ontology merging and alignment tool, OntoMerge, a tool for merging DAML ontologies, GLUE, a system based on machine learning techniques to semiautomatically create mappings.

Integration approaches may and usually do differ considerably due to a number of reasons. The first reason is the different intended use and objective an approach is designed for. Is it content explication, data integration, query support over heterogeneous sources, ontology evolution (versioning), or verification? The intended use imposes constraints with which integration has to comply. There are also additional questions to be asked: Are the resource ontologies to be directly integrated? Is there a reference, upper, or preferred ontology to be considered as a target? How is such a constraint materialized or imposed?

The second reason is that approaches may rely on different semantic elements, either because these are simply the only ones available or because these are considered as semantically more important or reliable than others. Most approaches to ontology integration (Calvanese, De Giacomo, and Lenzerini, 2001a) just rely on term similarity to express mappings between concepts. This proves to be a simple but not always effectual mechanism; therefore, other approaches attempt to incorporate additional descriptive information. A common way is by exploiting available definitions. When definitions are not available or regarded as semantically imprecise, some approaches make use of structural associations and subsumption relations among concepts in existing schemata or hierarchies as semantically revealing. It is also common to use parts, functions, and attributes as the base for comparing concepts. Finally, some approaches may resort to an extensional mapping of concepts on the basis of their corresponding instances, when of course available. Attributes and instances are not considered semantically very rich and are usually employed by database-oriented approaches.

The third reason for difference is the core idea behind integration (Section 14.7.2), as well as the architecture (topology) chosen (Section 14.7.1). Concepts may be compared and matched (ad hoc or systematically) on the basis of the available semantic components assumed in the previous paragraph; that is, term comparison, relation/property/attribute comparison, or instance comparison. The semantics used may be "deep" or "shallow"; while sometimes explication (syntactic-level) characteristics are used. The comparison may conclude as to whether two concepts are equivalent, different, related, etc. The semantics of top-level ontologies are often used to associate two ontologies indirectly. Another way of dealing with semantic correspondences and concept matching is by establishing compound similarity measures among the compared concepts from different ontologies. Structural resemblance is used by some as an indication of ontological resemblance. Inter-ontology mappings may also take place in the context of query-processing mechanisms.

The fourth reason is the context of comparison. Ontologies and their concepts have been developed according to different semantic contexts (see Chapter 7). Therefore, it is expected that they present various conflicts ranging from conceptualization to explication. Concepts can be compared in different ways, and the result can be diverse. Therefore, one has to decide in advance the parameters/dimensions to be considered as the base/reference of comparison. If too many parameters are included in the comparison, it is less likely to determine any similarity. This is an additional reason why explication differences do not matter and should be left out.

The fifth reason for difference is the role experts or users play during integration. Are differences reconciled automatically (without user involvement), semiautomatically (resolving some heterogeneities automatically and presenting possible/available choices to select from for the rest), or completely manually (based on intuition, experience, or agreement)? Machine learning techniques are sometimes employed (Doan et al., 2003) to assist semantic mappings in a semiautomated interaction environment. These start with manual mappings in order to train learners; these learners are subsequently applied to match schemas of new sources (Doan et al., 2003).

14.3 INTEGRATION FRAMEWORKS

In order to understand ontology integration approaches with their semantics from a huge and disparate literature, a framework is usually employed (Visser et al., 1997; Tama and Visser, 1998; Pinto, Gomez-Perez, and Martins, 1999; Wache et al., 2001; Calvanese, De Giacomo, and Lenzerini, 2001a; Klein, 2001; Ding and Foo, 2002; Uschold and Gruninger, 2002; Bouquet et al., 2004). Some of these frameworks are outlined below.

Uschold and Gruninger (2002) have presented an interesting analysis about semantic integration on the web, in which various architectures are evaluated with respect to four questions. These questions address: (a) who establishes the mapping (agents or designers), (b) when (at what stage) mapping takes place (in advance or dynamically), (c) the "topology" of mapping (one-to-one, via mediation, etc.), and (d) degree of agreement. On this basis, five architectures are distinguished (Uschold and Gruninger, 2002). The closest to the ultimate semantic integration is *Ontology Negotiation* (Bailin and Truszkowski, 2002) supporting automated one-to-one mappings, with no constraints or predefined scenarios. Its fully manual version, specified by a human-agent, is called *Manual Mapping* (Fillion et al., 1995; Obrst, Wray, and Liu, 2001). A partially automated version requiring predefined agreements between each ontology and a standard ontology or interlingua is called *Interlingua* architecture (Ciocoiu, Gruninger, and Nau, 2001). *Global Ontology* is the trivial case of imposing a universal conformity of all ontologies to a single ontology, in which case no mappings are actually required. Finally, *Community Ontologies* assume that predefined one-to-one mappings have been established among a number of ontologies of some community.

Another interesting framework for coordinating ontologies in the context of open distributed systems (like the semantic web) is that of the knowledgeweb project (Bouquet et al., 2004). In this framework, the general notion used is that of *alignment,* which is defined as a set of correspondences-mappings between concepts (called therein "entities") of the ontologies involved. Alignment is used to overcome several forms of heterogeneity similar to those already introduced in Chapter 3; syntactic, terminological, conceptual, and semiotic/pragmatic. The syntactic level deals with representation formats. At the terminological level, mismatches are caused by naming differences. Finally, the semiotic/pragmatic level addresses the way an ontology is interpreted or used by different users/communities and in different contexts. From all four levels, the focus is on terminological and particularly, on conceptual differences. The framework, following the existing artificial intelligence literature (Benerecetti, Bouquet, and Ghidini, 2000), distinguishes three groups of conceptual differences according to coverage, granularity, and perspective. Notice that this distinction presents similarities to the typology of heterogeneities adopted in Chapter 3. The framework expresses formally the semantics of the necessary mappings in each heterogeneity case. An interesting characteristic of the approach is that the mappings of the alignment, which are expressed as relations (e.g., equivalence, specialization, generalization, etc.), are also associated with a degree of trust denoting the confidence on the appropriateness of the mapping. Also, in model theoretic terms, two kinds of semantic mappings are envisioned, crisp and fuzzy.

Wache et al. (2001) provided an in-depth survey of existing approaches to ontology-based integration of information. In their evaluation framework, they first adopt the distinction by Kim and Seo (1991), and Kashyap and Sheth (1996) of (a) *structural* or *schematic heterogeneity* addressing differences in the data structures, and (b) *semantic heterogeneity* (which they equate to *data heterogeneity*) referring to the content of information items together with their intended meaning. With respect to the latter, they further adopt Goh's (1997) classification of semantic heterogeneities into *confounding conflicts, scaling conflicts,* and *naming conflicts.* Confounding conflicts are caused by information items, which deceptively appear to have the same meaning although they actually differ. Scaling conflicts are caused by the use of different reference systems. Finally, naming conflicts are caused by different naming schemes (e.g., homonyms and synonyms). Notice that similar classifications have already been presented in Chapter 3 on semantic interoperability. Also, from the conceptualization perspective that we have set, it appears that confounding and, to a certain degree, naming conflicts are semantically richer than scaling differences. Usually, the last can be easily taken care of by some transformation.

From this perspective, and with a focus on ontology-based information integration, four main criteria are used (Wache et al. 2001) to evaluate available approaches. The first criterion is the *role* ontologies play in explicating content. The simplest case is that of assuming (and thus committing to) a single shared ontology for providing common semantics. Integration approaches falling under this category can be effective only if all information sources share the same or a very close view of a domain. Because this is not often the case and information sources obey their own ontologies, multiple-ontology approaches that do not require commitment to a shared ontology are necessary. However, the price of such flexibility and independence is paid by the necessity to develop and provide inter-ontology mappings. This is not an easy task; therefore, hybrid approaches are devised to overcome the deficiencies of single or multiple ontology approaches. Different ontologies are possible to use; however, they have to share a common vocabulary, which may also constitute an ontology. Some approaches employ ontologies not only to explicate information content, but also to act as a global query schema, where global queries are reformulated to subqueries according to each appropriate information source. Finally, ontologies are sometimes used as a verification mechanism (Wache et al., 2001) for the mappings.

The second criterion Wache et al. (2001) use is the *ontology representation* language employed. As this is clearly an explication issue, it is not further elaborated. The third criterion—the *use of mappings*—is more appropriate to our scope. Here they distinguish approaches according to two different uses of mappings: (a) mappings between an ontology and the corresponding information sources, and (b) inter-ontology mappings, which again, are more appropriate to the objective of this chapter (see also Section 14.7). The fourth and last criterion concerns another explication issue, *ontology engineering* (development, supporting tools, and evolution) which may be only practically connected to inter-ontology mapping.

Klein (2001) has presented another framework of issues applying to ontology combination. This relies on an elaborated or slightly altered classification of ontology mismatches introduced by Visser et al. (1997), which is also in accordance with the two perspectives presented in Section 1.5. Three sets of issues are distinguished:

1. The first and most important are *ontology mismatches*. Here approaches address mismatches at (a) the *representation language level* and (b) the *ontology level*; further refined by *conceptualization* mismatches and *explication* (i.e., *terminological, modeling style,* and *encoding*) mismatches.
2. Second, there are *ontology versioning issues,* addressing ontology evolution and dependencies on it.
3. Finally, there are *practical problems,* which ontology designers need assistance to solve.

This framework of three sets of issues, further divided into subissues, has been applied (Klein, 2001) to describe the capabilities of existing tools and techniques. A last characteristic of the framework is a description of how this capability is implemented: (a) automatically, (b) as a mechanism providing a suggestion to the user, or (c) as a mechanism for manually specifying a solution.

Pinto, Gomez-Perez, and Martins (1999), after realizing that the notion of integration is often abused, attempt in their framework to clarify it by distinguishing three types of integration (and associated approaches): (a) those that build a new ontology by reusing other available ones; (b) those that create a single unified ontology by merging knowledge (concepts, terminology, definitions, constraints, etc.) from existing ontologies on a given theme; and (c) those that integrate ontologies into applications. In each type, they survey the available *tools,* the *ontologies and applications* built, and the *methodology* used. Notice that according to the framework set in the introduction of this chapter, only type (b) addresses the objective of ontology integration.

In their framework, Ding and Foo (2002) make use of the following dimensions for evaluating ontology mapping approaches: *mapping rules, resulting ontology, application areas,* and *tools and systems* developed by the approach.

From the perspective of a semantic web, Calvanese, De Giacomo, and Lenzerini (2001a) have presented a different formal framework for ontology integration systems (OISs). In this context where query mechanisms play a dominant role, a core issue is the way mappings are specified between the global ontology and local ontologies. As in various other approaches under this category (Halevy, 2000, 2001; Pottinger and Halevy, 2001; Capezzera, 2003), concepts in one ontology correspond to views (queries) of the others. The framework further distinguishes two types of approaches, *global-centric* and *local-centric.* In *global-centric* approaches (also known as global-as-view: GAV), concepts in the global ontology are mapped into queries over the local ontologies. GAV is more common to contemporary data integration architectures (Calvanese, De Giacomo, and Lenzerini, 2001a; Lenzerini, 2002; Bouquet et al., 2004). In *local-centric* approaches (also known as local-as-view: LAV) (Ullman 1997, Calvanese et al. 2000, Halevy 2001), concepts in the local ontologies are mapped into queries over the global ontology. A hybrid approach called *unrestricted mapping* has also been proposed (Calvanese, De Giacomo, and Lenzerini, 2001b) combining the advantages of the previous two. In this case, mappings between global and local ontologies are derived by relating views of the global ontology to views of the local ones. This third case, while presenting certain advantages, has not been thoroughly investigated yet. Another characteristic of the framework is that an indication of the level of the mappings between concepts and views is designated as *strict* (*sound, complete,* or *exact*) or *loose* (*loosely sound, loosely complete,* or *loosely exact*) (Calvanese, De Giacomo, and Lenzerini, 2001b).

Related to the above is a distinction of approaches to *tightly* versus *loosely coupled* (Arens, Knoblock, and Shen, 1996; Goh, 1997; Firat, 2003). In tightly coupled approaches, the user is left out of the heterogeneity problem, which is taken care of by the system (e.g., an administrator) that creates the global ontology either from the resource ontologies (bottom-up) or from existing knowledge of the domain (top-down). Heterogeneities between resources are resolved by mapping conflicting data items to a common view. This is usually accomplished by creating a common supertype. In loosely coupled approaches, the system provides users with the necessary tools to reconcile heterogeneities themselves. Tightly coupled approaches are not flexible, as they cannot accommodate multiple views; neither are they transparent. Creating and maintaining a global ontology/schema is very costly and suffers from the scalability problem. On the other hand, loosely coupled approaches provide better scalability, flexibility, and some transparency, as the user decides on the views and writes the mapping functions for every query. Their success, however, relies heavily on the complexity of the task and the ability/expertise of the user who undertakes the burden of carrying it out.

Tama and Visser (1998) have presented a framework for the comparison of techniques used for the integration of heterogeneous sources, placing emphasis on the role of ontologies. The framework is materialized by nineteen useful questions, addressing the role of ontologies in integration, grouped into three sets:

1. *General architectural features* refer to general design characteristics such as: (a) the heterogeneous sources, (b) the types of heterogeneity to deal with, (c) the role of agents in the process, (d) the operational level and the availability of the approach, and the like.
2. *Intra-ontology features* address issues such as: (a) whether the approach uses existing ontologies or builds its own, (b) the specification language used, (c) whether a top-level ontology is used for integration, (d) the accommodation of constraints, (e) the organization of the content (e.g., hierarchies, lattices, graphs, etc.), (f) the types of relations supported (e.g., is-a, part-of).
3. *Inter-ontology features* pertain to issues between ontologies such as: (a) the ability to support multiple inheritance, (b) the number of shared ontologies used, (c) the number of semantic translation steps, (d) the partitioning principle behind the ontologies involved (i.e., the ontology types), and (e) the relation types between multiple ontologies (subsumption, extension, mapping, etc.).

The brief review of the frameworks presented above shows the following:

- There is no general agreement on terminology or, more importantly, on the issues and dimensions of ontology integration. Understanding and mapping among the notions used in different approaches are similar problems but at the metalevel.
- There is significant partiality on the issues presented according to the interest/perspective/scope of the evaluator or establisher of the framework. Notice that this is a different issue from the interest/perspective/scope of the approach itself.
- From a different perspective each time, they address some conceptualization issues but their main arsenal is directed towards explication of mismatches of ontologies. These include syntactic, structural, and relation

differences, methodological issues, the level of agreement sought, and the role of designers and agents in the process.

- One common issue to most frameworks is the integration methodology employed, also called mapping rules. A lot of work originates from the database integration domain.
- Some differences emanate from different presumptions about the types of heterogeneity and how these can be resolved. The issue of how semantic information is derived and its effect on the heterogeneity reconciliation is not properly addressed.
- The dimensions of existing frameworks make it difficult to understand and modularize existing approaches, which would be helpful in combining methodologies to undertake a complex ontology integration task.
- Existing frameworks are only partially suitable to explain and overall systematize the problem of ontology integration. Because our interest is in the geographic domain, and according to the objectives and perspectives set out in Part 1, the integration of geographic ontologies entails additional issues. These are:
  - The way concepts are semantically defined and how such semantic information can be derived from existing sources.
  - The reconciliation of existing forms of semantic heterogeneities, as these have been already identified in Chapter 3.
  - The selection of the most appropriate type of ontology association according to the available semantic information, the objectives set, and the constraints of the integration endeavor.

This last overall realization has led to the extended framework presented in the next section.

14.4 AN EXTENDED FRAMEWORK

Attempting to develop an overarching framework for every aspect of integration would be an extremely difficult task, possibly of limited practical value. It is more rational to establish a framework from a specific perspective—in our case, an ontological one—and understand what semantic approaches have to offer. In order to do this, it is important to address the following preliminary questions:

1. Which semantic elements affect integration? Action: Define the semantic elements/components that shall prevail integration.
2. Where does semantics emanate from? Action: Determine the available sources of semantics.
3. How is semantics derived? Action: Use the appropriate approach to extract the semantic components from the available sources.
4. How are concepts and ontologies compared? Action: Define which concepts and ontologies are to be compared, and the basis (context) for their comparison.
5. How is similarity/heterogeneity among concepts determined? Action: Decide how the comparison takes place and what the possible/acceptable outcome of the comparison is.

6. How is heterogeneity resolved/reconciled? Action: Decide how heterogeneities among concepts are reconciled and what the outcome of reconciliation is.
7. What type of integration is preferred? Action: Define what resource ontologies are to be integrated, their role during and after integration, if one prevails over the others, if a target ontology is used to guide integration, etc.
8. Is user/expert involvement essential in the process? Action: Determine the degree of automation or interaction/expert involvement needed or preferred in resolving complex cases.
9. How is the result evaluated? Action: Determine the basis for assessing the result of integration, how objective/subjective the result may be, what kind of inconsistencies are expected, should be avoided, etc.

These questions are set from the user's point of view and are very helpful. Not only do they contribute to an understanding of what the overall objective is, but also which intermediate decisions are to be taken. Therefore, formalization, explication, and implementation details are not included. It is also important to see that the above questions can be further grouped into three sets, which define corresponding processes. The first set (questions 1 to 3) refers to *semantic extraction*, the second set (questions 4 to 6) refers to *comparison and reconciliation*, and only the third set (questions 7 to 9) refers specifically to the *integration* process. It proves useful to purport this distinction of the three (sub) processes as the basis for examining the available integration approaches and the way they deal with semantics in each (sub) process.

Figure 14.1 shows that the road map to the semantic integration process of geographic ontologies passes through two prerequisite processes: (a) semantic information extraction, and (b) identification and reconciliation of heterogeneities. These two preprocesses are often treated independently in the literature and so is the third process of integration. A comprehensive integration methodology of practical value necessitates some systematic collaboration of the processes, at least to the extent that it relates to their outcome. This is because the components used in each process must be compatible with the ones available or produced by the previous one. There have been different attempts, systems, or projects addressing various issues of these processes (see Chapter 15). The three processes, however, continue to be distinct having their own objectives. In order to make them comprehensible, six principal characteristics, which play a principal typifying role, can be identified:

1. *Assumptions made.* In order to understand each process, we need to know the assumptions made about the source of semantics and what the objective of the process is.
2. *Semantic level addressed.* Each process addresses certain "semantics." It is clear that there are approaches addressing fewer semantic elements such as terms or attributes, while others exploit richer semantics such as parts, functions, and finally semantic properties and relations (Green, Bean, and Myaeng, 2002; Kokla and Kavouras, 2005). The existence of richer semantics allows for "deeper conversation" among the associated ontologies.

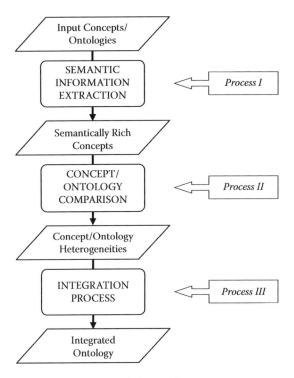

FIGURE 14.1   Processes toward semantic integration.

3. *Input (source)/output components.* The input/output to/from the process is also a critical factor for it may not satisfy the requirements of the initial objective or the method employed.
4. *Method used.* The method or strategy used to achieve the objective of the process is also important. Various methods may be available in each process. Extraction methods attempt to exploit differently the source components. Some comparison methods handle concepts/properties/ontologies separately; others treat them all together. Conflict resolution also differs. In other resource ontologies, the association may be horizontal, or via a top-level, target, or shared ontology.
5. *Ontology change.* The degree of alteration/change that the process has caused to the resource ontologies is also a critical factor for it may not allow backward interoperability.
6. *User involvement*: Talking about user involvement, we distinguish three types: automatic, semiautomatic, and manual. Here the importance is not in the degree of automation itself, for this is a concern of the second (design/implementation) perspective. We are rather interested in how objective or subjective the process is. Very often, "expert/domain" user manual involvement and "brainstorming" in ontology mapping causes additional conflicts very difficult to resolve afterwards.

The above principal characteristics are essential to a unified framework for ontology integration and the associative processes. A number of important issues related to each process are exemplified below.

14.5 PROCESS I: EXTRACTION

The first process is employed in order to extract semantic information from the available sources. Source components can be quite different, resembling ontologies somewhat as the latter are defined loosely or rigorously. In general, sources can consist of free text, corpora, thesauri, specialized text (e.g. definitions), terms, nomenclatures, data dictionaries, hierarchical classifications, database schemata, etc. (see Chapter 2). In geography, we encounter terminological ontologies more often than axiomatized or formal ontologies. In this wealth of source information, it is imperative to decide what constitutes semantic information so that it can be extracted with the appropriate extraction process. This is not a trivial task.

Semantic information in ontologies can be decomposed into the following semantic elements/components:

- Ontology domain/context
- Concepts
- Terms
- Semantic properties
- Semantic relations
- Subsumption relations
- Parts
- Functions
- Axioms—constraints
- Instances

These elements are usually contained explicitly or implicitly in the concept definition, when available, and in the ontology hierarchical structure as well. It is not always the case to have all these semantic components present in the available sources. There may be just terms, that is, class/category names, and possibly a hierarchy. In such a case, all the rest is implied from the subsumption relations and the ontology domain/context. At times, semantics are not derived from the definition, but from an existing database schema (Chang and Katz, 1989; Kashyap and Sheth, 1996; Hakimpour and Geppert, 2001). It is not always certain, however, whether and to what extent such information is indeed semantic.

Given the above source components, there are empirical ad hoc approaches attempting to formalize the concepts involved and design the associated databases. An advanced and systematic way, however, to extract semantics is by some semantic information extraction (SIE) approach based on natural language understanding or processing (NLU/NLP), the central notions in the area of computational linguistics and artificial intelligence (Soderland, 1997; Appelt, 1999; Cowie and Wilks, 2000). These approaches (see Chapter 12) are characterized by different levels of sophistication and automation. A proper application of such techniques in geographic ontologies can reveal salient semantic information (properties and relations) (Kokla and Kavouras, 2005; Tomai and Kavouras, 2004).

14.6 PROCESS II: COMPARISON, IDENTIFICATION, AND RECONCILIATION OF HETEROGENEITIES

Once semantic information becomes available, it is necessary to be able to compare the concepts of the ontologies involved (resource and target). There are various ways to perform a comparison (Klein, 2001; Hovy, 2002; Kavouras, Kokla, and Tomai, 2005). A comparison shall reveal and somehow measure similarities or *heterogeneities*. Heterogeneities are also known as *conflicts* or *mismatches*. Again here, various terms have been used (Sheth and Kashyap, 1992; Ding and Foo, 2002) to address the issue of semantic comparison, such as *equivalence, relevance, resemblance, compatibility, discrepancy, reconciliation,* and *relativism.* Comparison and similarity measures first reveal/depict how difficult integration (Process III) will be. But this is not enough. Process II also needs to resolve/reconcile heterogeneities (Kokla and Kavouras, 2005).

In order to comprehend and possibly engage in the issue of semantic heterogeneity between concepts, let us examine some common cases. The simplest one is to have/use only the concept terms of a taxonomy. Because terms are sometimes poorly understood or misleading, we may also consider additional descriptive information usually provided by concept definitions. In such a case, we are allowed to assume that the conceptualization of a real-world entity consists of a term T and a definition D (e.g., natural language definition); in other words, a concept C is represented by C = (T, D). Different combinations of these two elements result in a set of possible comparison outcomes between concepts as shown in Table 14.1 (with a self-explanatory notation). The clearest cases are those of "equivalence" (same term and definition) and "disjointness" (different terms and definitions). "Synonymy" occurs when two concepts are represented by different terms (e.g., "pond" and "pool") with the same definition ("small lake"). "Overlap" occurs when the terms are the same (e.g., "canal") but the definitions overlap (e.g., "artificial natural waterway used for irrigation and transportation" and "artificial waterway used for recreation and

---

TABLE 14.1
Different Combinations of Term (T) and Definition (D) Cases

	$T_1 = T_2$	$T_1 \neq T_2$
$D_1 = D_2$	Equivalence	Synonymy
$D_1 > D_2$	Further investigation required	IS-A
$D_1 \otimes D_2$	Overlap	Overlap
$D_1 \neq D_2$	Homonymy	Disjointness

transportation"). The case where the terms are the same but one of the two definitions is richer than the other can be differently interpreted and requires more thorough investigation. One case may be that the concepts are the same but the richer definition gives further explanatory information, not semantically essential. The other case may be that the richer definition identifies a more specific concept, that is there is an IS-A relation between the two concepts. This assertion is stronger in case the two concepts do not have the same terms, since this is further evidence for the differentiation of the two concepts.

This comparison can now be examined in more detail. In a systematic approach, with a hope for automation, definitions can not be directly compared. What is actually compared is the semantics contained. These are extracted by Process I (Section 14.5) in the form of semantic properties/relations and their values.

Selecting the right context has a crucial effect on the outcome of comparison. Context (see Section 7.2) also called view, perspective, basis, reference, framework, or situation, provides the basis for examining and interpreting things. But how should context be chosen? Each concept has its own local context. This can be defined as the set of its relations/properties. Some (Goh, 1997) call it the attribute vector. Thus, the vector of semantic relations for the concept of interest forms a semantic context. An ontology has its global context that applies to or emanates from all of its concepts. As a result, if semantic relations (Chapter 6) are to be used as the basis for concept comparison, then one can compare them with respect to:

- The global union (total) context of the two ontologies
- The global intersection (common) context of the two ontologies
- The local union context of the compared concepts
- The local intersection context of the compared concepts

Again, the wider the comparison context, the more differences are expected. A comparison on a narrower context results in more similarities, whereas overlap and inclusion cases are more easily resolved. We can and probably should compare concepts on the basis of their common semantic elements, given the same or similar term/name. The name has its own semantic weight and can be also considered as an important semantic property. Generally, certain semantic relations/properties are more essential to geographic concepts, especially those referring to physical bona fide entities. These are: name, is-a, embodiment (material-cover), role (purpose), meronyny, and spatiality.

14.7 PROCESS III: INTEGRATION

The process of ontology integration entails many issues. Most of them have already been addressed in the course of establishing the framework (Sections 14.1 through 14.4) with the principal characteristics for each process. Two issues remain to be explained in this section and are thus presented below: (a) the terminology of ontology integration according to the architecture established, and (b) the principle or idea behind ontology integration approaches.

14.7.1 INTEGRATION ARCHITECTURE TERMINOLOGY

As mentioned above, *integration* is used in the literature (Sowa, 2000; Klein, 2001; Uschold and Gruninger, 2002; Kavouras, 2005) as a general term to denote a number of related concepts, such as *association, coordination, combining, matching, mapping, translation, merging, partial compatibility, alignment, unification, fusion, mediation, true integration*, and the like. Most terms refer to the way ontologies are connected, describing thus the *architecture (or topology) of integration*. Since there is not an agreement on all terms used in the literature, some working definitions of the integration architecture are first presented here to be used thereinafter.

First, we embark on this presentation with the more general terms used. *Coordination, association,* and *combining* are probably the broadest terms just denoting the need to work in a task involving related but different ontologies. *Matching* is a term actually relating to the outcome of Process II (comparison and reconciliation) (Section 14.6). This procedure compares concepts in order to "match" those that are more similar in meaning according to a given context. *Mapping* is another widely used term to denote a general equivalence *matching* between concepts from different ontologies based on some metric (Klein, 2001). *Translation* is a characteristic such a process requires when ontologies are not expressed in the same formalism. During translation, the semantics remain unaltered. A term requiring some attention is that of *transformation*. Here, it is admitted that initial semantics may be intentionally somewhat altered in order to satisfy (or conform to) the semantics of a new application or domain (Klein, 2001). Such a distortion, no matter how mild or unavoidable, needs to be maintained or traced for it may lead to misunderstandings about the intended meaning. Transformation may also require translation.

The definition of the rest of the terms entails more complex characteristics. Most integration processes have two or more input ontologies, called *source* or *resource ontologies*, to integrate. Sometimes the integration process is performed in a forceful way, causing distortions that make the use of the initial ontologies impossible. The ability to allow usage of the resource ontologies is called *backward compatibility*. Another issue is whether the process of integration results in a single ontology or retains all resource ontologies. A third issue is whether the integration process relies on the resource ontologies only, or uses centrally a *target ontology*, acting as an *intermediate* or *shared* ontology. Depending on the resource ontologies and the integration objective, target ontologies can be general or domain ontologies. As a consequence, the outcome of the integration process also varies significantly. It can be a single integrated ontology, an alignment of resource ontologies without an intermediate ontology, an alignment between resource ontologies and a target or reference ontology as intermediate, an alignment between clusters of resource ontologies, and so on (Figure 14.2). The result of the integration process may be virtual or physical. Usually alignment and mapping result in virtual integrations, whereas unification and merging may also result in a physical integration. This, however, is an explication issue.

Based on the above characteristics, we are ready to clarify the most common complex types of integration architecture (Figure 14.3). Other mutations or combinations are of course possible.

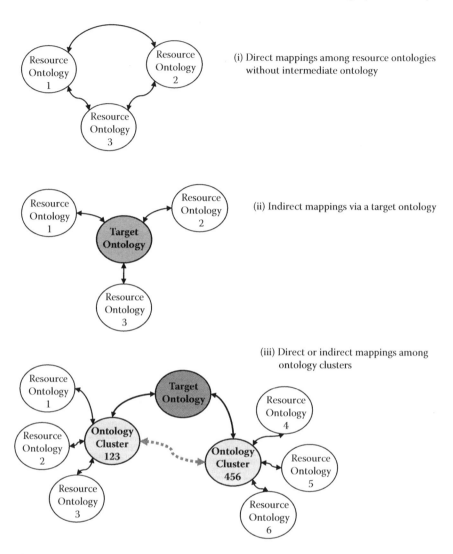

FIGURE 14.2 Direct and indirect ways of alignment/mapping among resource ontologies.

Alignment is a mapping between concepts of different ontologies bringing them into mutual agreement. It is specified by *articulation*, that is, a set of link nodes or semantic bridges (Ding and Foo, 2002), which articulate the two resource ontologies. Translation/conversion utilities are used to provide functionality. In alignment, no ontology is distorted. The user has still many ontologies to deal with. A target ontology may or may not be aligned with the resource ontologies.

Partial compatibility creates a merging of only those parts of ontologies that are considered more similar. The rest of the ontology parts are still necessary. The merged parts distort the initial common ontology parts. A target ontology may or may not be used for the merging of the common parts.

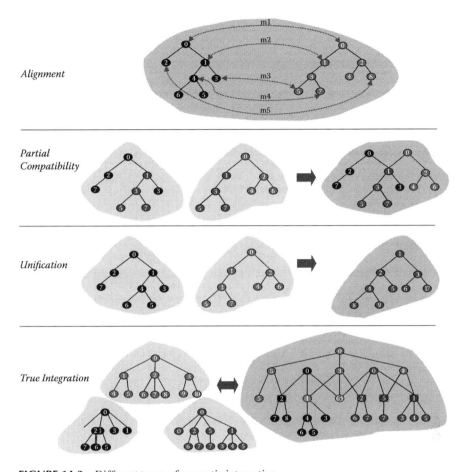

FIGURE 14.3 Different types of semantic integration.

Unification, sometimes also called *fusion*, extends partial compatibility to all ontologies and their concepts. The result is that each resource ontology is forced/distorted to become the same (and thus fully compatible) with the others. In other words, there is a single ontology at the end. A target ontology may or may not be used for defining the unified ontology. The user deals with one ontology. The initial (resource) ontologies are not usable anymore.

True integration creates a single integrated ontology whose parts are the resource ontologies including some additional concepts necessary for the association. The user deals with a single integrated ontology. The resource ontologies are not distorted, retaining their independence and usability. A target ontology may or may not be used in the integration.

Based on the above realizations, the methods for the integration process are distinguished with respect to three (almost orthogonal) dimensions.

TABLE 14.2
Classification of Ontology Integration Approaches

No.	D1	D2	D3	INTEGRATION TYPE	INTEGRATION CLASS
1	0	0	0	True integration	TRUE INTEGRATION
2	0	0	1	True integration targeted	
3	0	1	0	Alignment	ALIGNMENT
4	0	1	1	Alignment targeted	
5	1	0	0	Unification	UNIFICATION
6	1	0	1	Unification targeted	
7	1	1	0	Partial compatibility	PARTIAL COMPATIBILITY
8	1	1	1	Partial compatibility targeted	

- D1: The possible change/alteration/distortion caused by the process of integration. Two cases are distinguished here: NO/YES (0/1).
- D2: The number of ontologies resulting from the integration process. Two cases are distinguished here: SINGLE/MANY (0/1).
- D3: The possible use of a target/shared ontology in the integration process. Two cases are distinguished here: NO/YES (0/1).

As a result, each approach can be classified in one out of eight combinations (Table 14.2).

14.7.2 PRINCIPLE BEHIND AN INTEGRATION APPROACH

After having established a terminology about the architecture of ontology integration, the most important issue in selecting or developing an integration approach is the principle or core idea behind it. This pivoting matter and the direction taken, which determines to a great extent the complexity of the method and the quality of its result, have been already raised in Section 14.4. In the literature, there is a lot of confusion about the directions one may possibly take in an integration task. A fundamental objective of all approaches, no matter what methodology or architecture they subsequently employ, is more or less to compare the semantics of the given ontologies and determine the following:

- Whether the given ontologies are to some degree similar, related, or disjoint. *Similarity* is a general notion (see Chapter 13) denoting some overlapping between the two ontologies. *Disjoint* ontologies are those specifying unrelated domains. *Related* ontologies are those that are not overlapping but some of their concepts are related (e.g., by a subsumption relation) without being similar.
- How to compare concepts in overlapping or related ontologies, in order to identify the similar (overlapping), related, or disjoint concepts. This is a difficult but fundamental problem (Sheth, Ramakrishnan, and Thomas, 2005)

also known as *concept (or entity) identification/disambiguation*. This concept comparison step may be used without having to cover the previous ontology comparison step.
- How to associate the ontologies on the basis of the findings of the previous two steps and the possible architectures defined in Section 14.7.1.

Although the objectives seem clear, the context, perspective, and way the above issues are tackled differ a lot. Herein, some alternatives are presented that are commonly pursued in existing approaches. The objective is to clarify the principal directions and not to exhaust all variants—known or possible.

1. Conforming to a single central ontology
2. Manual ad hoc mappings
3. Intuitive mappings based on "light" lexical information
4. Intuitive mappings based on explication characteristics
5. Intuitive mappings based on structural similarity
6. Relating (grounding) to a single shared or top-level ontology
7. Direct mappings based on "deep" semantics
8. Integration by view-based query processing
9. Compound similarity measures
10. Extensional mappings based on common spatial reference

These directions are now discussed further.

1. *Conforming to a single global ontology.* Approaches following this principle attempt to establish a single global ontology that all users employ. This severely limited principle follows the old standardization paradigm—a rather procrustean way to enforce semantic interoperability, in which the need for mappings is entirely eliminated. Life has shown that such approaches are not successful, for various reasons. They only suffice for very narrow applications and small community needs for a limited time because they do not handle ontology evolution.
2. *Manual ad-hoc mappings.* This is a simple and commonly used approach, which lets the user/expert define arbitrary mappings between concepts of the two ontologies. The vast majority of mappings are still established manually. KRAFT (Visser et al., 1999; Preece et al., 2000) is a characteristic example of such an approach. Its major advantage is simplicity and user-controlled result. However, because it is an entirely subjective process, many inconsistencies will arise, while it is not certain that semantics are preserved (Wache et al., 2001). Being also laborious and error prone makes it highly inefficient to deal with many, large complicated ontologies with numerous overlapping concepts.
3. *Intuitive mappings based on "light" lexical information.* More refined (and of course, complex) approaches exploit basic (light) lexical information, such as terms (concept names) and their synonyms, to enable a more intuitive mapping between concepts. OBSERVER (Mena et al., 1998) is a

typical case of this. The advantage of approaches following this direction is that they are less subjective than the first one. As a result, some parts of the process can be semiautomated in the form of alternatives suggested to the user. The disadvantage is that mere term similarity (even with the use of vocabularies) is not sufficient to encapsulate the semantics of concepts. Therefore, the effectiveness of mapping approaches based only on such "shallow" semantics is limited.

4. *Intuitive mappings based on explication characteristics.* Some integration approaches, despite being called "semantic," attempt to solve explication problems resulting in a distortion of ontologies in order, for example, to make them computationally equivalent. These approaches resemble those from the field of database integration, where concepts (entities in this case) are compared/matched with respect to their syntactic similarity regarding explication characteristics (such as names, data types, and structures) of representation elements (attributes, relations, constraints, and instances). These approaches are very useful to integration at the explication level. Such syntactic information, however, is either inappropriate (leads to wrong conclusions) or insufficient to reveal semantic similarity, relation, or difference. Semantically equivalent concepts may differ at the explication level and vice versa. It is thus widely recognized (Sheth and Larson, 1990; Patridge, 2002) that databases do not contain sufficient information for semantic matching. The obscure assumption that syntactic similarity is strongly correlated to semantic similarity is rarely discussed (Doan, 2002) and certainly difficult to justify.

5. *Intuitive mappings based on structural similarity.* Some approaches originating from research on schema integration (Rahm and Bernstein, 2001), are developed on the following assumption: Similar ontologies also exhibit some structural (schematic) similarity. Along the same lines, if concepts in different ontologies match, then it is explored if the associated super/sub/side concepts also match. Another example would be the case of two concepts with a subsumption relation in the first ontology, matching two other concepts with another subsumption relation in the second ontology. In this case, it is possible that the in-between concepts also match (see the AnchorPROMPT approach by Noy and Musen, 2003). Since structures (schemas) can be treated as graphs, graph-matching techniques reveal the degree of structural similarity (Sheth, Ramakrishnan, and Thomas, 2005). Such approaches, no matter how logical, present various problems because ontology hierarchies are based on different contexts/domains, and it is not necessary that two equivalent concepts from two ontologies also have equivalent super or sub concepts. The validity of the assumption that schematic similarity is under conditions positively related to semantic similarity has been identified and investigated (Sheth and Kashyap, 1992; Kashyap and Sheth, 1996) as already presented in Chapter 13.

6. *Relating (grounding) to a single shared or top-level ontology.* Another family of approaches avoids the determination of direct correspondences between concepts from different ontologies by using a single common

top-level ontology. Each resource ontology inherits only superconcepts from the top-level ontology. This approach, known as *top-level grounding* (Wache et al., 2001), has some practical advantages, the most important being that the semantics of resource ontologies remain unaltered. However, the fact that only indirect correspondences are supported via more general superconcepts may create problems when exact correspondences are needed (Wache et al., 2001). An analogous principle is that of using a shared target ontology to map resource ontologies to it. Although mappings through a central ontology avoid the quadratic number of direct mappings (Weinstein and Birmingham, 1999), it is very difficult to specify central global ontologies that satisfy all anticipated uses (Hull, 1997).

7. *Direct mappings based on "deep" semantics.* Similarly to the semantic correspondences by Wache et al. (2001), the objective of this direction is to avoid (a) indirect mappings via a top-level ontology (direction 6), and (b) subjective direct mappings based on "shallow" semantic information (direction 3). In this family of approaches, in order to support direct mappings among concepts based on "deeper" semantics (e.g., semantic relations), it is essential that such information is derived from the available sources. *Linguistic techniques* such as NLP (see Process I in Section 14.5) are usually applied on *unstructured* data, while *constraint-based approaches* are more suitable in the case of *semistructured* data (Sheth, Ramakrishnan, and Thomas, 2005). Common vocabularies may be used to establish semantic correspondences between concepts from different ontologies. Wache et al. (2001) make use of semantic labels to compute correspondences between database fields. Stuckenschmidt et al. (2000) propose a description logic model of terms from different information sources, while relations between different terminologies can be established using subsumption reasoning. Formal Concept Analysis (Chapter 8), based on semantically "deep" properties, also establishes direct mappings between the concepts involved, in a concept lattice.

8. *Integration by view-based query processing.* This is a family of approaches in which querying mechanisms play a dominant role (Ullman, 1997; Calvanese et al., 2000; Halevy, 2000; Pottinger and Halevy, 2001; Capezzera, 2003), an issue already introduced as a framework for query-based approaches at Section 14.3. Integration in this context is expressed by mappings between a global ontology and the local ontologies. Such a service is usually provided by mediators (Wiederhold, 1992; Melnik, 2000) that usually offer abstract (nonmaterialized) integrated views over heterogeneous data sources.

9. *Compound similarity measures.* Concepts might be compared and matched on the basis of the available semantic components assumed in the previous paragraph; that is, term comparison, relation/property/attribute comparison, or instance comparison. This may conclude as to whether two concepts are equivalent, different, related, etc. Another way of dealing with semantic correspondences and concept matching is by establishing compound similarity measures among the compared concepts from different ontologies (Maedche and Staab, 2002; Maedche and Zacharias, 2002; Rodriguez and Egenhofer, 2003; Ehrig and Sure, 2004). The notion of similarity has been introduced

in Chapter 13. Such measures help, for example (Doan et al., 2003), to determine for each concept in one ontology, the most similar concept in another.
This can prove somewhat useful in approximate querying and data mining. The result of such approaches is usually a similarity (or dissimilarity)
matrix, which may additionally have other uses. An interesting use is that
of *spatialization* (W. Kuhn and Blumenthal, 1996; Skupin and Buttenfield,
1997; Kavouras, Kokla, and Tomai, 2005); that is, a concept mapping method
employing spatial metaphors and semantic distances for the visualization of
similarity and clustering. From the practical point of view, however, compound similarity values do not enlighten us about (a) how heterogeneities
must be reconciled, and (b) how to create an integrated ontology.

10. *Extensional mappings based on common reference.* Many of the above
families of approaches associate concepts relying on their intensional
information. However, there are approaches that associate concepts on the
basis of their extensional information, when of course available. The simple
assumption made here is that concepts having the same instances are likely
equivalent. An advantage of these approaches is that they contain an extensional base for comparison and reconciliation. There are, however, several
disadvantages: (a) extensional information may be unavailable, unknown,
incomplete, or circumstantial; (b) instances do not necessarily (or fully)
describe the semantic domain of a concept; and (c) the degree to which
extensional resemblance is directly and positively related to concept resemblance is not known or justified.

There is a case, however, which requires special attention for it relates to geographic space. Space is considered to act as a reference (container) where things
happen. The assumption here is that things closer in space are more associated. Furthermore, if two categories (concepts) of similar contexts/domains (ontologies) refer
to the same space (common reference), then it is likely that they are similar. If, for
example, a specific piece of land is classified as category a in land-cover ontology
O_1 and category b in land-cover ontology O_2, then the two categories are considered
as equivalent or connected via a subsumption relation. The same principle may be
applied for a general reference space, not necessarily geographic.

The way extensional type mappings are implemented is more complicated.
Here a mathematical theory of information flow known as *channel theory* (Barwise
and Seligman, 1997; Kalfoglou and Schorlemmer, 2003) is used to enable semantic interoperability between different communities that use their own ontologies, by
providing the necessary mappings across them. In addition, Kent (2001, 2004) has
developed the Information Flow Framework (IFF) for the standardization activity of
Standard Upper Ontology (SUO) and proposed a methodology for ontology merging. Interoperability in the context of information flow takes place both at the *type*
(category) and the *token* (instance) level as it can be understood from the notions
of *channel* and *infomorphism* (Chapter 10). Therefore, when generating mappings
between ontologies, the instances of the ontologies, as well as the included concepts,
are compared. In the field of geospatial integration, Tomai and Kavouras (2005)
have demonstrated how information flow can be used to associate different maps. An

extensional algebraic approach is also employed by Duckham and Worboys (2005), where the schema-level structure, necessary for the fusion, is inferred from instance-level information.

REFERENCES

Appelt, D. E. 1999. Introduction to information extraction. *AI Communications* 12: 161–172.

Arens, Y., C. Knoblock, and W. Shen. 1996. Query reformulation for dynamic information integration. *Journal of Intelligent Information Systems* 6(2/3): 99–130.

Bailin, S. and W. Truszkowski. 2002. Ontology negotiation between intelligent information agents. *Knowledge Engineering Review* 17(1): 7–19.

Barwise, J. and J. Seligman. 1997. *Information flow: The logic of distributed systems.* Cambridge Tracts in Theoretical Computer Science. Cambridge, UK: Cambridge University Press.

Batini, C., M. Lenzerini, and S. B. Navathe. 1986. A comparative analysis of methodologies for database schema integration. *ACM Computing Surveys* 18(4): 323–364.

Benerecetti, M., P. Bouquet, and C. Ghidini. 2000. Contextual reasoning distilled. *Journal of Experimental and Theoretical Artificial Intelligence* 12(3): 279–305.

Bouquet, P., J. Euzenat, E. Franconi, L. Serafini, G. Stamou, and S. Tessaris. 2004. Specification of a common framework for characterizing alignment, Deliverable D2.2.1 in KWEB EU-IST-2004-507482.

Busse, S., R. D. Kutsch, U. Leser, and H. Weber. 1999. Federated information systems: Concepts, terminology and architectures. Technical Report Nr. 99-9. Technische Universität, Berlin.

Calvanese, D., G. De Giacomo, and M. Lenzerini. 2001a. A framework for ontology integration. In *Proc. 1st Semantic Web Working Symposium SWWS 2001*, July 30–August 1, 303–316, Stanford University, Stanford, CA.

Calvanese, D., G. De Giacomo, and M. Lenzerini. 2001b. Ontology of integration and integration of ontologies. In *Description Logics 2001: Working Notes of the 2001 International Description Logics Workshop*, ed. C. Goble, R. Möller, and P. F. Patel-Schneider. Stanford, CA: Stanford University. http://CEUR-WS.org/Vol-49/ (accessed 1 May 2007).

Calvanese, D., G. De Giacomo, M. Lenzerini, and M. Y. Vardi. 2000. View-based query processing and constraint satisfaction. In *Proc. 15th IEEE Symposium on Logic in Computer Science LICS 2000*, June 26–29, 361–371, Santa Barbara, CA.

Capezzera, R. 2003. Query processing by semantic reformulation. Laurea thesis, *Faculty of Engineering at Modena,* Universitá di Modena e Reggio Emilia, Italy. http://www.dbgroup.unimo.it/tesi/capezzera.pdf (accessed 1 May 2007).

Chang, E. E. and R. H. Katz. 1989. Exploiting inheritance and structure semantics for effective clustering and buffering in an object-oriented DBMS. *ACM SIGMOD Record* 18(2): 36–45.Ciocoiu, M., M. Gruninger, and D. Nau. 2001. Ontologies for integrating engineering applications. *Journal of Computing and Information Science in Engineering* 1(1): 12–22.

Cowie, J. and Y. Wilks. 2000. Information extraction. In *Handbook of natural language processing*, ed. R. Dale, H. Moisl, and H. Somers, 241–260. New York: Marcel Dekker.

Denny, M. 2004. Ontology tools survey, revisited. http://www.xml.com/pub/a/2004/07/14/onto.html (accessed 1 May 2007).

Ding, Y. and S. Foo. 2002. Ontology research and development. II. A review of ontology mapping and evolving. *Journal of Information Science* 28(5): 375–388.

Doan, A. 2002. Learning to map between structured representations of data. Ph.D. thesis, University of Washington-Seattle.

Doan, A., N. Noy, and A. Halevy. 2004. Introduction to the special issue on semantic integration. *ACM SIGMOD record* 33(4): 11–33.

Doan, A., J. Madhavan, P. Domingos, and A. Halevy. 2003. Ontology matching: A machine learning approach. In *Handbook on ontologies in information systems*, ed. S. Staab and R. Studer, 397–416. Berlin: Springer Verlag.

Duckham, M. and M. F. Worboys. 2005. An algebraic approach to automated geospatial information fusion. *International Journal of Geographical Information Science* 19(5): 537–558.

Ehrig, M. and Y. Sure. 2004. Ontology mapping: An integrated approach. Bericht 427, Institut für Angewandte Informatik und Formale Beschreibungsverfahren, AIFB, Universität Karlsruhe (TH).

Fillion, F., C. Menzel, T. Blinn, and R. Mayer. 1995. An ontology-based environment for enterprise model integration. Paper presented at the IJCAI Workshop on Basic Ontological Issues in Knowledge Sharing, Montreal, Canada.

Firat, A. 2003. Information integration using contextual knowledge and ontology merging. Ph.D. thesis, Sloan School of Management, Massachusetts Institute of Technology, Boston.

Goh, C. H. 1997. Representing and reasoning about semantic conflicts in heterogeneous information systems. Ph.D. thesis, Massachusetts Institute of Technology, Boston.

Green, R., C. A. Bean, and S. H. Myaeng, eds. 2002. *The semantics of relationships: An interdisciplinary approach.* Dordrecht: Kluwer Academic Publishers.

Gruninger, M. and J. Kopena. 2003. Semantic integration through invariants. Paper presented at Workshop on Semantic Integration at ISWC-2003, Sanibel Island, FL.

Hakimpour, F. and A. Geppert. 2001. Resolving semantic heterogeneity in schema integration: An ontology based approach. *Proc. International Conference Formal Ontology in Information Systems FOIS-2001*, Ogunquit, Maine, 297–308. New York: ACM Press.

Halevy, A. 2000. Theory of answering queries using views. *ACM SIGMOD Record* 29(4): 40–47.

Halevy, A. 2001. Answering queries using views: A survey. *Very Large Database Journal* 10(4): 270–294.

Hameed, A., A. Preece, and D. Sleeman. 2003. Ontology reconciliation. In *Handbook on ontologies in information systems,* ed. S. Staab and R. Studer, 231–250. Berlin: Springer Verlag.

Hovy, E. 2002. Comparing sets of semantic relations in ontologies. In *The semantics of relationships: An interdisciplinary perspective*, ed. R. Green, C. A. Bean, and S. H. Myaeng, 91–110. Dordrecht: Kluwer Academic Publishers.

Hull, R. 1997. Managing semantic heterogeneity in databases: A theoretical perspective. In *Proc. 16th ACM SIGACT SIGMOD SIGART Symposium on Principles of Database Systems PODS-97*, Tucson, AZ, 51–61.

Kalfoglou, Y. and M. Schorlemmer. 2003. IF-Map: An ontology mapping method based on information flow theory. *Journal on Data Semantics* 1(1): 98–127.

Kashyap, V. and A. Sheth. 1996. Schematic and semantic similarities between database objects: A context-based approach. *International Journal on Very Large Data Bases* 5(4): 276–304.

Kavouras, M. 2005. A unified ontological framework for semantic integration. In *Next generation geospatial information: From digital image analysis to spatiotemporal databases*, ed. P. Agouris and A. Croitoru, 147–156. ISPRS Book series. London: Taylor & Francis.

Kavouras, M., M. Kokla, and E. Tomai. 2005. Comparing categories among geographic ontologies. *Computers & Geosciences* 31: 145–154.

Kent, R. E. 2001. The information flow framework. Starter document for IEEE, the IEEE Standard Upper Ontology Working Group. http://suo.ieee.org/IFF/ (accessed 1 May 2007).

Kent, R. E. 2004. The IFF Foundation for ontological knowledge organization. In *Knowledge organization and classification in international information retrieval*, 187–203, ed. N. J. Williamson and C. Beghtol. Cataloging and Classification Quarterly, Vol. 37(1,2). Binghamton, NY: Haworth Press.

Kim, W. and J. Seo. 1991., Classifying schematic and data heterogeneity in multidatabase systems. *IEEE Computer* 24(12): 12–18.

Klein, M. 2001. Combining and relating ontologies: An analysis of problems and solutions. In *Proc. IJCAI'01- Workshop on Ontologies and Information Sharing*, Seattle, WA, ed. A. Gomez-Perez, M. Gruninger, H. Stuckenschmidt, and M. Uschold. http://citeseer.ist. psu.edu/klein01combining.html (accessed 1 May 2007).

Kokla, M. and M. Kavouras. 2005. Semantic information in geo-ontologies: Extraction, comparison, and reconciliation. *Journal on Data Semantics* 3534: 125–142.

Kuhn, E., F. Puntigam, and A. Elmagarmid. 1991. Multidatabase transaction and query processing in logic. In *Database transaction models for advanced applications*, ed. A. Elmagarmid, 297–348. San Mateo, CA: Morgan Kaufmann Publishers.

Kuhn, W. and B. Blumenthal. 1996. Spatialization: Spatial metaphors for user interfaces. Technical Report, Department of Geoinformation, Technical University of Vienna, Vienna.

Lenzerini, M. 2002. Data integration: A theoretical perspective. In *Proc. 21st ACM SIGACT SIGMOD SIGART Symposium on Principles of Database Systems PODS 2002*, June 3–5, 233–246, Madison, WI.

Litwin, W. and A. Abdellatif. 1987. An overview of the multi-database manipulation language MDSL. *Proceedings of the IEEE* 75(5): 621–632.

Maedche, A. and S. Staab. 2002. Measuring similarity between ontologies. In *Proc. European Conference on Knowledge Acquisition and Management EKAW-2002*, Lecture Notes in Computer Science/Lecture Notes in Artificial Intelligence, 2473, 251–263. Berlin: Springer-Verlag.

Maedche, A. and V. Zacharias. 2002. Clustering ontology-based metadata in the semantic web in principles of data mining and knowledge. In *Proc. 6th European Conference PKDD 2002*, Lecture Notes in Computer Science, 2431, 348–360. Berlin: Springer-Verlag.

Melnik, S. 2000. Declarative mediation in distributed systems. In *Proc. International Conference on Conceptual Modeling ER'00*, Salt Lake City, UT, Stanford Database Group Publication Server.

Mena, E., V. Kashyap, A. Illarramendi, and A. Sheth. 1998. Domain specific ontologies for semantic information brokering on the global information infrastructure. In *Formal ontology in information systems*, Vol. 46, ed. N. Guarino, 269–283. Amsterdam: IOS Press.

Noy, N. F. 2004. Semantic integration: A survey of ontology-based approaches. *ACM SIGMOD Record, Special Issue on Semantic Integration* 33(4): 65–70.

Noy, N. F. and M. A. Musen. 2003. The PROMPT suite: Interactive tools for ontology merging and mapping. *International Journal of Human-Computer Studies* 59(6): 983–1024.

Obrst, L., R. Wray, and H. Liu. 2001. Ontology engineering for ecommerce: A real B2B example. In *Proc. International Conference Formal Ontology in Information Systems FOIS-2001*, Ogunquit, ME, 117–126. New York: ACM Press.

Parent, C. and S. Spaccapietra. 1998. Issues and approaches of database integration. *Communications ACM* 41(5): 166–178.

Partridge, C. 2002. The role of ontology in semantic integration. Paper presented at *Object-Oriented Programming, Systems, Languages, and Applications—11th Workshop on Behavioral Semantics*, Seattle, WA.

Pinto, H. S., A. Gomez-Perez, and J. P. Martins. 1999. Some issues on ontology integration. In *Proc. IJCAI99 Workshop on Ontologies and Problem Solving Methods: Lessons Learned and Future Trends*, Vol. 18, ed. V. R. Benjamins, B. Chandrasekaran, A. Gomez-Perez, N. Guarino, and M. Uschold, 7.1–7.12. Amsterdam: CEUR Publications.

Pottinger, R. and A. Halevy. 2001. MiniCon: A scalable algorithm for answering queries using views. *VLDB Journal* 10(2-3): 182–198.

Preece, A. D., K. J. Hui, W. A. Gray, P. Marti, T. J. M. Bench-Capon, D. M. Jones, and Z. Cui. 2000. The KRAFT architecture for knowledge fusion and transformation. *Knowledge Based Systems* 13(2-3): 113–120.

Rahm, E. and P. A. Bernstein. 2001. On matching schemas automatically. Technical Report MSR-TR-2001-17, Microsoft Research, Microsoft Corporation, Redmond, WA.

Rodríguez, M. A. and M. J. Egenhofer. 2003. Determining semantic similarity among entity classes from different ontologies. *IEEE Transactions on Knowledge and Data Engineering* 15(2): 442–456.

Sheth, A. and V. Kashyap. 1992. So far (schematically), yet so near (semantically). In *Proc. IFIPTC2/WG2.6 Conference on Semantics of Interoperable Database Systems DS-5*, IFIP Transactions A-25, 283–312. Amsterdam: North-Holland.

Sheth, A. P. and J. A. Larson. 1990. Federated database systems for managing distributed, heterogeneous, and autonomous databases. *ACM Computing Surveys* 22(3): 183–236.

Sheth, A., C. Ramakrishnan, and C. Thomas. 2005. Semantics for the Semantic Web: The implicit, the formal and the powerful. *International Journal on Semantic Web & Information Systems* 1(1): 1–18.

Shvaiko, P. and J. Euzenat. 2005. A survey of schema-based matching approaches. *Journal on Data Semantics* 4: 146–171.

Skupin, A. and B. P. Buttenfield. 1997. Spatial metaphors for display of information spaces. In *Proc. AUTO-CARTO 13*, Seattle, WA, 116–125.

Soderland, S. 1997. Learning text analysis rules for domain-specific natural language processing. Ph.D. thesis, University of Massachusetts, Department of Computer Science, Amherst, MA.

Sowa, J. F. 2000. *Knowledge representation: Logical, philosophical, and computational foundations*. Pacific Grove, CA: Brooks/Cole.

Stuckenschmidt, H., H. Wache, T. Vogele, and U. Visser. 2000. Enabling technologies for interoperability. In *Proc. Workshop on the 14th International Symposium of Computer Science for Environmental Protection*, ed. U. Visser and H. Pundt, 35–46. TZI, University of Bremen, Bonn, Germany.

Tamma, V. A. M. and P. R. S. Visser. 1998. Integration of heterogeneous sources: Towards a framework for comparing techniques. In *Proc. "6 Convegno dell'Associazione Italiana per l'Intelligenza Artificiale (AI*IA), Workshop su Strumenti di organizzazione ed accesso intelligente per informazioni eterogenee,"* Padua, Italy, 89–93.

Tomai, E. and M. Kavouras. 2004. From "Onto-GeoNoesis" to "Onto-Genesis": The design of geographic ontologies. *Geoinformatica* 8(3): 285–302.

Tomai, E. and M. Kavouras. 2005. Mappings between maps: Association of different thematic contents using situation theory. In *Proc. International Cartographic Conference ICC 2005*, A Coruna, Spain.

Ullman, J. D. 1997. Information integration using logical views. In *Proc. 6th International Conference on Database Theory ICDT'97*, Lecture Notes in Computer Science, 1186, 19–40. Berlin: Springer Verlag.

Uschold, M. and M. Gruninger. 2002. Creating semantically integrated communities on the World Wide Web. Invited talk in Workshop on the Semantic Web, Co-located with WWW 2002, Honolulu, HI. http://semanticweb2002.aifb.uni-karlsruhe.de/USCHOLD-Hawaii-InvitedTalk2002.pdf (accessed 1 May 2007).

Uschold, M. and R. Jasper. 1999. A framework for understanding and classifying ontology applications. In *Proc. IJCAI99 Workshop on Ontologies and Problem-Solving Methods(KRR5)*, ed. V. R. Benjamins. Amsterdam: CEUR Publications. http://sunsite.informatik.rwth-aachen.de/Publications/CEUR-WS/Vol-18/ (accessed 1 May 2007).

Visser, P. R. S., D. M. Jones, T. J. M. Bench-Capon, and M. J. R. Shave. 1997. An analysis of ontological mismatches: Heterogeneity versus interoperability. In *Proc. AAAI 1997 Spring Symposium on Ontological Engineering*, Stanford, CA.

Visser, P.R.S., D. M. Jones, M. D. Beer, T. J. M. Bench-Capon, B. M. Diaz, and M. J. R. Shave. 1999. Resolving ontological heterogeneity in the KRAFT project. Paper presented at the 10th International Conference and Workshop on Database and Expert Systems Applications DEXA '99, University of Florence, Italy.

Wache, H., T. Vogele, U. Visser, H. Stuckenschmidt, G. Schuster, H. Neumann, and S. Hubner. 2001. Ontology-based integration of information: A survey of existing approaches. In *Proc. IJCAI-01 Workshop: Ontologies and Information Sharing*, Seattle, WA, ed. H. Stuckenschmidt, 108–117.

Weinstein, P. C. and W. P. Birmingham. 1999. Comparing concepts in differentiated ontologies. In *Proc. 12th Workshop on Knowledge Acquisition, Modeling and Management, KAW-99*, Banff, Alberta, Canada.

Wiederhold, G. 1992. Mediators in the architecture of future information systems. *IEEE Computer* 25(3): 38–49.

15 Integration Approaches

15.1 CLASSIFICATION OF APPROACHES

There are an ever-growing number of approaches to integration of ontologies. Also, existing approaches and tools evolve constantly to improve on their capabilities and functionality. Hence, their exhaustive survey is outside the scope of a research textbook. More important and useful is to present and evaluate the distinguishing (i.e., more universal) characteristics of approaches. Based on this objective, some of the most known, interesting, or different approaches are selected and presented in the next sections. Approaches that are not presented here, or will be developed in the future, can still be understood, compared, or evaluated on the basis of the framework presented in the previous chapter, and in a manner very similar to the one presented below.

Most of the available general approaches are results of research projects. MOMIS, KRAFT, PROMPT, and Chimaera are some of them (see a survey on ontology tools by the OntoWeb Consortium, 2002). These approaches are outlined briefly and discussed in the first sections (15.2 to 15.9).

In the geospatial domain, there have been a limited number of approaches to semantic interoperability and ontology integration addressing different issues and focusing on diverse application needs. These can be generally called *geointegration* approaches and are classified into general directions (e.g., architectures to achieve integration, intensional, and extensional-based approaches) and discussed in Sections 15.10 to 15.13. Throughout the presentation of existing approaches, for reasons of homogeneity and comparison, an attempt has been made to follow the terminology introduced so far.

The last section of this chapter concludes with a comparative presentation and demonstrates how the unified framework presented above, based on three processes and their principal characteristics, helps to understand and evaluate the integration approaches. Six exemplary approaches are used in the summary tables.

15.2 PROMPT

PROMPT, formerly known as SMART, is a general algorithm and tool developed at Stanford University (Noy and Musen, 2000; 2003) in order to support ontology merging and alignment. By merging, it is implied that the result is a single consistent ontology, while alignment implies that the resource ontologies are made consistent with one another, but kept separately. It is based on the idea that semantic integration (merging, alignment or other) cannot be fully automated. Therefore, some form of user/expert intervention is inevitable. PROMPT aims at assisting and automating user choices while revealing resulting conflicts and suggesting remedies. With respect to the framework established above, PROMPT does not cover the process of information extraction, but rather addresses the other two processes, that is, ontology comparison with resolution of conflicts and ontology merging/alignment.

The PROMPT algorithm is developed as an interactive tool in Protégé-2000, a popular ontology-design and knowledge-acquisition environment (Gennari et al., 2003). The knowledge model underlying Protégé is frame-based and OKBC-compatible (Chaudhri et al., 1998). The basic and only knowledge model elements required by Protégé and PROMPT are classes, slots, facets, and instances. PROMPT works in a stepwise fashion, as follows:

1. In step 1, the two resource ontologies O_1 and O_2 are loaded by the user.
2. In step 2, a syntactic comparison takes place looking for identical or similar class names. Once matched, a list of suggestions is made to the user.
    * In the process of merging, if the concept names are identical, the suggestion is either a merging of classes or a removal of one. If concept names are not identical but similar, then a weaker merging of lower confidence is suggested with a link between the classes.
    * In the process of alignment, the user has first to distinguish which ontology between the two is more general (e.g., O_1). Classes in the less general ontology (e.g., O_2) are to be linked with subsumption (superclass-subclass) relations to those in O_1. If the same class name is identified in both ontologies, then the suggestion is to merge them. For a class in O_2, which does not have a match in O_1, the suggestion is to find its parent in O_1.
3. In step 3, the user selects an operation to perform, taking into consideration (but not being constrained by) the suggestions of step 2.
4. In step 4, PROMPT performs automatically all updates on the operations selected by the user, checks for possible resulting conflicts and suggests actions and remedies.
5. Steps 3 to 4 are reiterated until O_1 and O_2 are completely merged or aligned.

The operations supported by PROMPT applicable to merging and alignment involve all frame types and include: merge classes, merge slots, shallow copy of a class (just the class itself without associated elements—superclasses or slots), deep copy of a class (including all superclasses until the root of the hierarchy, and associated slots), remove parent, add parent, rename a frame, etc. The resulting conflicts in step 4, after an operation is performed, may include: naming conflicts (same name used for different frames), dangling references (relations to nonexistent frames), redundancies (multiple paths from classes to superclasses), slot-values violating class inheritance, and others.

The user may specify a preferred ontology, which plays the role of a target ontology. If there is such a preference expressed, all conflicts are resolved automatically in favor of the preferred ontology.

Recent developments report various augmentations to PROMPT, such as Anchor-PROMPT (Noy and Musen, 2001) and PROMPTDiff (Noy and Musen, 2002) in order to facilitate certain operations. AnchorPROMPT helps determine similar classes (and eventually slots) between two resource ontologies more efficiently. While PROMPT and Chimaera (Section 15.4) compare classes and slots directly associated (called local context), the use of some heuristics may help extend the analysis to the entire ontologies, which are represented as graphs of classes (graph nodes) and slots (graph edges). In order to associate the two graphs, it is necessary to establish, manually or automatically, some pairs of similar classes acting as control points or anchors. The basic assumption is that hierarchies about similar domains resemble both in terms and structure. Therefore, the intermediate classes between anchor matches are likely to exhibit similarity. PROMPTDiff is a heuristic algorithm intended to compare not different ontologies, but versions of the same ontology. While PROMPT focuses mainly on identifying similarities among ontologies, PROMPTDiff focuses on the differences, which constitute only a small fraction.

According to the framework established here, PROMPT treats integration more at the explication level and less at the semantic level. It makes the assumption, reasonable to database integration, that ontological knowledge is fully described by the four constituents of its knowledge frame-based model. Subsequently, PROMPT looks in the resource ontologies for classes with identical names, reasonably assuming that they represent the same (or similar) concept(s).

The important issue, however, also valid in geographic ontologies, is that mere class names, slots (as attributes), and subsumption relations describe only a small portion of a concept's semantics. High-level semantics, that is, meaning differences, are left for the expert to resolve. This is especially true for ontologies whose semantics are contained in lexical descriptions, such as definitions. PROMPT does not attempt to differentiate the meaning or importance of slots in relation to the concept described. Slots are mostly used as plane attributes, with whatever semantic importance they may carry. It remains to the ontology designer to include such semantic information in the knowledge model, something that is not (or cannot easily be) checked by the algorithm or even the average user. Tools like AnchorPROMPT may prove useful in certain cases where there is a lot of syntactic or structural resemblance among the resource ontologies. Their effectiveness, however, decreases when ontologies have been constructed differently, and this is a common case in geospatial ontologies. A proper association with semantic information extraction tools may increase the semantic value of the functionality offered by the tool.

15.3 KRAFT

KRAFT (Visser et al., 1998, 1999) is a project that aims at the reconciliation and integration of heterogeneous information systems, using an agent architecture. All three integration processes are performed: (1) information extraction using text-based analysis, (2) comparison, and (3) merging based on a shared ontology. KRAFT defines ontologies as hierarchies of categories with attributes and axioms. Heterogeneities between ontologies are classified according to the framework described in Visser et al. (1998). KRAFT has the ability to integrate constraints apart from data. The Constraint Interchange Format (CIF) is used for the definition of constraints. Its architecture is based on a set of smart programs, agents, and wrappers, which deal with the problems of information extraction and translation.

Most of the processes implemented by KRAFT are semiautomatic, that is, they are carried out with user interaction. The general aspect of KRAFT's workflow includes information extraction from different information sources (text, databases, ontologies, etc.), use of agents and wrappers to translate the information into the KRAFT language, and ontology creation and mapping processes, resulting in a common ontology with links to the original information sources.

Wrappers are used to extract information from multiple sources, whereas agents semiautomatically translate this information into the KRAFT language. KRAFT examines the original ontologies in order to identify categories with similar terms. Once a match is found, the categories' attributes are examined to confirm or falsify the match. All these matching and nonmatching categories are used later in the mapping process. KRAFT does not define direct mappings between the original ontologies. Instead, it creates a shared ontology and the mapping takes place between the shared ontology and the original ones. For the construction of the shared ontology, a top-level ontology is used as a frame, and then the categories of the original ontologies are linked to it.

The shared ontology is created according to the following steps:

1. In the beginning, a text-based analysis of corpora related to the domain is performed in order to extract information for the domain and select the terms that fully describe it. These terms are examined relative to their valid explication at the top-level ontology. The terms with valid explication at the top-level ontology (supported seed terms) will be used in the shared ontology. The terms that are not correctly explained in the top-level ontology (unsupported seed terms) are not discarded, as they will still be used in the next step. At this step, a decision is taken on whether a term will be represented as a concept, an attribute, or a value.
2. The next step consists in the ontology definition. The top-level ontology is enriched with concepts corresponding to the supported seed terms. Once this process is completed, an effort is made to relate the unsupported seed terms to concepts of the shared ontology. Then attributes and constraints are created for all concepts that have been added to the shared ontology.
3. The final step towards the construction of the shared ontology is the creation of extra concepts, attributes, and constraints. These are special elements that are created in a way that they can be linked to the original ontologies, thus creating an association between the shared ontology and the information sources.

Once the shared ontology is created, the mapping process takes place. For every category of the original ontologies, a similar term in the shared ontology is identified. This is done using not only the concept names, but also—in case the concept names are not enough to define a link—the attribute names, or even the attribute values.

The mapping process is completed after the above-described workflow, resulting in a shared ontology and the links that connect its components to the original ontologies.

15.4 CHIMAERA

Chimaera (McGuinness, 2000; McGuinness et al., 2000a, 2000b) is a set of tools for ontology integration and merging that began as a supporting project for the Ontolingua server, an online ontology editor developed from the Knowledge Systems Laboratory at the University of Stanford. Chimaera's basic functionality includes ontology merging and problem resolution in a single ontology or in multiple ontologies, like class reorganization, correction of multiple appearances of same objects, and ontology transformation from one format to another.

Chimaera is a completely web-based application and provides interactivity with the user through HTML interface. Like its "partner," Ontolingua, Chimaera has all the advantages that derive from the OKBC (Open Knowledge Base Connectivity) architecture such as: functioning through the Internet, concurrent use of an ontology library by many users around the world, administration of multiple ontologies, and interaction among multiple users on a common knowledge base. Chimaera no longer needs Ontolingua as a requirement to function; it can easily work with other ontology editors as well.

During the merging process of ontologies, the first action Chimaera takes is resolving name conflicts between classes of the ontologies involved. This is a semi-automated process, as the system compares the classes that are being used in terms of nomenclature and presents them to the user. The user then has to confirm or falsify the possible relationships between classes and take the appropriate action: merge two classes, redefine the place of a class in the hierarchy, or assign a class as superclass or subclass of a related class.

Once the first step described above is completed, Chimaera uses a browsing interface to direct the user to the points where the new ontology is likely to display overlapping problems. When classes from different ontologies merge as subclasses of another class in the new ontology, it is a common problem to have semantic overlap between them. On the other hand, they could actually prove inadequate to fully describe their superclass, in which case more information is needed (i.e., another subclass). The system points out the areas where these kinds of problems might emerge, and the user proceeds by defining which classes actually overlap and which do not.

After the correction of the issues that concern the classes of the merged ontologies, Chimaera processes the slots that contain the attributes of the classes and determines the cases in which slots of merged classes are related and thus have to be unified as well. As in the previous steps, the user has to make the corrections that he or she feels fit to each occasion.

15.5 MOMIS

MOMIS (Mediator envirOnment for Multiple Information Sources) (Bergamaschi et al., 2001; Beneventano et al., 2003; Beneventano and Bergamaschi, 2004) was originally developed by the University of Modena and Reggio Emilia along with the University of Milano and Brescia. MOMIS is a system designed to exploit the advantages of formalized information with the purpose of mining it from its source.

It has also a semiautomatic mechanism that allows it to integrate information from different sources.

MOMIS performs the information extraction using ODL-I3, an object-oriented language, which checks for any common objects between the different sources and translates the original information into its own language schemata using description logic processes. The integration of the heterogeneous information systems is continued by merging their schemata to a single common ontology, enabling a total review of the information and the schemata involved. In this process, clustering techniques are also applied. These are provided by ARTEMIS (Analysis and Reconciliation Tool Environment for Multiple Information Sources), a tool designed specifically to analyze the semantic part of the information source schemata and identify objects with similar characteristics. The analysis is followed by a structuring phase, during which the schemata that contain similar objects are grouped together, and small structures like hierarchies are created. The functioning of ARTEMIS is based on affinity coefficients that are calculated for objects coming from different information sources in order to determine possible relations between them.

During the activity of the above last steps, the unified ontology undergoes additional corrections and refinements by the SI-Designer, the graphic environment that guides the integration process. The SI-Designer tool confirms the relations found by the ARTEMIS environment while creating new ones using ODB Tools. WordNet is used to draw information about the objects of the newly created ontology and propose additional semantic relations based on the lexical relevance of the objects.

The integration process also involves a percentage of interactivity between the system and the user. The core of the MOMIS project is realized as a mediator that has partial control over the steps taken and the results produced. The user is called on to interact both in the first schema integration part and in the final refinement activities through WordNet. In the first case they use the ODB Tools to add manually relations between objects that have not been found automatically, while in the second part they exploit the advantages of WordNet to provide even more detailed information about relations between objects that appear to have similar lexical attributes.

The activities that MOMIS implements during its workflow, in order to achieve integration, can be summarized in the following:

- Identification of similarities between objects from the information sources
- Information extraction from one or different sources
- Transformation of the information in its own language form
- Creation of a total overview of the newly built ontology that maintains the original schemata
- Organization of hierarchies and other structural formations
- Interaction with the WordNet ontology in order to identify more similarities between objects based on their semantic relations
- Interaction with the user to confirm the above-mentioned propositions and provide additional relations between classes or attributes

15.6 GLUE

GLUE (Doan, Domingos, and Halevy, 2001; Doan et al., 2002) is a tool that deals with the problem of ontology mapping. Machine learning techniques are applied in order to semiautomatically identify semantic mappings between ontologies, which are represented as concept taxonomies. The aim of the approach is the following: for each concept node in one taxonomy, the most similar concept node in the other taxonomy is identified. An interesting feature of GLUE is the multicriteria use of similarity measures between concepts from different ontologies. Furthermore, GLUE uses a technique originating from the computer vision and image processing fields called *relaxation labeling*, in order to incorporate domain constraints and general heuristics in the mapping process.

Regarding its architecture, GLUE consists of three main parts:

1. The Distribution Estimator
2. The Similarity Estimator
3. The Relaxation Labeler

The *Distribution Estimator* takes as input two taxonomies, O_1 and O_2, and their instances and calculates joint probability distributions for pairs of concepts using computer-learning techniques. For any pair of concepts A (A $\in O_1$) and B (B $\in O_2$), their joint probability distribution comprises four probabilities: $P(A,B)$, $P(A,\overline{B})$, $P(\overline{A},B)$, $P(\overline{A},\overline{B})$, where the term $P(A,\overline{B})$ denotes the probability that an instance belongs to concept A, but not to concept B. Because not every instance in the world is known, the assumption made by GLUE is that the set of instances of each original ontology is representative of the whole number of instances in the world that are described by the corresponding ontology. GLUE calculates $4|O_1||O_2|$ numbers, where $|O_1|$ is the number of concepts in the ontology O_1.

GLUE follows a multistrategy learning approach, in order to maximize its efficiency in estimating the joint probability distribution. It incorporates two base learners, the *Content Learner* and the *Name Learner* and a metalearner to combine the results of the base learners. The *Content Learner* builds a prediction hypothesis based on the frequencies of words in what is called the *textual content* of an instance (i.e., the name, the set of attributes and attribute values of the instance) using the Naive Bayes learning technique. The Name Learner follows a similar approach as the Content Learner, but instead of the textual content of an instance, it deals with its *full name*, that is, the sequence of concept names from the root of the taxonomy to the instance. The metalearner combines the similarities identified by the Content and Name Learners by manually applying weights to each one and computes their weighted sum.

The *Similarity Estimator* takes as input the numbers computed by the Distribution Estimator regarding the probability distributions and calculates a similarity value for each concept pair (A $\in O_1$, B $\in O_2$) based on a similarity function specified by the user. The output is a similarity matrix that contains the similarity values between the concepts of the two ontologies.

The *Relaxation Labeler* takes as input the similarity matrix derived from the Similarity Estimator and applies a series of techniques, like heuristic knowledge and domain-based constraints, to complete the mapping process. Relaxation labeling is a technique for the attachment of labels to nodes of a graph. This technique relies on the position that the features of a node's neighborhood affect the node's label. These features are constraints, which narrow the possibilities of matching candidates for a concept and thus increase the quality of the mapping process. They are classified into (a) domain-independent and (b) domain-dependent constraints. Domain-independent constraints consist of general rules of interrelations between concepts in a taxonomy. For example, the condition that two concepts with the same subconcepts and superconcepts are probably equivalent is a typical domain-independent constraint. Domain-dependent constraints deal with interrelations between concepts in the taxonomies that depend on the specific application or domain. For example, if we take an administrative example, it is obvious that the instances of the ontology concepts cannot override the real-world hierarchy that exists. For example, a prefecture cannot be a subclass of a municipality.

To sum up, the concept comparison approach performed by GLUE is based on instances, concept terms, attributes and their values, as well as taxonomical relations. Taxonomical relations represented as domain-independent and domain-dependent constraints play an important role in the approach. Thus, concept similarity is greatly influenced by the existence of common subconcepts and superconcepts in the hierarchy. Therefore, the approach is more suitable for the integration of ontologies with a common categorization perspective–view. Ontologies with different categorization perspectives are likely to differ in the hierarchical structure, although they may include common concepts. Similarity is based on the existence of common concept terms, attributes, and attributes values. An extensional direction is also incorporated, since the approach can be applied in case instances exist. Concerning the integration approach, GLUE performs ontology mapping, attempting to match similar concepts, that is, finding the proper pair for every concept in the taxonomy.

15.7 ECOIN

ECOIN (Extended Context Interchange) (Firat, 2003; Firat, Madnick, and Manola, 2005) is an extension of the COIN framework, which addresses the issues of contextual heterogeneity. COIN (Bressan et al., 1997; Goh, 1997) models the information used in the mapping process in a domain model, which describes the part of the physical world being studied in a more generic way. The mapping process is performed based on the source databases or ontologies and the relations between the sources and the domain model. This results in the creation of categories that form a new ontology.

ECOIN, as an extension of COIN, uses the same principles but introduces a set of new features for the ontology heterogeneity representation and methods for reasoning. It is actually a framework for information integration and ontology merging. The ECOIN framework is a generic logic-based data model, which includes four components: (a) the ontology, (b) the source declarations, (c) the context (instances), and (d) the mapping functions between contexts.

Although it is primarily developed for application merging, it involves the process of ontology merging as well. The ontology consists of the domain and context model. The domain model is the common knowledge of the domain, that is, the general concepts that describe the domain. The specializations of these concepts as defined by different information sources represent the context model of the ontology. ECOIN deals with relational sources, although nonrelational sources (e.g., XML web sites) may be transformed to relational ones via proper wrappers. Contexts define specializations for data values originating from representational or semantic differences. Mappings are conversion functions defined in order to appropriately link different contexts.

Besides ontology mapping and merging, ECOIN also includes information extraction processes. These are implemented through the use of a special tool developed for this purpose, a wrapper engine called Caméléon. This engine performs data extraction from web pages based on specific extraction rules defined by regular expressions. Caméléon uses specifications created with XML terms to identify similar information sources and then extracts information requested by the user.

Initially ECOIN treats heterogeneity at three different levels: contextual, ontological, and temporal. The main idea behind the approach of ECOIN is that it translates the heterogeneity among ontologies into heterogeneity among contexts. Therefore, ECOIN requires defined context frames for the ontologies to be merged.

ECOIN offers two possibilities for the integration process: merging and mapping. More specifically, it enables the creation of a new ontology based on the original ontologies but also performs mapping actions between the related terms of the original ontologies. As has already been mentioned, ECOIN uses the context of the two ontologies to define the mappings between their categories and to create conversion functions for some of the ontologies' elements.

ECOIN creates a new context for the new ontology and then builds the new categories, based on the related categories of the original ontologies. It searches for semantically similar elements in the original ontologies and then maps these elements to the newly created object in the new ontology. To define semantic similarities between ontologies' elements, context values and semantic types of categories are examined. ECOIN utilizes a custom constraint definition language that is called CHR (Constraint Handling Rules) (Firat, 2003) in order to address the heterogeneities between the original ontologies. Furthermore, ECOIN automatically creates multiple data views by combining elements from the original ontologies and applications. This process can be further refined by user intervention, which, however, is not mandatory.

15.8 FCA-MERGE

FCA-Merge (Stumme and Maedche, 2001) is a method for merging ontologies based on natural language processing techniques and Formal Concept Analysis (see Chapter 8). The method is performed bottom-up and is claimed to "offer a structural description of the merging process" (Stumme and Maedche, 2001, 225). The FCA-Merge method performs the first and third integration processes. Semantic information extraction is performed based on linguistic analysis techniques. However, the aim of the extraction process is to extract instances from a set of documents, overlooking any semantic element of concepts. Integration is performed with a combination of Formal Concept Analysis algorithms and a number of queries, which support the ontology engineer in generating the merged ontology.

The merging process is performed in three steps. The first step extracts instances from domain-specific documents and generates the formal context corresponding to each of the original ontologies (see Chapter 8). This step takes as input the original ontologies and a set of natural language documents, which include the concepts of the original ontologies. The document set is considered as the object set and the set of concepts as the attribute set. A binary relation I is held when document g contains concept m. In order to produce satisfactory results, the input documents should meet the following specifications:

- The documents should be relevant to the source ontologies.
- The documents should cover all concepts of the source ontologies.
- The documents should properly differentiate the concepts of the source ontologies, because, if two concepts always appear in the same documents, they are represented as one concept in the final ontology.

Linguistic techniques are applied to extract domain-specific information. More specifically, the input documents are analyzed in order to associate words or expressions with concepts from the ontologies. The selection of the appropriate words or expressions is based on a domain lexicon.

The second step derives the merged context of the formal contexts generated in the previous step. At this step, the concepts are indexed in order to allow a concept with the same name but different semantics to exist in more than one ontology, but to be treated differently. After generating the merged context using Formal Concept Analysis, the method applies an algorithm called TITANIC to derive the *pruned concept lattice*. The pruned concept lattice is so called because it contains only those formal concepts that resulted from original concepts, as well as their superconcepts; subconcepts are removed from (pruned down) the concept lattice.

The third step derives the merged ontology from the pruned concept lattice. Unlike the two previous steps, which are fully automatic, this step is performed semiautomatically, because it requires domain knowledge provided by an expert. More specifically, a number of queries support the ontology engineer in generating the concepts and relations of the final ontology based on the pruned concept lattice and the relations between concepts of the original ontologies. First, formal concepts, which are generated by exactly one ontology concept from one of the source ontologies, are included in the final ontology, without any contribution by the ontology engineer. Second, formal concepts, which are generated by two or more ontology concepts, are dealt with. In that case, the concepts are integrated into one concept in the final ontology, and the ontology engineer decides its name. At this stage, the final ontology contains all concepts from the source ontologies. Also, the method copies all relations from the source ontologies into the final ontology, and the ontology engineer is asked to resolve possible conflicts. Third, the ontology engineer deals with formal concepts, which are generated by at least two concepts from the source ontologies. These are added either as ontology concepts or relations in the final ontology. Finally, the is-a relation between concepts is automatically derived from the pruned concept lattice.

15.9 IF-MAP

Kalfoglou and Schorlemmer (2003) developed a method for ontology mapping, called IF-Map, based on the mathematical theory of Information Flow, Channel Theory (Barwise and Seligman, 1997) (see Chapter 10), and on the Information Flow Framework (Kent, 2004, 2005) (see Section 4.5). The work also draws from Schorlemmer's approach to aligning ontologies, from Kalfoglou's heuristics for examining probable mappings between ontologies, and from the FCA-Merge method (Stumme and Maedche, 2001).

The assumption made by the approach is that original ontologies commit to a reference ontology, which provides a common basis for information sharing. The method aims at providing mappings between the original ontologies via the reference ontology. These ontology mappings, together with the link between original ontologies via the reference ontology, constitute the global ontology, which is the final output of the method. Based on channel theory, the method provides a formalization of (a) ontologies as *local logics* and (b) ontology mappings as *logic infomorphisms*.

The architecture of the approach consists in the following steps:

1. Ontology harvesting, that is, ontology acquisition from different sources (e.g., ontology libraries, ontology editors, etc.)
2. Translation of the original ontologies to Prolog, which is the representation language of the IF-Map approach
3. Infomorphism generation, that is, provision of mappings between original ontologies
4. Display of results in RDF format

As was mentioned in Chapter 10, information flow between the components of a distributed system occurs when there are regularities between the connections of these components. Moreover, the connections between the tokens allow information flow both at the token and the type levels. According to this assumption, in case two communities willing to share information generate links between the instances that populate their corresponding ontologies, then it is possible to draw inferences about the relations among the concepts of the ontologies.

In order to generate ontology mappings, the IF-Map approach also formalizes the notions of ontology, ontology morphism, and ontology mapping using the formal notions of local logic and logic infomorphism described in Chapter 10. The approach assumes that local logics represent ontologies while logic infomorphisms are used to relate local logics with each other. In this way, IF-Map achieves the definition of mappings between different ontologies and more especially between an unpopulated ontology, corresponding to the reference ontology, and a populated ontology, corresponding to a local ontology. The application of the method is illustrated in the case of mapping ontologies of computer science departments from different universities.

15.10   ARCHITECTURE-BASED APPROACHES
TO GEOINTEGRATION

In the geospatial domain, there are a number of geointegration approaches, which mainly focus on architectural issues. Three approaches belonging to this family are presented below.

ISIS (Interoperable Spatial Information System) (Leclercq, Benslimane, and Yétongnon, 1999; Jouanot, Cullot, and Yétongnon, 2003) is a research project that presents a spatial object-oriented data model and a mediation architecture to achieve GIS interoperability. The project focuses on the dynamic resolution of semantic heterogeneities in order to provide autonomy, flexibility, and extensibility.

The project is based on a multi-agent architecture. Six different types of agents are used: (1) Wrapper Agent, (2) Cooperation Agent, (3) Ontology Agent, (4) Semantic Router Agent, (5) Global Query Processor Agent, and (6) User Interface Agent.

The Wrapper Agent expresses each local GIS and aims at processing queries derived from the respective Cooperation Agent. For each local GIS, the Cooperation Agent defines the semantics of objects included in the wrapper schema. The Ontology Agent aims at facilitating communication between Cooperation Agents by defining the common knowledge of the domain. Ontological agreements specify the mappings between the concepts of the ontology and the objects of the cooperation schema. These agreements are stored by the Semantic Router Agent and are further used by the Global Query Processor Agent to identify relevant sources and plan the execution of a query posed by a Cooperation Agent. The User Interface Agent provides the interface to the user for submitting queries to a Cooperation Agent and receiving the result.

The AMUN data model aims at providing a set of concepts for: (1) the representation of thematic and spatial information, (2) the definition of semantic contexts, (3) the resolution of semantic heterogeneities, and (4) the conversion and transfer of objects among different systems.

In order to represent semantic information, ISIS introduces the notion of context, which is further classified into two types: (1) reference context and (2) cooperation contexts. The reference context specifies the common knowledge of a domain as mediation classes, which are described by static properties, behavior-methods, and semantics. The semantics of mediation classes are defined as (a) a domain value (i.e., the admissible set of values for an attribute), (b) a semantic value (i.e., the description of the meaning of an attribute), and (c) a logic expression of an assertion or a constraint.

On the other hand, the cooperation contexts define the reference context in regard to the local schemata. They consist of cooperation classes, which provide different versions (mediation roles) of mediation classes on different local schemata. Each mediation role is linked to a context transformation, that is, a mapping function from a local value to the corresponding cooperation value.

Cruz et al. (2002) propose an approach to handling heterogeneities in distributed geographic databases. The approach supports the development of a central ontology and the specification of agreements (declarative transformations) between local schemata and the central ontology. It is tested under the Wisconsin Land Information

System (WLIS), which is a distributed web-based system with heterogeneous data lying on local and state servers. Under the WLIS, semantic heterogeneities mainly arise due to the existence of different land use classifications in the local databases.

The system architecture is based on that proposed by Leclercq, Benslimane, and Yétongnon (1999) and comprises five components:

1. GIS: this component includes the local database (represented as an XML document) and the respective agreement file.
2. Ontology: this component consists of the common knowledge of the domain and acts as the logical database schema; end user queries are formulated according to the categories of the ontology. The ontology is created by a domain expert.
3. Semantic registry: this component is the registry of the local databases in the system and is used by the query processor component to perform a sub-query on a local database.
4. Query processor: this component recomposes the end user query into subqueries according to the ontology and the agreement files of the local databases.
5. User interface: this component consists of two subcomponents. One helps end users to query local databases. The other helps an expert responsible for a local database to define the agreement between that local database and the ontology.

In contrast to ISIS, which uses procedural methods in the cooperation agent, the current approach uses a declarative data access mechanism to ease the addition of new local databases to the system. The agreement file is an XML document which specifies the mappings of the categories of the local databases to those of the shared ontology. It is created by a local expert familiar with the categories of the local database as well as their relation to those of the shared ontology, whenever a new local database is incorporated in the system. The domain expert specifies one-to-one mappings between attribute names in the ontology and those in the local databases. For attribute values, however, there are five different types of mappings (between attribute values in the ontology and those in the local databases) supported by the system: (1) one-to-one, (2) one-to-many, (3) many-to-one, (4) one-to-null, and (5) children.

The approach proposes an architecture to incorporate multiple views and support queries over related heterogeneous data. Semantic heterogeneities are not really resolved, but are reconciled by a domain expert, who creates the agreement file for each local database, that is, the expert defines mappings between the local databases and the shared ontology based on the domain knowledge. An advantage of the approach is that the final ontology preserves the original classifications, although these may be defined according to different semantic contexts. Therefore, local databases can be mapped to the ontology without loss of semantic resolution. For example, the concept "manufacture" may be a subconcept of "industrial areas" in one classification and a subconcept of "secondary sector" in another. In contrast to some mediation approaches, the current approach preserves both views (contexts, classifications) in the final ontology.

Ram et al. (2002) present an information integration approach, called GeoCosm, which is based on (a) a spatiotemporal semantic model to define the content of the geospatial sources and (b) on a conflict resolution ontology to resolve semantic heterogeneity. The approach uses metadata in order to process spatiotemporal queries and to evaluate the data sources. These metadata are classified into three categories: semantic, context, and geospatial metadata. Semantic metadata define the representation and organization of geospatial data; context metadata specify logical relations between data in distributed systems; and geospatial metadata provide information on data acquisition, processing, and accuracy, in order to facilitate the appropriate use and evaluation of geospatial data.

The approach is based on the integration of local schemata of distributed data sources into a global schema based on a shared ontology. Although claimed to be semantic-based, the approach mainly resolves schematic heterogeneities between local schemata. The approach extends the integration methods developed by Hearst et al. (1998) and Knoblock et al. (2001) in the geospatial domain. The architecture has two basic components: (a) metadata management and (b) query plan generation.

For metadata management, the approach adopts the Spatio-Temporal Unifying Semantic Model (ST USM) (Khatri, Ram. and Snodgrass, 2001; Khatri et al., 2002). ST USM is used as a canonical model in order to define spatiotemporal concepts, that is, position, geometry, spatial resolution, event, state, temporal granularity, etc. Ontologies are used to store context metadata (i.e., concepts and functions between concepts) of the shared domain. The correlation between local schemata and the shared ontology is called "context mapping" and is used to resolve schematic heterogeneities between local schemata, for example, different measurement units for the same concepts. In order to assist users in the evaluation of the relevance and the usefulness of the available data sources, geospatial metadata are also incorporated in the developed architecture. The approach relies on the Content Standard for Digital Geospatial Metadata (FGDC, 1998) to include elements such as identification information, data quality, spatial reference, citation, time period, and contact information.

The query plan generation is completed in three steps: (a) preplanning for dynamic factors, (b) rule-based plan rewriting, and (c) interleaving of plan reformulation and plan execution.

The developed architecture has three basic components: metadata manager, mapping manager, and query manager. The metadata manager is a mechanism for the representation of semantic, context, and geospatial metadata. The mapping manager includes the semantic mapper for the association of local schemata to the global schema, and the context mapper for the association of local schemata to the shared ontology. The mapping manager focuses on the agreement of all distributed data sources with a common model in order to perform query processing. The query manager is based on a mediator and performs three processes: (a) selection of appropriate data sources and creation of an initial query plan, (b) application of rewriting rules to the initial query plan and decomposition into subplans, and (c) execution of subplans and iteration of the previous steps in order to reach desired results.

Based on this approach, a web-based prototype has been developed called Geo-Cosm. GeoCosm allows (a) experts to register a source by defining the appropriate metadata and (b) users to query distributed data sources, compare them, and choose the appropriate one through a single interface.

15.11 A CONCEPT COMPARISON APPROACH

Although concept comparison does not consist of a complete integration approach, it comprises an important part of the integration process and therefore a similarity-based approach focused on geographic concepts is described in this section.

Rodríguez and Egenhofer (2003) developed a model to compute semantic similarity between spatial concepts (or entity classes, as referred to in the approach) from different ontologies. The assumption made by this approach is that the input ontologies include concepts, their between semantic relations, as well as distinguishing features for describing concepts. This model can be used as a systematic tool to detect similar concepts among different ontologies, thus facilitating the comparison process.

The basic components used for concept representation are: (a) a set of synonym words, (b) a set of semantic relations (is-a and part-whole relations) between concepts, and (c) a set of distinguishing features to differentiate one concept from another. Distinguishing features are further classified as functions, parts, and attributes. These components are used for the specification of a similarity measure, which consists of the weighted sum of the similarity of each component (Equation 15.1). The functions S_w, S_u, and S_n express the similarity between synonym words (S_w), features (S_u) and semantic neighborhoods (S_n) between concept a of ontology p and concept b of ontology q; w_w, w_u, and w_n are the corresponding weights.

$$S(a^p, b^q) = w_w S_w\left(a^p, b^q\right) + w_u S_u\left(a^p, b^q\right) + w_n S_n\left(a^p, b^q\right)$$

$$\text{for } w_w, w_u, \text{ and } w_n \geq 0 \tag{15.1}$$

In order to be able to apply the similarity measure, concept representations must have some common components on which the comparison will be based.

The similarity measure is based on the normalization of Tversky's model (Tversky, 1977) (see also Chapter 13) and the set-theory functions of intersection ($A \cap B$) and difference (A/B) (Equation 15.2). A and B represent components (synonym sets, semantic neighborhoods, and distinguishing features) of concepts a and b, respectively; $\|$ is the cardinality of a set; a is a function that defines the relative importance of the noncommon elements.

$$S(a,b) = \frac{|A \cap B|}{|A \cap B| + a(a,b)|A/B| + (1 - a(a,b))|B/A|} \quad \text{for } 0 \leq a \leq 1 \tag{15.2}$$

According to the model, similarity is the result of not only the common but also the different characteristics between concepts. Furthermore, similarity is not always a symmetric relation, as the similarity of a class with its superclass (e.g., "bus" and "vehicle") may be greater than the similarity between a superclass and the class (e.g., "vehicle" and "bus"). Taking this into account, the approach measures the distance from these concepts to the immediate superconcept that subsumes them. In case the concepts belong to different ontologies, the approach assumes that the two ontologies are connected to a hypothetical concept "anything." Therefore, the function a is calculated in relation to the function depth () (Equation 15.3), which represents the shortest path from the concept to the hypothetical concept "anything."

$$a\left(a^p,b^q\right)=\begin{cases}\dfrac{depth(a^p)}{depth(a^p)+depth(b^q)}\, depth(a^p)\le depth(b^q)\\[2ex]1-\dfrac{depth(a^p)}{depth(a^p)+depth(b^q)}\, depth(a^p)> depth(b^q)\end{cases} \tag{15.3}$$

This matching model is used to determine similarity measures for: (a) synonym words (word matching), (b) distinguishing features (feature matching), and (c) semantic relations between concepts (semantic neighborhood matching). Word matching compares terms of concepts. Therefore, A and B in Equation 15.2 represent terms for concepts a and b, respectively.

Feature matching compares the sets of distinguishing features A and B of concepts a and b, respectively (Equation 15.2). Equation 15.4 is used to compute feature similarity in case both ontologies include a further categorization of features into parts (S_p), functions (S_f), and attributes (S_a); w_p, w_f, and w_a are weights for S_p, S_f, and S_a, respectively:

$$S_u(a^p,b^q)= w_p\cdot S_p\left(a^p,b^q\right)+w_f\cdot S_f\left(a^p,b^q\right)+w_a\cdot S_a\left(a^p,b^q\right) \tag{15.4}$$

$$\text{for } w_p,\ w_f,\ \text{and } w_a \ge 0 \text{ and } w_p+w_f+w_a=1.0$$

Semantic neighborhood matching identifies semantic neighborhoods based on sets of synonyms or distinguishing features. Equation 15.5 calculates semantic neighborhood similarity (S_n) with radius r between concepts a^p and b^q of ontologies p and q correspondingly; the result depends upon the cardinality $(||)$ of the semantic neighborhoods (N) and the approximate cardinality of the set intersection $(\cap_n)$ between these semantic neighborhoods. The intersection among semantic neighborhoods is given by Equation 15.6, where $S()$ is the semantic similarity of concepts, a^p_i and b^q_j are concepts in the semantic neighborhood of a^p and b^q, respectively, and n, m are the numbers of concepts in the corresponding semantic neighborhoods.

$$S_n(a^p,b^q,r)=$$

$$\dfrac{a^p\cap_n b^q}{a^p\cap_n b^q +a\left(a^p,b^q\right)\cdot\delta\left(a^p,a^p\cap_n b^q,r\right)+\left(1-a\left(a^p,b^q\right)\right)\cdot\delta\left(b^q,a^p\cap_n b^q,r\right)}$$

$$\text{with } \delta\left(a^p,a^p\cap_n b^q,r\right)=\begin{cases}\left|\left|N\left(a^p,r\right)\right|-\left|a^p\cap_n b^q\right|\right|\, if\left|N\left(a^p,r\right)\right|>\left|a^p\cap_n b^q\right|\\[1ex]0 \quad \text{otherwise}\end{cases} \tag{15.5}$$

$$\left|a^p \cap_n b^q\right| = \left[\sum_{i \le n} \max_{j \le m} S\left(a_i^p, b_j^q\right)\right] - \varphi S\left(a^p, b^q\right), \text{ where}$$

$$\varphi = \begin{cases} 1 \text{ if } S\left(a^p, b^q\right) = \max_{j \le m} S\left(a_i^p, b_j^q\right) \\ 0 \quad \text{otherwise} \end{cases} \tag{15.6}$$

$$S\left(a_i^p, b_j^q\right) = w_l S_l\left(a_i^p, b_j^q\right) + w_u S_u\left(a_i^p, b_j^q\right) \text{ with } 0 < w_l + w_u \le 1$$

According to the radius of the semantic neighborhood, the approach specifies and compares superclasses, subclasses, parts, and wholes among concepts. An application of the similarity measure is given in Section 16.2.

To sum up, the method is effective in case the input ontologies are represented in the above-mentioned way, that is, concepts are defined by a set of synonymous terms and distinguishing features (parts, functions, and attributes) and are related to each other with is-a and part-of relations. When one or more of these elements are missing from the representation of one (or both) of the compared concepts, the comparison is confined to the available elements, thus resulting in a simplification of the similarity measure. An extreme example would be an ontology including only attributes for the representation of concepts, compared to another including only parts. In that case, feature similarity is zero. Furthermore, the similarity model does not incorporate values for attributes, but only examines whether attributes have the same name. This may result in misleading conclusions, because an attribute may take a wide range of values, or even contradictory ones (e.g., natural versus artificial, flowing versus stagnant, high versus low, wet versus dry, etc.).

The inclusion of semantic neighborhood matching in the similarity model implies that the hierarchical structure and the neighbor concepts are essential characteristics of a concept. However, the way concepts are hierarchically related may be due to a different perspective or context (or ontology type—scope, e.g., WordNet versus CORINE Land Cover), which does not necessarily influence the semantics of the concepts internally ("essentially"). For example, the concept "canal" may be a sub-concept of "watercourse" in a land cover ontology, and a subclass of "transportation network" in a transportation ontology. Despite their different hierarchical relations, the concepts "canal" may still be semantically equivalent.

Generally, the result of similarity measures for concept comparison in integration endeavors needs further evaluation, that is, which is the minimum value (threshold) from which two concepts are considered similar? A solution to strengthening the decision of how similar two concepts are would be to incorporate an extensional basis in the similarity assessment. In case similarity is not very high, the inclusion of possibly available instances may facilitate the comparison. If the instances–objects referring to the two concepts coincide, then there is a higher possibility that the concepts are also similar. Last, but not least, any final number as an expression of similarity is not by itself very informative, unless it is accompanied by a formal description of the comparison context.

15.12   INTENSIONAL–BASED GEOINTEGRATION APPROACHES

Bishr (1997, 1998) developed an approach primarily focused on the data modeling direction of information sharing. It aims at providing a model to capture the semantics of geospatial objects and develop a semantic translator to achieve their identification and sharing in a heterogeneous distributed geographic information system.

The model developed in this approach is called Semantic Formal Data Structure (SFDS) and is based on the notion of *context*. More specifically, the SFDS formalizes the Proxy Context, which is a mediator between two or more contexts sharing knowledge on the domain. The formalism is designed in three layers, in order to deal with the respective types of heterogeneity: (a) the syntactic, (b) the schematic, and (c) the semantic layer.

The syntactic layer defines the geometric and thematic characteristics of geospatial objects. The schematic layer defines a reference model and a set of functions for the representation of federated schemas. The elements of the reference model are: proxy context, proxy hierarchy, proxy class, proxy attribute. The functions are designed to help the designer deal with the elements of the reference model. The semantic model defines the context information necessary to enable information sharing. Context information is defined for all contexts as well as the proxy context and consists of three types: (a) context information for categories definition, (b) context information for class intension definition, and (c) context information for geometric description.

A semantic translator implements the information-sharing process and is based on the combination of a federated schema, context information, and a common ontology of the domain. The resolution of heterogeneities is performed in two phases: the first phase computes the semantic similarity between the elements of a context and those of a proxy context, whereas the second phase resolves schematic heterogeneity between similar elements. A domain expert performs the mapping between each context and the proxy context. A shared ontology of the domain is used as a framework for the association of the federated and export schemata. Concerning the resolution of schematic heterogeneity, the approach proposes a classification of possible conflicts and an action to resolve each one. Schematic heterogeneity is classified as differences in classes, attributes, and hierarchies. The process is completed with the mapping among the export and the federated schemata.

The approach uses the notion of context in order to formally define the semantics of geospatial objects and map heterogeneous information sources. Like most approaches in the geographic domain, it is performed manually by a domain expert or database designer. Information loss is not avoided, because the common ontology determines the shared information. The case in which a local database includes a category or attribute absent from the common ontology leads to intensional information loss.

In the context of the MIGI (Metadata Integration and Geodata Integrity) project, a semiautomatic approach to schema integration based on formal ontologies is presented (Hakimpour and Geppert, 2001a, 2001b, 2002; Hakimpour and Timpf, 2002). Explicit and formal definitions of semantics of categories provided by formal ontologies are used to deal with semantic heterogeneity between different schemas during schema integration. Description Logic is used for the formalization and representation of ontologies. This formalism constitutes the basis of a reasoning system for the identification and possible resolution of semantic conflicts between two different ontologies.

The approach proceeds in three phases. The first phase uses a reasoning system to merge the underlying ontologies. The result of the ontology merging process is used in the second phase for the generation of the global schema from the local schemas. The third phase deals with the specification of mappings between the global and local databases.

The assumption made by this approach is that semantics refers to the meaning of schema elements and is context-dependent, in contrast to syntax, which is context-independent. Furthermore, class and attribute names in local schemas are based on terms defined in the underlying ontologies. Therefore, in order to identify similarities and heterogeneities between local schemes, the approach compares intensional definitions of terms in formal ontologies. Intensional definitions are definitions of concepts using logical axioms. For example, the intensional definition of the relation "Faculty" is as follows (Hakimpour and Geppert, 2001a):

$$\text{"t[Faculty(x)]} = \text{Person(x)} \wedge (\exists y: \text{Course(y)} \wedge \text{teaches(x, y))"}$$

The terms themselves are not taken into consideration for ontology merging. Only intensional definitions are used to infer similarity relations between schema elements, that is, definitions of terms in the ontologies using logical axioms.

More specifically, the first phase merges ontologies available for local schemas. The term "merge" is used to imply that the underlying ontologies are not integrated, that is, the method does not attempt to identify or resolve heterogeneities between them. The merging process is performed by a reasoning system (such as Power Loom). This process uses a common reference ontology in order to compare intensional definitions as specified by the underlying formal ontologies. The notions "concept" and "relation" refer to intensional relations with arity 1 and intensional relations with arity greater than 1, respectively. Four similarity relations are used to associate the corresponding terms: equality, specialization, overlapping, and disjoint. These similarity relations emanate from the four-intersection model (Egenhofer and Herring, 1991):

- Disjoint definitions: in case the conjunction of two intensional definitions is false, then the corresponding concepts or relations are considered disjoint.
- Overlapping definitions: in case the conjunction of two intensional definitions cannot be proven to be false, then the corresponding concepts or relations are considered overlapping. The result of the merging process is a new concept or relation, which represents the conjunction of the original overlapping ones.
- Specialized definitions: in case the intensional definition of a concept or relation is a specialization of the intensional definition of another concept or relation, then there is a specialization relation between the two concepts

or relations. The merging process defines a subconcept or subrelation similarity between the two original concepts or relations.
• Equal definitions: in case two intensional definitions are the same, then the corresponding concepts or relations are considered equal. The result of the merging process is a single intensional definition related to both original terms.

The second phase of the approach generates the global schema using the previously defined similarity relations. A schema integrator generates the global schema, as well as mappings between the global schema and the local schemes. The process is performed in two steps. At the first step, classes and their between hierarchical relations are generated, whereas at the second step, classes are ascribed attributes. For the generation of the global classes, the process is carried top-down, that is, superclasses in the local schemas are the first to be added to the global schema, followed by their subclasses and their between hierarchical relations. Each category of similarity relations is dealt with differently. Equal concepts or relations are inserted in the global schema once. In case of a specialization relation between two concepts or relations, a hierarchical relation is defined between them. For overlapping concepts or relations, a new class is generated representing their conjunction together with the hierarchical relations to both overlapping concepts or relations. In order to generate meaningful results, this step requires supervision by an expert. For the global attribute generation, for each attribute of a class in the local schema, an attribute of the corresponding class in the global schema is defined.

The third phase performs the mapping of class instances from the local databases to the corresponding ones from the global schema using a reasoning system.

This approach can be applied in case intensional definitions thoroughly describe the semantics of concepts and relations. However, in the geospatial domain, there are many cases where there are no intensional definitions for concepts, or the semantics of geographic concepts are rather complicated, thus impeding the straightforward applicability of the approach.

ONTOGEO is a project focused on the semantic integration of geographic ontologies (Kokla and Kavouras, 2001, 2002, 2005; Kavouras and Kokla, 2002). The project aims at identifying and resolving heterogeneities among geographic ontologies developed according to different semantic contexts: thematic, detail, temporal. These differences in the conceptualization and categorization of geographic concepts may not be easily identified or formalized due to their weak semantic content. Therefore, they raise problems when ontologies from heterogeneous contexts are to be integrated. In order to unambiguously define the integrated ontology, it is important to identify and resolve these heterogeneities in an explicit and objective way.

The ONTOGEO project aims at the development of a methodology able to deal with such crucial aspects of the integration process. It is oriented towards true integration, because it creates a single integrated ontology without distorting the resource ontologies, which retain their independence and usability. The source ontologies are usually terminological concerning their level of formality. This implies that they include a hierarchy of geographic concepts described by natural language definitions. The methodology is based on the view that definitions are the primary and usually the only descriptions of concept terms in geographic ontologies. The ONTOGEO methodology performs all three processes: semantic information extraction, category comparison and reconciliation, and semantic integration.

The first process deals with the semantic information extraction from geographic concept definitions. The process is implemented using natural language processing techniques. These techniques are based on the realization that definitions are an important source of scientific knowledge stated in natural language. Definitions describe concepts' meaning and therefore contain sufficient information to disambiguate similar concepts. Geographic information sources use natural language definitions to describe the essential features of their concepts. By analyzing geographic concept definitions, the aim of the project is to identify and represent the knowledge contained in an explicit form, in order to identify similarities and heterogeneities between similar concepts. The techniques employed to extract semantic information from definitions take advantage of their special structure and content.

The process of extracting semantic information from definitions is performed in two stages. The first stage involves the syntactic analysis (parsing) of definitions in order to determine the form, function, and syntactical relations of each part of speech. The second stage involves the application of a set of heuristic rules that examine the existence of syntactic and lexical patterns, that is, words and phrases in definitions systematically used to express specific semantic elements (see Chapter 12).

The ONTOGEO project distinguishes two types of semantic elements derived from definitions:

- Semantic properties describe characteristics of the concept itself (internal characteristics)
- Semantic relations describe characteristics of the concept relative to other concepts (external characteristics)

In the literature, there is no complete list of semantic information that can be extracted from definitions because they vary according to the dictionary from which they are extracted (Barriere, 1997). In Chapter 12, different geographic ontologies, standards, and categorizations (e.g., Cyc Upper Level Ontology, WordNet, CORINE Land Cover, DIGEST, SDTS, etc.) were analyzed in order to identify patterns that are systematically used to express specific semantic elements (properties and relations) and formulate the corresponding rules. The ones most commonly used are shown in Table 12.1 and Table 12.2. The analysis resulted in the identification of general semantic elements (e.g., PURPOSE, CAUSE, TIME, etc.), as well as "geographically oriented" ones, for example information on location (on, below, above, etc.), topology (e.g., adjacent-to, surrounded-by, connected-to, etc.), proximity (e.g., near, far, etc.), and orientation (e.g., north, south, towards, etc). Context-specific semantic elements were also identified. For example, concepts relative to hydrography are described by semantic elements such as nature (natural or artificial) and flow (flowing or stagnant). Concepts relative to agriculture are described by semantic elements such as crop rotation, tillage, irrigation, vegetation type, etc.

The first process is implemented with a tool called GeoNLP (see also Section 16.2) and is performed with minimal user involvement. The output of this process is the formalization of definitions into a set of semantic elements and corresponding values. These are used subsequently to provide an objective way to perform the second process, concept comparison.

Concept comparison consists in the identification of similarities and heterogeneities between similar concepts. The components used to perform the comparison are the terms, as well as semantic elements and their values. Different combinations of these elements lead to four possible comparison cases between two concepts: equivalence, difference, subsumption, and overlap. The comparison is performed for those concepts that are semantically similar, that is, exhibit similarity either in terms (synonymous terms) or in the set of semantic elements (common semantic elements and values).

As long as similarities and heterogeneities are identified, the ONTOGEO methodology proceeds with heterogeneity reconciliation. Reconciliation is implemented by a conceptual analysis procedure known as Semantic Factoring (Section 8.4) and ends up in a set of nonoverlapping, nonredundant concepts, which will be the building blocks of the integrated ontology. Equivalent concepts are merged into one concept, subsumption relations are defined for superordinate-subordinate concepts, and overlapping concepts are split into their nonoverlapping parts.

The third process builds the integrated ontology from the set of nonredundant, nonoverlapping building blocks, called semantic factors. Semantic factors inherit the semantic elements and values of the original concepts from which they were derived. An algorithm based on Formal Concept Analysis (FCA) (Chapter 8) is used to derive the integrated ontology, which is represented as a concept lattice. The concept lattice is formed by formal concepts and their between hierarchical relations. Formal concepts consist of objects (semantic factors), attributes (semantic properties and relations), and values. The algorithm produces relations, not only between concepts with common objects, but also between concepts with common attributes. The final categories resulting from the application of FCA are not pre-defined, but are generated directly by the algorithm. Moreover, the algorithm derives formal concepts resulting from the fusion or division of original concepts and reveals hierarchical relations that were not initially obvious.

The rationale behind the application of FCA for the integration process is that it is a conceptual classification theory, which reveals all possible combinations between similar concepts and requires minimal user involvement. The structure of a concept lattice allows a concept to belong to more than one superordinate concept, which is a basic demand in a geographic ontology covering multiple subdomains. The integration process results in one integrated ontology. The original ontologies are not altered, neither are they forced to comply with a target ontology.

ONTOGEO takes into account concept semantics immanent in definitions for performing the three processes (extraction, comparison, and integration) in contrast to other approaches, which are usually performed empirically based on features that do not completely define category semantics, such as terms and attributes; this may introduce errors in the integrated ontology. The methodology may be applied to terminological geographic ontologies, which are usually defined as hierarchies of concept terms with their definitions, without additional features like properties, axioms, functions, etc.

ONTOGEO formalizes the process of semantic heterogeneity identification and resolution, which is usually performed superficially during ontology integration.

The four possible comparison cases (i.e., equivalence, difference, subsumption, and overlap) explicitly describe all possible relations between two concepts. Overlap is a complicated case, which is usually neither identified nor resolved in ontology integration approaches.

Furthermore, the approach minimizes human intervention during the semantic integration in order to maximize explicitness and objectiveness. Concept comparison does not depend on experts' subjective decisions, or on superficially defined features. It rather relies on an explicit process based on semantic elements identified with specific rules.

15.13 EXTENSIONAL–BASED GEOINTEGRATION APPROACHES

Duckham and Worboys (2005a, 2005b) developed an extensional approach to information fusion based on algebra and first-order logic. The approach aims at inferring semantic information from geospatial information. It is particularly relevant to the geographic information domain, because geospatial location provides the common basis for the fusion process. A simplified example of the basic idea behind this approach is that of two land cover maps, where all regions classified as "lake" in one map are classified as "inland water bodies" in the other. In that case, we may infer that the category "lake" is a subclass of the category "inland water bodies" (intensionally). The authors stress the following advantages of using instance-level information in the integration process:

1. Instance-level information may provide useful indications of the actual use of information.
2. Instance-level information provides a wealth of examples that can be used for the partial or full automation of the integration process.

The process of deriving general rules from specific cases is called *inductive inference*. A risk of applying inductive inference to the extensional integration approach is met at the occurrence of deductive invalidity, that is, the case when the premises are true, but the final conclusion is not. Deductive invalidity may be avoided through the use of large data sets, which include a wide range of spatial relations between the classes present in the data set. In case intensional information is also available, it can be used in the integration process.

The extensional approach operates under the following assumptions: (1) the input data sets refer to related thematic domains, and (2) the data sets are spatially coincident. Furthermore, the approach should be cautiously applied in case the extensional information is inaccurate or imprecise, because the reliability of the inductive inference process decreases. In order to deal with categories for which there is no knowledge of their semantics, the extensional approach to information fusion considers only the extensional form of categories, that is, the set of all instances that belong to each category.

The implementation of the approach is based on first-order logic and algebra. The basic idea behind the development of an automated information fusion system is the formulation of rules for the inductive inference process and their implementation within an automated reasoning system. A geographic data set is formalized as a set S that is a partition of a spatial region, a taxonomy $(C, \leq)$, and a function e: $C \rightarrow 2^S$. A taxonomy is defined as a partially ordered set $(C, \leq)$, where C is a set of categories and $\leq$ is the subsumption relation between these categories. The function e is called extension function and relates each category in the taxonomy to a unique set of elements from the spatial partition S.

For example, for two data sets $G_1 = (S_1, (C_1, \leq_1), e_1)$ and $G_2 = (S_2, (C_2, \leq_2), e_2)$ the first-order rule "for all $x \in C_1$ and $y \in C_2$, $e_1(x) \subseteq e_2(y)$ implies $x \leq y$" means that where the spatial extent of a category y includes the spatial extent of a category x, then x is a subcategory of y. These rules were implemented within the description logic reasoning system RACER (Haarslev and Moeller, 2001).

This approach has two main weaknesses: (1) that the partial order is too general a structure to represent a geographic ontology, and (2) that the fusion process is not associative or commutative, that is, the system may produce different results, based on the insertion order of the data. In order to overcome these weaknesses, a modification of the approach is presented in Duckham and Worboys (2005b). A geographic data set—represented as a partition of space, a lattice, and an extension function—is combined with the fusion operator form, in order to develop a *fusion algebra,* which ensures that the integration process is closed, associative, and commutative.

Tomai and Kavouras (2005) compare the information content of two different thematic maps of the same phenomenon by generating mappings between them. In this research, maps and their thematic content are formalized based on channel theory (Barwise and Seligman; 1997). This theory has been applied to areas such as ontological knowledge organization (Kent, 2004), and ontology mapping (Kalfoglou and Schorlemmer, 2003; Kent, 2005) (for further details see Section 15.9).

To demonstrate the approach, two thematic maps (map_1 and map_2) showing population density in a part of Europe at different periods of time (years 1996 and 1999) are used. These maps are considered as classifications whose types correspond to the legend symbols and tokens correspond to the regions of the maps bearing these symbols. Since the two maps are based on different classifications of population density, there is no possibility to infer population density changes over time. In order to draw inferences about population density change, a source of information is needed that is able to account for changes in population density across time. For this purpose, a third map (map_3) is used that shows total population change rates between 1996 and 2000 in the same European regions. In terms of channel theory, map_3 is used as a channel within which the information between the two classifications can flow.

According to the following steps, the approach transforms the population density classification of map_1 into the corresponding classification of map_2:

- Calculation of population density in 1999 for the given regions with respect to map_3 and map_1.
- Comparison of the results to the population density deduced from map_2, ending up with its true value for every region for the year 1999.
- Generation of mappings between the two classifications.

The resulting relations between the tokens of the two classifications can be described in terms of inclusion, overlap, extension, and refinement. Inclusion is met in cases where a population density class of the first classification is included in a population density class of the second one. Overlap holds when a part of a class of the first classification is included in a class of the second one. Extension regards expansion of the limits of the initial class, while refinement involves the opposite procedure when the limits of the initial class are confined.

In general, extensional approaches offer an additional basis for concept comparison and integration. They also overcome limitations of intensional approaches, as, for example, in cases of ill-defined ontologies. However, extensional approaches raise several problematic issues (see also Chapter 14). They are based on the position that there is differentiation between a concept's definition and use (probably due to human cognitive, cultural, scientific, etc., diversity) but this position virtually implies that there is subjectivity in both definition and use of information. Based on this assumption, the adoption of an approach that relies on the use of information may increase the uncertainty and unreliability of the results. Major concerns also arise relative to the applicability of extensional approaches, especially in domains where instances are not clearly defined or do not constitute a close set (e.g., like months or days of a week).

Furthermore, ill-defined ontologies usually result in ill-defined instances, since vagueness at the intensional level (e.g., categories, hierarchical structure, etc.) is usually transferred to the extensional level as well (Tomai, 2006). Last but not least, relations between instances do not necessarily stand among concepts; therefore the results at the extensional level cannot be transferred directly to the intensional level (Tomai, 2006).

15.14 COMPARATIVE PRESENTATION

The difficulty in evaluating and comparing integration approaches impedes usability as well as methodological interoperability. In order to facilitate understanding, the key issues of semantic integration in a unified framework were presented in Chapter 14. The framework proves to be a useful prescriptive tool to a better understanding/comparison of the available approaches and, as a result, to making better choices in an integration endeavor on the basis of (a) what semantics are available, (b) what kind of integration is sought, and (c) what level of interaction is possible/needed. The presentation pursued in the present chapter facilitates the comparison of the approaches according to the framework presented in Chapter 14, as well as the identification of their principal characteristics, their advantages, and disadvantages.

Furthermore, it is useful for the selection of the appropriate approach in an integration endeavor according to specific needs. For example, if there is a need to perform ontology mapping between large informal ontologies whose semantics are basically defined by their concept terms, relying on user intervention, then a user-assisting tool may be used (e.g., PROMPT, KRAFT, etc.). In case the original ontologies are ill-defined, but there is reliable extensional information, an extensional-based

TABLE 15.1

Comparative Presentation of Approaches for Process I: Semantic Information Extraction

	Involvement-Interaction	Method	Information Extracted	Source Components
PROMPT	—	—	—	—
KRAFT	Semiautomatic	Text-based analysis	Terms that describe the domain	Text, databases, ontologies, etc.
MOMIS	Semiautomatic	ODL-I3 (an object-oriented language, with an underlying DL)	Terminological relations	Schema-/lexicon derived relations, designer-supplied relations, inferred relations
Rodríguez and Egenhofer	—	—	—	—
MIGI	—	—	—	—
ONTOGEO	Minimal user involvement.	NLP techniques to extract semantic information from definitions	Semantic elements and their values	Definitions of terms

method may be applied (Stumme and Madche, 2001; Kalfoglou and Schorlemmer, 2003; Duckham and Worboys, 2005a, 2005b; Tomai and Kavouras, 2005). These comparison capabilities prove to be essential to the next chapter, where guidelines to perform integration are provided. A summary of the principal characteristics with respect to the framework is illustrated in Tables 15.1 to 15.3 for six representative approaches. For example, Table 15.1 shows that for the process of information extraction, MOMIS follows a semiautomatic method to extract terminological intensional and extensional relationships. From Table 15.2, it is apparent that KRAFT uses a semiautomatic method to identify similarities and heterogeneities between input ontologies, using as components for the comparison concept terms and attributes. Table 15.3 shows that for the integration process ONTOGEO uses minimum user involvement and the method used is based on Formal Concept Analysis in order to generate the integrated ontology in the form of a concept lattice.

TABLE 15.2

Comparative Presentation of Approaches for Process II: Concept/Ontology Comparison and Reconciliation

	Involvement-Interaction	Method	Components Used	Heterogeneities Identified/Resolved
PROMPT	Semiautomatic	Syntactic comparison for term similarity (any term matching algorithm)	Terms of classes and slots, class hierarchy, slot attachment to classes	Name conflicts, dangling references, redundancies, slot-values that violate class inheritance
KRAFT	Semiautomatic	Linguistic comparison of terms, then comparison of attributes	Terms and attributes	Mismatches according to Visser et al., 1998
MOMIS	Semiautomatic	ARTEMIS (tool based on affinity-based clustering techniques)	Terms and attributes	Semantically related classes are clustered together
Rodríguez and Egenhofer	May be automated	Similarity-based approach	Terms, is-a and part-whole relations, attributes, parts, and functions	Identification of a similarity value
MIGI	Semiautomatic	Use of a common reference ontology—no identification or resolution of heterogeneities between original ontologies	Intensional definitions of terms in formal ontologies	Similarity relations between intensional definitions: disjoint, overlapping, specialized, equal
ONTOGEO	Minimal user involvement	Comparison of terms, semantic elements, and their values	Terms, semantic elements and values	Semantic heterogeneities (equal, different, subsumed, overlapping concepts)—resolution using Semantic Factoring

TABLE 15.3

Comparative Presentation of Approaches for Process III: Integration

	Involvement-Interaction	Method	Ontology Change	Number of Resulting Ontologies	Use Target
PROMPT	Semiautomatic	Operations: merge classes, merge slots, deep/shallow copy of class, remove/add parent, rename frame	Yes	One	Yes (preferred ontology may be used)
KRAFT	Semiautomatic	Construction of a shared ontology and definition of mappings between resource ontologies and shared ontology	No	One	Yes
MOMIS	Semiautomatic	Unification of affinity clusters (combination of clustering techniques and DL)	No	Many plus a mediator schema	No (plan for the future)
Rodríguez and Egenhofer	—	—	—	—	—
MIGI	Semiautomatic	Generation of a global schema and mappings between global schema and local schemata	No	One	Yes
ONTOGEO	Minimal user involvement	Construction of a concept lattice using FCA	No	One	No

REFERENCES

Barriere, C.. 1997. From a children's first dictionary to a lexical knowledge base of conceptual graphs. Ph.D. thesis, School of Computing Science, Simon Fraser University.

Barwise, J. and J. Seligman. 1997. *Information flow: The logic of distributed systems.* Cambridge Tracts in Theoretical Computer Science 44. Cambridge, UK: Cambridge University Press.

Beneventano, D. and S. Bergamaschi. 2004. The MOMIS methodology for integrating heterogeneous data sources. Paper presented at IFIP World Computer Congress, Toulouse France, August 22–27. http://dbgroup.unimo.it/prototipo/paper/ifip2004.pdf (accessed 3 May 2007).

Beneventano, D., S. Bergamaschi, F. Guerra, and M. Vincini. 2003. Synthesizing an integrated ontology. *IEEE Internet Computing Magazine* 7(5): 42–51.

Bergamaschi, S., S. Castano, D. Beneventano, and M. Vincini. 2001. Semantic integration of heterogeneous information sources. *Data and Knowledge Engineering, Special Issue on Intelligent Information Integration* 36(1): 215–249.

Bishr, Y. 1997. Semantic aspects of interoperable GIS. Ph.D. thesis, Wageningen Agricultural University and ITC, Netherlands.

Bishr, Y. 1998. Overcoming the semantic and other barriers to GIS interoperability. *International Journal of Geographical Information Science* 12(4): 229–314.

Bressan, S., K. Fynn, C. H. Goh, M. Jakobisiak, K. Hussein, H. Kon, T. Lee, S. E. Madnick, T. Pena, J. Qu, A. Shum, and M. Siegel. 1997. The context interchange mediator prototype. In Proc. ACM SIGMOD/PODS Joint Conference, Tuczon, AZ, 525–527.

Chaudhri, V. K., A. Farquhar, R. Fikes, P. D. Karp, and J. P. Rice. 1998. OKBC: A programmatic foundation for knowledge base interoperability. In *Proc. Fifteenth National Conference on Artificial Intelligence (AAAI-98)*, 600–607. Madison, WI: AAAI Press.

Cruz, I. F., A. Rajendran, W. Sunna, and N. Wiegand. 2002. Handling semantic heterogeneities using declarative agreements. *In Proc. 10th ACM International Symposium on Advances in Geographic Information Systems*, 168–174. New York: ACM Press.

Doan, A., P. Domingos, and A. Halevy. 2001. Reconciling schemas of disparate data sources: A matching-learning approach. In *Proc. ACM SIGMOD Conference*, Santa Barbara, CA, 509–520.

Doan, A., J. Madhavan, P. Domingos, and A. Halevy. 2002. Learning to map between ontologies on the semantic web. In *Proc. 11th International WWW Conference (WWW2002)*, Honolulu, HI, 662–673.

Duckham, M. and M. F. Worboys. 2005a. Rosetta: Automated geographic information fusion and ontology integration. Paper presented at GISPlanet 2005, Lisbon, Portugal.

Duckham, M. and M. F. Worboys. 2005b. An algebraic approach to automated information fusion. *International Journal of Geographic Information Science* 19(5): 537–557.

Egenhofer, M. J. and J. R. Herring. 1991. *Categorizing binary topological relations between regions, lines and points in geographic databases.* Technical Report, Department of Surveying Engineering, University of Maine, Orono.

FGDC (Federal Geographic Data Committee). 1998. Content standard for digital geospatial metadata, FGDC-STD-001-1998. ed. National Spatial Data Infrastructure. http://www.fgdc.gov/standards/projects/FGDC-standards-projects/metadata/base-metadata/index_html (accessed 1 May 2007).

Firat, A. 2003. Information integration using contextual knowledge and ontology merging. Ph.D. thesis, Massachusetts Institute of Technology.

Firat, A., S. E. Madnick, and F. Manola. 2005. Multi-dimensional ontology views via contexts in the ECOIN semantic interoperability framework. Working Paper No. 2005-05, MIT Sloan Working Paper No. 4543-05, CISL.

Gennari, J. H., M. A. Musen, R. W. Fergerson, W. E. Grosso, M. Crubezy, H. Eriksson, N. F. Noy, and S. W. Tu. 2003. The evolution of Protege: An environment for knowledge-based systems development. *International Journal of Human-Computer Studies* 58(1): 89–123.

Goh, C. H. 1997. Representing and reasoning about semantic conflicts in heterogeneous information systems. Ph.D. thesis, Sloan School of Management, Massachusetts Institute of Technology.

Haarslev, V. and R. Moeller. 2001. Description of the Racer system and its applications. In *Proc. International Description Logics Workshop (DL 2001)*, ed. C. Goble, D. L. McGuinness, R. Moeller, and P. F. Patel-Schneider, 131–141. Stanford, CA, August 1–3 2001. Available online as *CEUR Workshop Proceedings*, Vol. 49.

Hakimpour, F. and A. Geppert. 2001a. Ontologies: An approach to resolve semantic heterogeneity in databases. Technical Report. http://arvo.ifi.unizh.ch/dbtg/Projects/MIGI/publication/ontoreport.pdf (accessed 1 May 2007).

Hakimpour, F. and A. Geppert. 2001b. Resolving semantic heterogeneity in schema integration: An ontology based approach. In *Proc. International Conference on Formal Ontology in Information Systems FOIS-2001*, Vol. 2001, 297–308.

Hakimpour, F. and A. Geppert. 2002. Global schema generation using formal ontologies. In *Proc. 21st International Conference on Conceptual Modeling ER2002*. Lecture Notes in Computer Science, Vol. 2503, 307–321. Berlin: Springer-Verlag.

Hakimpour, F. and S. Timpf. 2002. A step towards geodata integration using formal ontologies. In *Proc. 5th AGILE Conference on Geographic Information Science*, Palma, Balearic Islands, Spain.

Hearst, M. A., A. Y. Levy, C. Knoblock, S. Minton, and W. Cohen. 1998. Information integration. *IEEE Intelligent Systems* 13(5): 12–24.

Jouanot, F., N. Cullot, and K. Yétongnon. 2003. A solution for contextual integration based on the calculation of a semantic distance. In *Proc. 5th International Conference on Enterprise Information Systems (ICEIS 2003)*, France, April 22–26, 483–486.

Kalfoglou, Y. and M. Schorlemmer. 2003. IF-Map: An ontology mapping method based on information flow theory. *Journal on Data Semantics* 1(1), 98–127.

Kavouras, M. and M. Kokla. 2002. A method for the formalization and integration of geographical categorizations. *International Journal of Geographical Information Science* 16(5): 439–453.

Kent, R. E. 2004. The IFF foundation for ontological knowledge organization. *Cataloging and Classification Quarterly* 37(1/2): 187–203.

Kent, R. E. 2005. Semantic integration in the information flow framework. In *Proc. Dagstuhl Seminar 04391Semantic Interoperability and Integration*, ed. Y. Kalfoglou, M. Schorlemmer, A. Sheth, S. Staab, and M. Uschold. Internationales Begegnungs- und Forschungszentrum (IBFI), Dagstuhl, Germany. http://drops.dagstuhl.de/opus/frontdoor.php?source_opus=41 (accessed 2 May 2007).

Khatri, V., Ram, S., and Snodgrass, R.T., 2001, ST-USM: Bridging the semantic gap with a spatio-temporal conceptual model, TIMECENTER, Technical Report TR-64.

Khatri, V., S. Ram, R. T. Snodgrass, and G. M. O'Brien. 2002. Supporting user-defined granularities in a spatiotemporal conceptual model. *Annals of Mathematics and Artificial Intelligence, Special Issue on Spatial and Temporal Granularity*, 36 (1-2): 195–232.

Knoblock, C. A., S. Minton, J. L. Ambte, N. Ashish, I. Muslea, A. G. Philbot, and S. Tejada. 2001. The Ariadne approach to web-based information integration. *International Journal of Cooperative Information Systems Special Issue on Intelligent Information Agents: Theory and Applications* 10(1-2): 145–169.

Kokla, M. and M. Kavouras. 2001. Fusion of top-level and geographic domain ontologies based on context formation and complementarity. *International Journal of Geographical Information Science* 15(7): 679–687.

Kokla, M. and M. Kavouras. 2002. Extracting latent semantic relations from definitions to disambiguate geographic ontologies. Paper presented at GIScience2002, Boulder, CO.

Kokla, M. and M. Kavouras. 2005. Semantic information in geo-ontologies: Extraction, comparison, and reconciliation. *Journal on Data Semantics*, Lecture Notes in Computer Science, 3534, 125–142.

Leclercq, E., D. Benslimane, and K. Yétognon. 1999. ISIS: A semantic mediation model and an agent based architecture for GIS interoperability. In *Proc. International Symposium Database Engineering and Applications (IDEAS)*, 87–91. IEEE.

McGuinness, D. L. 2000. Conceptual modeling for distributed ontology environments. In *Proc. Eighth International Conference on Conceptual Structures Logical, Linguistic, and Computational Issues (ICCS 2000)*, Darmstadt, Germany, August 14–18. Lecture Notes in Computer Science, Conceptual Structures: Logical, Linguistic, and Computational Issues. Volume 1867/2000, 100–112.

McGuinness, D. L., R. Fikes, J. Rice, and S. Wilder. 2000a. An environment for merging and testing large ontologies. In *Proc. Seventh International Conference on Principles of Knowledge Representation and Reasoning (KR2000)*, Breckenridge, CO, ed. A. G. Cohn, F. Giunchiglia, and B. Selman 483–493.

McGuinness, D. L., R. Fikes, J. Rice, and S. Wilder. 2000b. The Chimaera ontology environment. In *Proc. Seventeenth National Conference on Artificial Intelligence (AAAI 2000)*, Austin, TX, July 30–August 3, 1123–1124.

Noy, N. F. and M. A. Musen. 2000. PROMPT: Algorithm and tool for automated ontology merging and alignment. In *Proc. Seventeenth National Conference on Artificial Intelligence (AAAI-2000)*, Austin, TX, July 30–August 3, 405–455.

Noy, N. F. and M. A. Musen. 2001. Anchor-PROMPT: Using non-local context for semantic matching. In *Proc. Workshop on Ontologies and Information Sharing at the Seventeenth International Joint Conference on Artificial Intelligence (IJCAI-2001)*, Seattle, WA, 63–70.

Noy, N. F. and M. A. Musen. 2002. PROMPTDiff: A fixed-point algorithm for comparing ontology versions. In *Proc. Eighteenth National Conference on Artificial Intelligence (AAAI-2002)*, Edmonton, Alberta, Canada, 744–750.

Noy, N. F. and M. A. Musen. 2003. The PROMPT suite: Interactive tools for ontology merging and mapping. *International Journal of Human-Computer Studies* 59(6): 983–1024.

OntoWeb Consortium. 2002. A survey on ontology tools. Ontology-based information exchange for knowledge management and electronic commerce, Deliverable 1.3, 31 May 2002, IST-2000-29243. http://ontoweb.aifb.uni-karlsruhe.de/About/Deliverables/D13_v1-0.zip (accessed 2 May 2007).

Ram, S., V. Khatri, L. Zhang, and D. D. Zeng. 2002. GeoCosm: A semantics-based approach for information integration of geospatial data. In *ER 2001 Workshops*, H. Arisawa and Y. Kambayashi, 152–165. Lecture Notes in Computer Science, 2465. Berlin: Springer-Verlag.

Rodríguez, A. and M. Egenhofer. 2003. Determining semantic similarity among entity classes from different ontologies. *IEEE Transactions on Knowledge and Data Engineering* 15(2): 442–456.

Stumme, G. and A. Madche. 2001. FCA-Merge: Bottom-up merging of ontologies. In *Proc. 7th Intl. Conf. on Artificial Intelligence (IJCAI '01)*, Seattle, WA, 225–230.

Tomai, E. 2006. Towards intensional/extensional integration between ontologies. In *Proc. Joint ISPRS Workshop on Multiple Representation and Interoperability of Spatial Data*, Hannover, Germany, February 22–24. http://www.ikg.uni-hannover.de/isprs/workshop2006/Paper/6036/tomai_hannover.pdf (accessed 2 May 2007).

Tomai, E. and M. Kavouras. 2005. Mappings between maps: Association of different thematic contents using situation theory. In *Proc. International Cartographic Conference*, A Coruna, Spain, July 6–16.

Tversky, A. 1977. Features of similarity. *Psychological Review* 84(4): 327–352.

Visser, P. R. S., D. Jones, T. J. M. Bench-Capon, and M. J. R. Shave. 1998. Assessing heterogeneity by classifying ontology mismatches. In Formal ontology in information systems, ed. N. Guarino, 148–162. Amsterdam: IOS Press.

Visser, P. R. S., D. Jones, T. J. M. Bench-Capon, M. J. R. Shave, and B. M. Diaz. 1999. Resolving ontological heterogeneity in the KRAFT Project. In *Proc. 10th International Conference on Database and Expert Systems Applications*, Florence, Italy, August 30–September 3, 668–677.

16 Integration Guidelines

16.1 GENERAL GUIDELINES AND SCENARIOS

Semantic integration is not a one-way process because it is affected by a set of factors or characteristics as outlined in Chapter 14. Given the complexity of the issue, the user needs to know what his or her choices are and what is involved in each choice. According to the framework introduced in Chapter 14, this translates to knowing in advance what to expect from the integration process, considering bottlenecks, impossibilities, costs, and eventually to selecting the optimum solution. The first critical issues (questions to be answered) are:

1. *Objective-assumptions*—assumptions made about the source of semantics and what the desired objective is. The type of the process (extraction, comparison, integration) or their combination must also be specified here.
2. *Semantic level addressed*—each process addresses certain "semantics." It is clear that not all approaches use (or are capable of using) all semantic elements.
3. *Input (source)/output components*—here, the user must specify what the input to the approach is and the desired output.
4. *Ontology change*—approaches also differ in the way they alter the resource ontologies. The user may specify, for example, whether individual ontologies will continue to be used or they will be abandoned after merging.
5. *User involvement*—whether user/expert involvement is desirable in order to make critical choices or whether an automated approach is preferable instead.

Answering these five questions is very useful if not essential. It is obvious, however, that, although they address different aspects of the integration process, they are not completely orthogonal. For instance, the user may desire to address the deep level of semantics, but the available sources may be inappropriate for achieving that. Actually, not all options are possible, especially when essential components of the approach are missing and cannot be replaced. For example, a method requiring attributes cannot be used if such information is not available.

Another issue of some importance is how the user expresses what is desirable. The issue of conceptual sameness or similarity is critical here. Two concepts may be similar with respect to certain properties and dissimilar with respect to others. The user should know what is preferable. This resembles the well-known issue of the two error types in statistics and which type is preferable to commit to. Type I error would be to classify as similar two concepts that are not. Type II error could be to classify as different two concepts that are in fact the same. It somewhat also relates to the *lumper and splitter* problem. Sometimes the user prefers to lump things together (in fewer concepts/categories) rather than to end up with more concepts; therefore, similarity is weighted much more than dissimilarity. Splitters do quite the opposite. The right answer often lies in the middle. Any hope for systematization, automation, or simply objectivity requires that the context of comparison and the desired outcome are well understood before they are specified. Also, the more characteristics one includes, the more likely it is to find dissimilarity and end up with splitting.

The present chapter describes a set of guidelines to perform semantic integration according to the principal characteristics of Chapter 14. These guidelines are based on the following real case scenarios and cover various typical integration tasks:

Scenario 1: Concept comparison focusing on the ontologies' structure
Scenario 2: Concept comparison based on their similarity
Scenario 3: Mapping to a shared ontology
Scenario 4: Advanced ontology mapping
Scenario 5: Ontology merging
Scenario 6: Advanced ontology merging
Scenario 7: True integration
Scenario 8: Extensional integration
Scenario 9: Integration based on natural language documents
Scenario 10: Temporal integration

Because this is a highly complex problem domain, in some scenarios there are solutions available—even alternatives at times. However, there are also cases where no ideal solution exists because it is still an open research problem.

16.1.1 SCENARIO 1: CONCEPT COMPARISON FOCUSING ON THE ONTOLOGIES' STRUCTURE

Objective-assumptions: comparison of two ontologies represented as hierarchies of terms. The basic assumption is that hierarchies of similar domains resemble both in terms and in structure.

Semantic level addressed: low since only concept terms and their hierarchical relations are taken into account for the concept comparison.

Input (source) components: concept terms and their hierarchical relations.

Output components: identification and mapping of similar concepts.

Ontology change: original ontologies are not altered since no merging-integration is performed

User involvement: user involvement may be required.

SUGGESTION-GUIDELINE: A concept comparison approach that enforces the effect of taxonomical relations is suitable in the current scenario. For example, taxonomical relations represented as domain-independent and domain-dependent constraints play an important role in the approach implemented by GLUE (Section 15.6).

An alternative solution would be to use a tool such as AnchorPROMPT (Section 15.2). In order to associate two ontologies represented as graphs, AnchorPROMPT establishes manually or automatically some pairs of similar concepts acting as control points or anchors. Therefore, the intermediate concepts between anchors are likely to exhibit similarity. Another alternative would be to apply a similarity measure that is based on hierarchical relations between concepts, for example, the network or edge-counting similarity measure or the information content similarity measure (see Chapter 13) or a similarity measure based on the concepts' semantic neighborhood (Section 15.11).

16.1.2 Scenario 2: Concept Comparison Based on Similarity

Objective-assumptions: the objective is to incorporate similarity measures in the concept comparison subprocess.

Semantic level addressed: usually medium; it depends on the availability and quality of the elements used in the similarity measures. The semantic level increases in case there are available elements that reinforce concepts' semantic description such as properties, functions, and parts.

Input (source) components: concept terms, attributes and values, taxonomic relations, or other elements such as parts and functions may be used as input components to the similarity measures.

Output components: a value reflecting the similarity between concepts from different ontologies.

Ontology change: original ontologies are not altered since no merging-integration takes place.

User involvement: no user involvement is desirable. However, an expert should elaborate the result of the similarity assessment in order for this to be applicable to the integration subprocess.

SUGGESTION-GUIDELINE: in the current scenario we may use either a method that exclusively performs concept comparison (Section 15.11) or a tool such as MOMIS (Section 15.5) or GLUE (Section 15.6), which incorporates similarity measures in the concept comparison subprocess.

16.1.3 Scenario 3: Mapping to a Shared Ontology

Objective-assumptions: the objective is to map original ontologies to a top-level ontology. The assumption made by the current scenario is that ontologies consist of hierarchies of concepts defined by attributes and constraints. Different kinds of source elements are available, for example, corpora, databases, ontologies, etc. Therefore, an information extraction process is also necessary in order to exploit these information sources. A further requirement of the current scenario is to ground the mapping process on a top-level ontology.

Semantic level addressed: low-medium; semantics is based on concept terms and attributes.

Input (source) components: corpora, databases, and ontologies defined as hierarchies of concepts with attributes and constraints as well as a top-level ontology which comprises the basis for the mapping process.

Output components: a shared ontology and mappings between the shared ontology and the original ones; no direct mappings are established between original ontologies.

Ontology change: original ontologies are not distorted.

User involvement: user-expert involvement is required.

SUGGESTION-GUIDELINE: Because there are a variety of information sources available, it is advisable to use an approach capable of performing information extraction from free text. Information (i.e., concepts, attributes, and values) derived from these sources may then be associated to a top-level ontology in order to enrich it with the necessary information elements. This step results in the shared ontology, which should then be mapped to the original ones. In this way, no direct mappings between the original ontologies are established; instead, the shared ontology acts as a basis for the association. A user-assisting tool may be used in this case, such as KRAFT (Section 15.3). KRAFT follows a semiautomatic workflow, resulting in a shared ontology and links that connect its components to the original ontologies.

16.1.4   Scenario 4: Advanced Ontology Mapping

Objective-assumptions: the objective is to perform concept comparison and ontology mapping, that is, for each concept in one ontology the most similar concept in the other ontology should be identified. Except from the original ontologies, which are represented as taxonomies of concept terms with attributes and attribute values, instances are also available for the current scenario and should be utilized.

Semantic level addressed: medium; concept semantics are defined by concept terms, attributes, attribute values, taxonomical relations, and instances.

Input components: original ontologies represented as concept taxonomies with attributes and attribute values, as well as instances.

Output components: mappings between original ontologies.

Ontology change: no alteration of original ontologies.

User involvement: user assistance is required.

SUGGESTION-GUIDELINE: an ontology-mapping tool may be used in the current scenario. An example is GLUE (Section 15.6), which applies machine-learning techniques in order to semiautomatically identify semantic mappings between ontologies. GLUE also incorporates multicriteria use of similarity measures between concepts from different ontologies.

16.1.5   Scenario 5: Ontology Merging

Objective-assumptions: the objective is to merge two informal ontologies represented as hierarchies of concepts defined by their terms and attributes. These elements (i.e., terms and attributes) are assumed to adequately represent the concepts' semantics. Usually, this type of integration scenario does not comprise the first process (i.e., information extraction), as the comparison process is based on word-matching algorithms, which identify similar terms.

Semantic level addressed: low because semantics is based solely on concept terms, that is, homonymous terms define equivalent concepts. However, attributes may also contribute to the decision on whether two concepts are the same and should be merged. In that case, the semantic level of the approach depends on whether attributes represent the concepts' meaning or are superficially defined.

Input (source) components: concept terms, attributes, and hierarchical relations between concepts.

Output components: a merged ontology.

Ontology change: original ontologies are changed.

User involvement: user-expert involvement is required.

SUGGESTION-GUIDELINE: a user-assisting tool may be used in this case, for example PROMPT (Section 15.2) or Chimaera (Section 15.4). A detailed implementation of such a scenario, using PROMPT, is shown in Section 16.2.1.

16.1.6 SCENARIO 6: ADVANCED ONTOLOGY MERGING

Objective-assumptions: the objective is to perform complete merging of two ontologies represented by concept terms and attributes; the scenario is based on the assumption that concept terms and attributes are the source of semantics. A further requirement is to perform all three integration subprocesses. More specifically, the extraction process should identify semantic relations between concepts of the original sources; the comparison process should be based on similarity measures defined between concepts and the integration process should result in a single shared ontology.

Semantic level addressed: medium; concepts' semantics is based on terms and attributes.

Input (source) components: concept terms, attributes, and hierarchical relations between concepts. Other information sources may also be used to enrich the semantics of the integrated ontology, for example general ontologies such as WordNet.

Output components: a merged ontology represented by a hierarchy of concept terms and attributes ascribed to them.

Ontology change: complete alteration of the original ontologies because they will cease to be used after the merging process.

User involvement: expert involvement required.

SUGGESTION-GUIDELINE: a user-assisting tool may be used in the current integration scenario. MOMIS (Section 15.5) is an example of such a tool. It performs all three integration subprocesses. The extraction process semiautomatically discovers semantic relations between concepts of the original sources (synonyms, hypernyms, hyponyms, and related terms); these are further enriched with semantic relations extracted from WordNet. The comparison process is based on the definition of similarity measures taking into account terms and attributes of related concepts. Similar concepts are grouped together into semantic clusters. The integration process results in a shared ontology defined by the representative concepts (global classes) of each cluster.

16.1.7 SCENARIO 7: TRUE INTEGRATION

Objective-assumptions: the objective is to perform true integration of two ontologies represented as hierarchies of concept terms with definitions in natural language. The main assumption relies on definitions as a rich source of semantics; therefore, an extraction process is required to derive and formalize semantic information latent in definitions.

Semantic level addressed: high, based on semantic elements (properties and relations) and their values, which represent the meaning of concepts.

Input (source) components: the input components are the original ontologies that are comprised of concept terms and their definitions in natural language as well as hierarchical relations between concepts.

Output components: the expected output component is an integrated ontology consisting of concept terms defined by semantic elements and the hierarchical relations between them.

Ontology change: no alteration of the original ontologies is desirable because these may also be used independently of the integrated ontology.

User involvement: requirement for minimum user involvement in order to achieve maximum objectivity.

SUGGESTION-GUIDELINE: in order to ensure a high semantic level in the integration result, it is necessary to use an approach consisting of all three semantic integration subprocesses (extraction, comparison, integration). Furthermore, it is also necessary that each subprocess utilizes and preserves the components derived by the previous process to avoid semantic information loss. Therefore, it is recommended that an information extraction process be employed in order to derive the semantic elements that will be used subsequently by the comparison and integration processes. OntoGEO is an example of such an approach whose detailed implementation is presented in Section 16.2.3.

16.1.8 SCENARIO 8: EXTENSIONAL INTEGRATION

Objective-assumptions: the objective is to perform extensional-based integration relying on the assumption that extensional information is available and is considered accurate and precise.

Semantic level addressed: low, based on concept terms and on the classification of instances according to them.

Input (source) components: two databases with concept terms and hierarchical relations as well as instances.

Output components: a global ontology consisting of ontology mappings together with the link between original ontologies via the reference ontology.

Ontology change: original ontologies are not altered because only mappings between them and the reference ontology are generated.

User involvement: user assistance is required.

SUGGESTION-GUIDELINE: use of an extensional approach such as IF-Map (Section 15.9), which aims at providing mappings between the original ontologies via the reference ontology. Extensional-based geointegration approaches (Section 15.13) provide another alternative in case the input data sets refer to related thematic domains and the data sets are spatially coincident.

16.1.9 SCENARIO 9: INTEGRATION BASED ON NATURAL LANGUAGE DOCUMENTS

Objective-assumptions: the objective of the current scenario is to perform integration based not on the semantic elements of the original ontologies but on a set of natural language documents describing the application domain. The assumption made is that these documents accurately and sufficiently describe the semantics of the application domain; they will be the basis for extracting the concepts describing the domain and for performing integration. Hence, an adequate information extraction process is required in order to derive all the necessary concept terms.

Semantic level addressed: low, because only concept terms are used in the integration process.

Input (source) components: original ontologies and a set of natural language documents including the original ontologies' concepts.

Output components: an integrated ontology.

Ontology change: no alteration of original ontologies.

User involvement: user assistance is required.

SUGGESTION-GUIDELINE: use of FCA-Merge (Section 15.8), an approach that performs the first and third integration subprocesses. Semantic information extraction is performed based on linguistic analysis techniques, whereas Formal Concept Analysis (see Chapter 8) is applied for integration.

16.1.10 SCENARIO 10: TEMPORAL INTEGRATION

Objective-assumptions: classification of real-world instances at different time periods based on different schemata. In order to monitor evolution through time, an association of the classification schemata is also necessary.

Semantic level addressed: varies according to the elements used and their ability to describe concepts' semantics (e.g., concept terms, attributes, relations, etc.).

Input (source) components: two classification schemata referring to different time periods.

Output components: an integrated ontology as well as the association to the original ones in order to extract information about temporal changes.

Ontology change: the original ontologies should not be altered because they will be used in parallel with the integrated ontology to derive results about evolution through time.
User involvement: user involvement may be required.

SUGGESTION-GUIDELINE: in the present scenario it is necessary to use an integration method that does not alter the original ontologies because they will continue to be in use after the integration process. OntoGEO is an example of such an integration approach (Section 16.2.3). In case there are no different ontologies to be integrated but different versions of the same ontology, a tool for comparing different ontology versions such as PROMPTDiff (Section 15.2) may be used.

16.2 INTEGRATION SCENARIOS AND WORKING EXAMPLES

The present section describes a set of guidelines needed to perform semantic integration according to the principal characteristics of Chapter 14. The guidelines involve different integration types, deal with some or all of the integration processes and address different semantic levels. They are based on three real case scenarios and cover the most typical integration tasks.

Furthermore, the same example is used throughout the three scenarios in order to facilitate their understanding and their comparison. The example includes excerpts of two typical geographic ontologies described by a hierarchy of concepts (Figure 16.1 and Figure 16.2). Concepts are basically defined by their terms and definitions (Table 16.1). However, because the scenarios are based on different approaches, which address different semantic levels, there is a slight modification of the semantic elements used to describe the concepts of the input ontologies according to the approach followed. For example, in the first scenario, a semiautomatic tool uses concept terms, attributes, and hierarchical relations to perform ontology comparison and merging. In the second scenario, a similarity measure is applied which, apart from the elements used in the first scenario, also uses concepts' parts. The third scenario involves a semantic information extraction process that extracts semantic

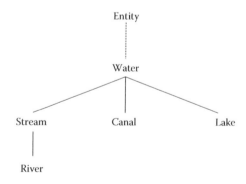

FIGURE 16.1 Hierarchical structure of Ontology A.

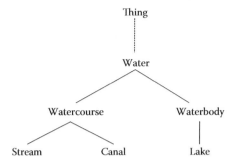

FIGURE 16.2   Hierarchical structure of Ontology B.

TABLE 16.1

Definitions of Ontologies A and B Terms

Ontology A

Water: area covered with water

Stream: natural flowing body of fresh water

River: natural stream of water, normally of a large volume

Lake: body of water surrounded by land

Canal: artificial waterway created to be path for boats, or for irrigation

Ontology B

Water: the part of the Earth's surface covered with water

Watercourse: natural or artificial channel through which water flows

Stream: natural flowing watercourse

Canal: man-made or improved natural waterway used for transportation

Waterbody: inland body of water

Lake: body of water surrounded by land

---

elements and values from definitions of geographic concepts. These are further used in the comparison and integration processes.

The example is small due to space limitations but includes all kinds of semantic conflicts that may occur between two concepts in order to test the ability of different approaches to identify and resolve the conflicts. Table 16.2, Table 16.3, Table 16.4, and Table 16.5 show examples of different semantic relations between two geographic concepts. The most complicated case to identify and resolve is that of overlapping concepts. In Table 16.5, Ontologies A and B have overlapping definitions for the concept "canal" because for Ontology A a canal is an artificial waterway used for transportation or irrigation, whereas for Ontology B a canal is an artificial or improved natural waterway used only for transportation.

In order to increase the semantic value of the approaches, semantic properties instead of superficially defined attributes have been ascribed to concepts. If these

TABLE 16.2
Example of Equivalent Concepts

Equivalence	Ontology A	Ontology B
Term	Lake	Lake
Definition	Body of water surrounded by land	Body of water surrounded by land

TABLE 16.3
Example of Different Concepts

Difference	Ontology A	Ontology B
Term	Lake	Stream
Definition	Body of water surrounded by land	Natural flowing watercourse

TABLE 16.4
Example of Subsumed Concepts

Subsumption	Ontology A	Ontology B
Term	Lake	Waterbody
Definition	Body of water surrounded by land	Body of water

TABLE 16.5
Example of Overlapping Concepts

Overlap	Ontology A	Ontology B
Term	Canal	Canal
Definition	Artificial waterway created to be path for boats or for irrigation	Man-made or improved natural waterway used for transportation

elements are not included in the ontologies, they may be extracted using a semantic information extraction process as shown in the third scenario.

The first scenario deals with ontology merging under the guidance of a semi-automatic tool. This tool aims at assisting the user in the ontology comparison and merging processes by revealing conflicts and suggesting suitable operations for their resolution. The tool performs concept comparison to identify identical or similar concept terms. Based on these findings, the tool guides the user in performing a series of operations, such as concept merging, attribute merging, and removal or addition of a superconcept, which will result in a single merged ontology.

The second scenario deals with the second integration process, that is, concept comparison. A similarity measure is applied in order to draw conclusions about similar geographic concepts. The semantic elements taken into account are concept terms, attributes, and parts, as well as hierarchical and part-of relations. This scenario does not include an integration process; however, the result of the comparison may be used as input for a subsequent integration process. The approach is useful when concepts are described by the above semantic elements (i.e., terms, attributes, functions, parts, as well as hierarchical and part-of relations).

The third scenario involves the semantic integration of two terminological ontologies. A terminological ontology consists of a hierarchy of concepts defined by natural language definitions. It is the most commonly encountered type of existing geographic metadata source. The third scenario covers all three integration processes. Its purpose is to analyze geographic concept definitions and extract immanent semantic information, which will subsequently be used to identify similarities and resolve heterogeneities between original concepts. A requirement is to perform the process with minimum human interaction in order to ensure maximum objectivity. The result is a single ontology; however, original ontologies are not altered.

16.2.1 SCENARIO 1: ONTOLOGY MERGING

The first scenario deals with a simple integration task, that is, merging two informal geographic ontologies. These ontologies are basically defined as hierarchies of concepts; each concept is described by a term and is assigned a set of attributes. The purpose of this scenario is to perform ontology merging with the guidance of a user-assisting tool. This scenario is suitable in the following cases: (a) when light integration is preferred against more sophisticated integration types, (b) when original ontologies do not include other semantic elements except concept terms and attributes, and (c) when these elements are considered suitable and sufficient as a basis for the ontology merging process.

For this purpose, PROMPT (Section 15.2) was chosen as a freeware, typical tool, which is based on the popular ontology design environment of Protégé (Gennari et al., 2003). PROMPT does not address the process of information extraction but rather performs the other two processes, that is, ontology comparison with conflict resolution and ontology merging. However, as already mentioned, in order to strengthen the result of the integration process, an information extraction process such as that implemented in Section 16.2.3 has been used to derive semantic properties and relations instead of using superficial attributes.

PROMPT aims at assisting and automating user choices while revealing resulting conflicts and suggesting appropriate operations. Concept comparison is based on terms of concepts and their attributes (slots). Other semantic information such as definitions or values of attributes is not taken into account. However, in order to increase the semantic value of the approach, semantic properties and relations derived from a semantic information extraction process may be used, instead of superficially defined attributes.

Ontologies are defined according to the knowledge model supported by Protégé, which includes the following knowledge elements: classes, slots, facets, and instances. For the purpose of the current scenario, instances were not included in the representation of the ontologies. Figure 16.3 and Figure 16.4 show the representation of ontologies A and B correspondingly in the knowledge model of Protégé. The upper-left part of each figure (Class Relationship pane) shows the hierarchy of

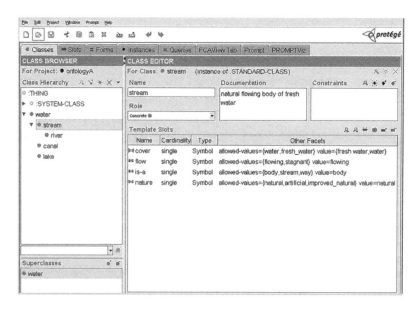

FIGURE 16.3 Representation of Ontology A in Protégé.

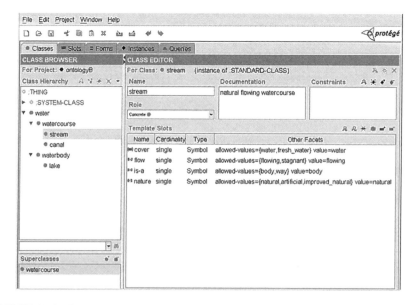

FIGURE 16.4 Representation of Ontology B in Protégé.

classes, which constitute the Ontologies A and B correspondingly. When a single class from the hierarchy is selected, the lower-left part (Superclasses pane) shows the immediate superclass of the selected class, while the right part shows the available information about the class: name, documentation, slots, facets, etc. For Ontology A (Figure 16.3), a stream is a "natural flowing body of fresh water" and is assigned four slots (cover, flow, is-a, and nature).

The merging process takes place in a new project; each operation selected by the user results in the creation of classes and corresponding slots of the new merged ontology. The original ontologies are claimed to be kept separately and not to be affected by the merging process. However, they are altered by the process and cannot be reused in combination with the merged one.

As already described in Section 15.2, PROMPT follows a stepwise procedure. At the first step, the user loads the original ontologies (Figure 16.5). At this step, the user has the ability to choose one of the two ontologies as a preferred ontology, which plays the role of target ontology. For the present scenario, none of the two ontologies was chosen as a preferred one; therefore, they both have the same weights for the merging process. Furthermore, at the same step, a case-sensitive comparison may be selected.

At the second step, a syntactic comparison is performed to identify identical or similar class names. Once matched, a list of suggestions is presented to the user.

- If the concept terms are identical, then the user has to choose between a concept merge or a concept removal.
- If the concept terms are not identical but similar, then a weaker concept merge of lower confidence is suggested to the user.

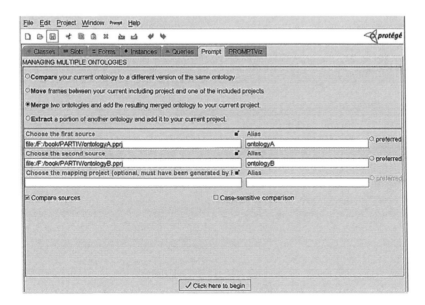

FIGURE 16.5   PROMPT: loading of original ontologies.

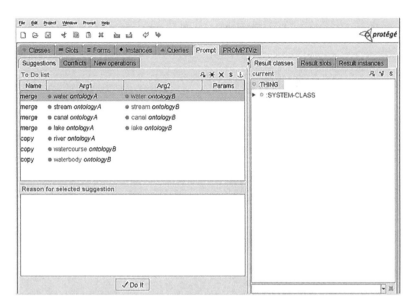

FIGURE 16.6 PROMPT: suggestions to the user after term comparison.

Figure 16.6 shows the list of suggestions presented to the user according to the term comparison (upper-left part) as well as the reason for the selected suggestion (lower-left part). More specifically, four concepts with identical terms in both ontologies were identified (water, stream, canal, and lake) and it is suggested that they are merged. Furthermore, PROMPT suggests the copy of three classes (river, watercourse, and waterbody), which belong to only one of the two ontologies.

The user performs an operation taking into account (without being constrained by) the suggestions of the previous steps. The operations supported by PROMPT are: (a) class merge, (b) slot merge, (c) shallow copy of a class (copy only of the class without associated elements like superclasses or slots), (d) deep copy of a class (with associated slots and superclasses until the root of the hierarchy), (e) parent removal, (f) parent addition, and (g) frame renaming. On the right part, the Result Window shows the resulting classes, slots, and instances of the merging process. For example, after selecting the merge of class "water" as defined by Ontologies A and B, the new class is added to the merged ontology (Figure 16.7). The same procedure is followed for the slots. Figure 16.8 shows the merge of slots "is-a" for class "stream." The user is asked to select facets for the merged slot. The resulting slots are shown on the right part of the window.

PROMPT automatically updates the operations selected by the user, identifies resulting conflicts and suggests actions for their resolution. The resulting conflicts are shown at the Conflicts Tab and include: naming conflicts (same name for different frames), dangling references (relations to nonexistent frames), redundancies (multiple paths from classes to superclasses), slot values violating class inheritance, etc. For example, the merge of class "lake" results in a redundancy, as shown in Figure 16.9, because the class has two superclasses and, therefore, a cycle is created in the merged ontology. The user is asked to remove one of the two superclasses (parents).

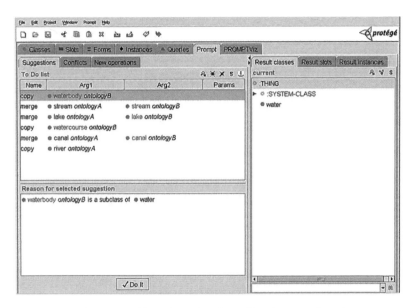

FIGURE 16.7   PROMPT: addition of the class "water" to the merged ontology.

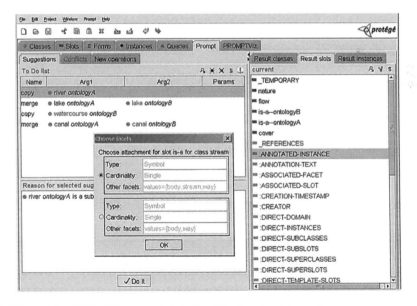

FIGURE 16.8   PROMPT: merge of slots "is-a" for the class "stream."

Steps 3 and 4 are repeated until the two ontologies are completely merged. At the New Operations Tab, the user is allowed to perform operations that are not suggested by PROMPT. For the present example, a conflict occurred due to the merge of class "stream," which belongs to two superclasses in the merged ontology ("water" and "watercourse") (Figure 16.10). This conflict had to be manually identified

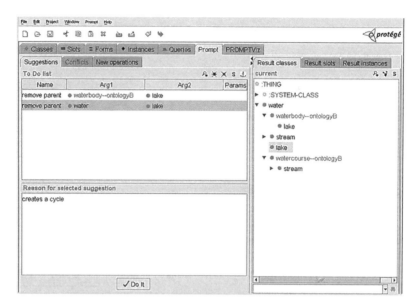

FIGURE 16.9   PROMPT: automatic identification and resolution of a conflict for the class "lake."

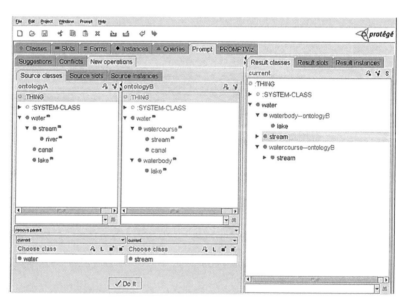

FIGURE 16.10   PROMPT: manual identification and resolution of a conflict for the class "stream."

and resolved. An important conflict that can neither be identified nor resolved by PROMPT has occurred for the concepts "canal." Although the two concepts have the same name, their definitions are overlapping. For Ontology A, a canal is "an artificial waterway created to be paths for transportation, or for irrigation," whereas for

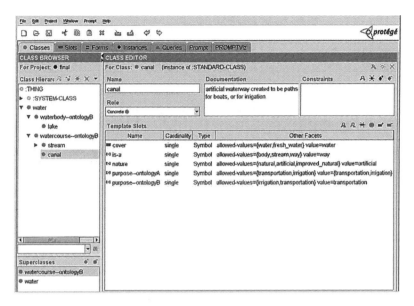

FIGURE 16.11   PROMPT: final merged ontology.

Ontology B, a canal is "a man-made or improved natural waterway used for transportation." The overlapping between these concepts is also depicted in the values of the slots "nature" and "purpose." However, because PROMPT identifies similarities between concepts based only on their terms, the two concepts are considered equivalent, and the suggestion made to the user is that of a class merge. Therefore, this semantic conflict cannot be really resolved; it may only be reconciled either by keeping both concepts in the merged ontology or by merging them and keeping both overlapping slots. The final result of the ontology merging is shown at Figure 16.11; the left part shows the hierarchy of the final ontology, whereas the right part shows the information (e.g., name, documentation, slots, etc.) for each class (in the specific case, "canal").

16.2.2 SCENARIO 2: CONCEPT COMPARISON

The second scenario deals with the second integration process, that is, concept comparison. It consists of applying a similarity measure for comparing concepts. For that purpose, the similarity measure developed by Rodríguez and Egenhofer (2003, 2004) is employed (Section 15.11). The semantic elements taken into account are concept terms, attributes, and parts, as well as hierarchical and part-of relations (Table 16.6). These elements are important in order to apply the similarity measure and can be derived from a semantic information extraction process in case they are not recorded in the original ontologies. For the purpose of the current scenario, these elements are extracted with the method shown at the third scenario.

This scenario does not include an integration process; however, the result of the comparison may be used as input for a subsequent integration process. The approach is useful when concepts are described by the above semantic elements (i.e., terms, attributes, functions, and parts, as well as hierarchical and part-of relations).

TABLE 16.6

Concepts and their Semantic Elements for Concept Comparison Scenario

Ontology A

Water:	Part-of: the Earth's surface	Attributes: {cover}
Stream:	Parts: {ford, meander, midstream}	Attributes: {{cover}, {nature}, {flow}}
River:	Parts: {estuary, rapid, waterfall}	Attributes: {{cover}, {nature}, {size}, {flow}}
Lake:	Parts: {inlet}	Attributes: {{cover}, {surroundness}}
Canal:	Parts: {lock, lockage}	Attributes: {{cover}, {nature}, {purpose}}

Ontology B

Water:	Part-of: the Earth's surface	Attributes: {cover}
Watercourse:		Attributes: {cover}
Stream:	Parts: {midstream, riverbank}	Attributes: {{cover}, {nature}, {flow}}
Canal:	Parts: {lock}	Attributes: {{cover}, {nature}, {purpose}}
Waterbody:		Attributes: {cover}
Lake:	Parts: {inlet}	Attributes: {{cover}, {surroundness}}

For the present scenario we assume that:

- Attributes are inherited from superconcepts to subconcepts.
- The comparison takes place between homonymous or synonymous concepts in order to simplify the process.
- All types of distinguishing features (parts, functions, and attributes) contribute evenly to the representation of concepts and therefore, receive equal weights ($w_p = w_f = w_a = 0.33$).

From an examination of concepts, we conclude that the similarity measure should be applied between concepts "water," "stream," "canal," and "lake," which are included in both ontologies A and B. The method assumes that the two compared ontologies are connected to an imaginary root "anything" in order to be able to derive the level of generalization between concepts (Figure 16.12). The comparison starts bottom-up, that is, from the most detailed to the most general concepts and comprises three similarity measures, which are defined according to Equation 15.2:

1. Word matching
2. Feature matching
3. Semantic neighborhood matching

In order to determine the semantic similarity between the homonymous concepts "stream" as defined by Ontology A (stream^A) and Ontology B (stream^B), first, the depth of each concept is determined as the shortest path from the concept to the hypothetical root (concept "anything"). According to Figure 16.12, since depth (stream^A) = 3 < depth (stream^B) = 4, the function **a** is defined as follows (Equation 15.3):

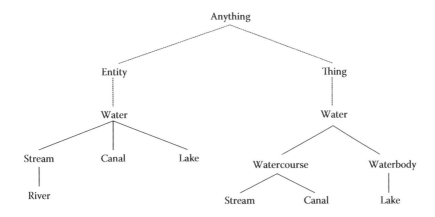

FIGURE 16.12   Connection of Ontologies A and B to an imaginary root "anything."

$$a(stream^A, stream^B) = \frac{depth(stream^A)}{depth(stream^A) + depth(stream^B)} = \frac{3}{3+4} = \frac{3}{7} = 0.43$$

The three similarity measures for the concepts stream^A and stream^B are specified as follows:

1. Word similarity is (Equation 15.2):

$$S_w(stream^A, stream^B) = \frac{\left|\{stream\}\right|}{\left|\{stream\}\right| + 0.43\left|\{\ \}\right| + 0.57\left|\{\ \}\right|} = \frac{1}{1} = 1$$

2. Feature similarity is (Equation 15.4):

$$S_u(a^p, b^q) = w_p \cdot S_p\left(a^p, b^q\right) + w_f \cdot S_f\left(a^p, b^q\right) + w_a \cdot S_a\left(a^p, b^q\right)$$

$$\text{for } w_p, w_f, \text{ and } w_a \geq 0 \text{ and } w_p + w_f + w_a = 1.0$$

No functions are defined for the concepts stream^A and stream^B, and parts and attributes are considered to contribute evenly to concept representation ($w_p = w_a = 0.50$). Part and attribute similarity are defined according to Equation 15.2, where X represents the common elements (parts or attributes correspondingly), Y represents the elements that belong to stream^A and not to stream^B, and Z represents the components that belong to stream^B and not to stream^A. For example, for part similarity:

$X = $ stream^A.parts $\cap$ stream^B.parts $= \{midstream\}$
$Y = $ stream^A.parts - stream^B.parts $= \{\{ford\}, \{meander\}\}$
$Z = $ stream^B.parts $-$ stream^A.parts $= \{riverbank\}$

$$S_p(stream^A, stream^B) = \frac{|X|}{|X| + 0.43|Y| + 0.57|Z|} = \frac{1}{1 + 0.43 \cdot 2 + 0.57 \cdot 1}$$

$$= \frac{1}{2.43} = 0.41,$$

Therefore, feature similarity is:

$$S_u(stream^A, stream^B) = w_p \cdot S_p(stream^A, stream^B) + w_a \cdot S_a(stream^A, stream^B) =$$

$$= 0.50 \cdot 0.41 + 0.50 \cdot 1 = 0.705$$

3. Semantic neighborhood similarity is determined according to Equations 15.5 and 15.6.

For radius $r = 2$, the following semantic neighborhoods (N) are identified:

$$N(stream^A, 2) = \{stream, water, entity, river\}, n = 4$$

$$N(stream^B, 2) = \{stream, watercourse, water\}, m = 3$$

The semantic similarity of one of the compared concepts to the semantic neighborhood of the other $S(a, b_j^q)$ is:

For $j = 1$, $S(stream^A, stream^B) = w_w S_w + w_u S_u = 0.50 \cdot 1 + 0.50 \cdot 0.705 = 0.85$
For $j = 2$, $S(stream^A, watercourse^B) = 0.50 \cdot 0 + 0.50 \cdot 0.50 = 0.25$
For $j = 3$, $S(stream^A, water^B) = 0.50 \cdot 0 + 0.50 \cdot 0.55 = 0.275$

The intersection over semantic neighborhoods is:

$$\left| a^p \cap_n b^q \right| = \left[\sum_{i \leq n} \max_{j \leq m} S\left(a_i^p, b_j^q \right) \right] - \varphi S\left(a^p, b^q \right) = 1.85 - 1 \cdot 0.85 = 1$$

The semantic-neighborhood similarity is:

$$S_n(stream^A, stream^B, 2) = \frac{1}{1 + 0.50 \cdot 3 + 0.50 \cdot 3} = \frac{1}{4} = 0.25$$

because

$$\left| N\left(a^p, r \right) \right| > \left| a^p \cap_n b^q \right|$$

and

$$\delta\left(a^p, a^p \cap_n b^q, r \right) = \left| N\left(a^p, r \right) \right| - \left| a^p \cap_n b^q \right| = 4 - 1 = 3$$

Finally, the semantic similarity between stream^A and stream^B is defined as follows:

$$S(stream^A, stream^B) = w_w S_w(stream^A, stream^B) + w_u S_u(stream^A, stream^B) +$$

$$w_n S_n(stream^A, stream^B) = 0.33 \cdot 1 + 0.33 \cdot 0.705 + 0.33 \cdot 0.25 = 0.615$$

This value means that stream^A is 61.5% similar to stream^B.

The semantic similarity between the homonymous concepts "canal" as defined by Ontology A (canal^A) and Ontology B (canal^B) is calculated accordingly.

1. Word similarity is (Equation 15.2):

$$S_w(canal^A, canal^B) = \frac{|\{canal\}|}{|\{canal\}| + 0.43|\{\ \}| + 0.57|\{\ \}|} = \frac{1}{1} = 1$$

2. Feature similarity is (Equation 15.4):

$$S_u(canal^A, canal^B) = w_p \cdot S_p(canal^A, canal^B) + w_a \cdot S_a(canal^A, canal^B)$$

$$= 0.50 \cdot 0.54 + 0.50 \cdot 1 = 0.77$$

3. Semantic neighborhood similarity is (Equations 15.5 and 15.6):

$$S_n(canal^A, canal^B, 2) = \frac{1}{1 + 0.43 \cdot 2 + 0.57 \cdot 2} = \frac{1}{3} = 0.33$$

Finally, the semantic similarity between canal^A and canal^B is defined as follows:

$$S(canal^A, canal^B) = w_w S_w(canal^A, canal^B) + w_u S_u(canal^A, canal^B)$$

$$+ w_n S_n(canal^A, canal^B)$$

$$= 0.33 \cdot 1 + 0.33 \cdot 0.77 + 0.33 \cdot 0.33 = 0.693$$

This value means that canal^A is 69.3% similar to canal^B.

The semantic similarity between the homonymous concepts "lake" as defined by Ontology A (lake^A) and Ontology B (lake^B) is calculated as follows.

1. Word similarity is (Equation 15.2):

$$S_w(lake^A, lake^B) = \frac{|\{lake\}|}{|\{lake\}| + 0.43|\{\ \}| + 0.57|\{\ \}|} = \frac{1}{1} = 1$$

2. Feature similarity is (Equation 15.4):

$$S_u(lake^A, lake^B) = w_p \cdot S_p(lake^A, lake^B) + w_a \cdot S_a(lake^A, lake^B)$$

$$= 0.50 \cdot 1 + 0.50 \cdot 1 = 1$$

3. Semantic neighborhood similarity is (Equations 15.5 and 15.6):

$$S_n(lake^A, lake^B, 2) = \frac{1}{1 + 0.43 \cdot 2 + 0.57 \cdot 2} = \frac{1}{3} = 0.33$$

Finally, the semantic similarity between lake^A and lake^B is defined as follows:

$$S(lake^A, lake^B) = w_w S_w(lake^A, lake^B) + w_u S_u(lake^A, lake^B) + w_n S_n(lake^A, lake^B)$$

$$= 0.33 \cdot 1 + 0.33 \cdot 1 + 0.33 \cdot 0.33 = 0.77$$

This means that lake^A is 77% similar to lake^B.

The semantic similarity between the homonymous concepts "water" as defined by Ontology A (water^A) Ontology B (water^B) is calculated as follows.

1. Word similarity is (Equation 15.2):

$$S_w(water^A, water^B) = \frac{\left|\{water\}\right|}{\left|\{water\}\right| + 0.5\left|\{\ \}\right| + 0.5\left|\{\ \}\right|} = \frac{1}{1} = 1$$

2. Feature similarity is (Equation 15.4):

$$S_u = S_a(water^A, water^B) = \frac{|X|}{|X| + a|Y| + (1-a)|Z|} = \frac{1}{1 + 0.5 \cdot 0 + 0.5 \cdot 0} = \frac{1}{1} = 1$$

3. Semantic neighborhood similarity is (Equations 15.5 and 15.6):

$$S_n(water^A, water^B, 2) = \frac{4.7}{4.7 + 0.50 \cdot 3.3 + 0.50 \cdot 3.3} = \frac{4.7}{8} = 0.59$$

Finally, the semantic similarity between water^A and water^B is defined as follows:

$$S(water^A, water^B) = w_w S_w(water^A, water^B) + w_u S_u(water^A, water^B)$$

$$+ w_n S_n(water^A, water^B)$$

$$= 0.33 \cdot 1 + 0.33 \cdot 1 + 0.33 \cdot 0.59 = 0.85$$

This value means that water^A is 85% similar to water^B.

The result of the concept comparison process using a similarity measure is similarity values between the compared concepts. These similarity values need to be further evaluated to be useful in an integration process. For the revelation of semantic similarity and the visualization of the comparison result, a spatialization technique, as those introduced in Chapter 13, may be utilized. These techniques uncover semantic similarity between concepts by forming semantic clusters.

16.2.3 SCENARIO 3: TRUE INTEGRATION

The third scenario involves all three integration processes. It aims at the semantic integration of two terminological ontologies with minimum human interaction in order to ensure maximum objectivity. The first process analyzes geographic concept definitions and extracts immanent semantic information, which is further used by the second process to identify similarities and resolve heterogeneities between original concepts. The third process generates the final integrated ontology.

16.2.3.1 Semantic Information Extraction

The first process is performed by a tool developed by the ONTOGEO group called GeoNLP (Kokla and Kavouras, 2005; Kokla, 2005; Mourafetis, 2005). This tool aims at the automatic extraction of semantic elements from geographic concept definitions. The tool is based on a methodology for analyzing definitions and extracting immanent semantic information in the form of semantic elements (e.g., LOCATION, PURPOSE, IS-PART-OF, etc.) introduced by Jensen and Binot (1987), and further pursued by Vanderwende (1995) and Barriere (1997). The tool identifies the main semantic properties and relations recorded in geographic concept definitions as outlined in Chapter 12. This approach is based on:

- Definition parsing (syntactic analysis)
- Application of rules that locate certain syntactic and lexical patterns (or defining formulas) in definitions.

Parsing determines the structure of a definition, that is, the form, function, and syntactical relationships of each part of speech. An appropriate tool called parser performs syntactic analysis. The result is usually presented as a parse tree. GeoNLP performs parsing using DIMAP-4 (CL Research, 2001), a software for creating and maintaining dictionaries for use in natural language and language technology applications. DIMAP-4 provides functionality for parsing free text and definitions, and for identifying basic semantic information, especially IS-A relations, and generates a parse tree.

The parsing result is subsequently used by a set of heuristic rules. These rules examine the existence of syntactic and lexical patterns, that is, words and phrases in definitions systematically used to express specific semantic information (Chapter 12). For example, the PURPOSE semantic property is determined by specific phrases containing the preposition "for" (e.g., for (the) purpose(s) of, for, used for, intended for) followed by a noun phrase, present participle, or infinitival clause. With some slight but necessary modifications, the rule for extracting this semantic property from definitions is the following (Vanderwende, 1995):

> If the verb used (intended, etc.) is post-modified by a prepositional phrase with the preposition "for," then there is a PURPOSE semantic property with the head(s) of that prepositional phrase as the value.

The PROPERTY-DEFINED LOCATION semantic property implies that an action or activity (provided by the value of the semantic property) takes place in the defined concept. With some slight but necessary modifications, the rule for extract-ing the semantic property PROPERTY-DEFINED LOCATION is the following (Dolan, Vanderwende, and Richardson, 1993):

> If the genus term is in the set {place, area, space, ...} and there is a relative clause and the relativizer is in the set {where, in which, on which}, then there is a LOCATION relation between the headword and the verb of the relative clause (along with any of its arguments).

The HAS-PART semantic relation is determined by phrases such as "consist of," "comprised of," "composed of," and "made of." The rule to extract this semantic relation is formulated as following:

> If the verb consist (comprise, compose, etc.) is post-modified by a prepositional phrase with the preposition "of," then there is a HAS-PART semantic relation with the head(s) of that prepositional phrase as the value.

GeoNLP utilizes the output of DIMAP-4 to identify patterns and extract seman-tic elements from definitions. Every semantic element is identified with a specifically developed rule. GeoNLP uses its own programming language and enables the user to specify new rules or to change the existing ones. This means that new rules may be generated for dealing with domain or application ontologies with specific semantic properties and relations. GeoNLP offers two possibilities. The first one is to load a whole ontology and generate an XML file that contains the values of the semantic elements of each ontology concept. The second possibility is to analyze single defini-tions in order to identify specific semantic elements. Figure 16.13, Figure 16.14, and Figure 16.15 show the identification of the semantic relation IS-A for the concepts stream, river, and lake with GeoNLP. The definition of each concept is given at the upper-left window. The right window shows the output of the parsing process using DIMAP-4. The symbols used in the parse trees are illustrated in Table 16.7. The middle-left window shows the rule for extracting a specific semantic element, whereas the bottom-left window shows the value of the semantic element.

Table 16.8 shows the set of semantic elements and values identified for the con-cepts of Ontology A and B. Thus, each geographic concept definition is replaced with a set of semantic elements and their values. However, in order to perform concept comparison, it is necessary to process the output of Table 16.8 and determine syn-onyms and hypernyms for concept terms and values. Reference ontologies, diction-aries, or thesauri may provide this information; however, human intervention may also be necessary at this phase. For the purpose of our running example, WordNet and *Merriam-Webster* online were used. For example, value "man-made" is replaced by the synonymous value "artificial" (Table 16.9). Methodologies for interpreting compound nouns are also valuable at this stage. For example, WordNet defines the term "waterway" as "a navigable body of water." Therefore, this noun compound can be decomposed into two values:

- "way" for the IS-A semantic relation and
- "water" for the MATERIAL-COVER semantic property.

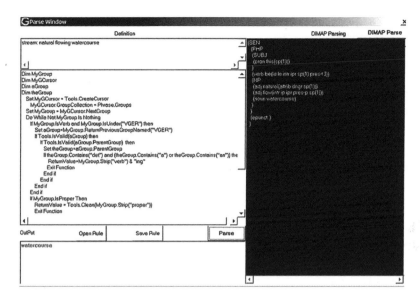

FIGURE 16.13   Identification of the IS-A relation for the concept "stream."

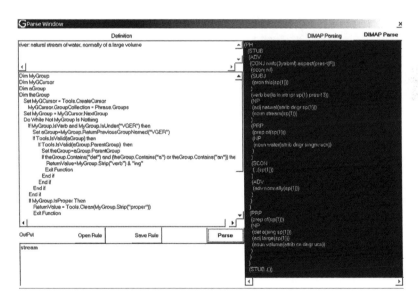

FIGURE 16.14   Identification of the IS-A relation for the concept "river."

16.2.3.2   Concept Comparison

Concept comparison consists in the identification of similarities and heterogeneities between similar concepts. This process relies on available elements, which describe concepts' semantics, such as terms and definitions. According to the previous section, definitions can be further analyzed into semantic elements and values. Therefore, if

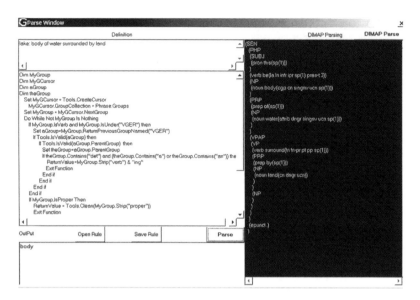

FIGURE 16.15   Identification of the IS-A relation for the concept "lake."

we assume that a concept definition is analyzed into a set of semantic elements and their corresponding values, then a concept C_i is represented by the triple $<T_{Ci}, E_{Ci}, V_{eiCi}>$, where T_{Ci} is the term, E_{Ci} the set of semantic elements, and V_{eiCi} the set of corresponding values, that is:

$$E_{C_i} = \left\{ e_{1C_i}, \ e_{2C_i}, \ \ldots, \ e_{nC_i} \right\},$$

$$V_{e_iC_i} = \left\{ v_{e_1C_i}, \ v_{e_2C_i}, \ \ldots, \ v_{e_nC_i} \right\}.$$

Different combinations of T_{Ci}, E_{Ci}, and V_{eiCi} lead to four possible comparison cases (expressing the degree of equivalence) between two concepts:

- Equivalence, when the concepts are identical in meaning.
- Difference (nonequivalence), when the concepts have different meanings.
- Subsumption (partial equivalence), when one concept has broader meaning than the other.
- Overlap (inexact equivalence), when concepts have similar but not precisely identical meanings.

In order to cover all possible cases, we assume that terms may be either the same (or synonymous) ($T_{C1} = T_{C2}$) or different ($T_{C1} \neq T_{C2}$). Two sets of semantic elements may be:

- Equal ($E_{C1} = E_{C2}$).
- Different ($E_{C1} \neq E_{C2}$).

TABLE 16.7
Symbols Used by DIMAP-4

Symbol	Explanation
ADJP	Adjective phrase
ADV	Adverb (adv) or interjection
CONJ	Conjunctive phrase, which may consist of any other type of phrase
NP	Noun phrase
PHN	Phrases after "as," "than," the complementizer of a relative clause, a subordinating conjunction introducing full subordinate clauses
PRP	Prepositional phrase
SCON	Conjunction phrase
SUBC	Subordinate clause
SUBJ	A NP or noun-equivalent that has been relabeled as the subject of a sentence or clause
VP	Verb phrase
VPAP	Verb phrase in which the verb is a past participle
WTHAT	"that" appositive clause
adj	Adjective
adv	Adverb
det	Determiner
fsuc	Subordinating conjunction introducing full subordinate clauses
noun	Noun
oconj	Ordinary conjunctions: and, but, nor, not, or
prep	Preposition
pron	Pronoun
v-be	Auxiliary verb be
verb	Verb

- The one a subset of the other ($E_{C1} \supset\subset E_{C2}$), that is, one concept has more semantic elements than the other.
- Overlapping ($E_{C1} \oplus E_{C2}$), that is, concepts have some semantic elements in common.

Correspondingly, two sets of values may be:

- Equal ($V_{eC1} = V_{eC2}$), when concepts have the same values for all semantic elements.
- Different ($V_{eC1} \neq V_{eC2}$), when concepts have different values for all common semantic elements.
- The one a subset of the other ($V_{eC1} \supset\subset V_{eC2}$), when the value of at least one semantic element of one concept is more general than that of the other.
- Overlapping ($V_{eC1} \oplus V_{eC2}$), when concepts have overlapping values for a semantic element, for example, V_{eC1} = "irrigation or transportation" and

TABLE 16.8
Example of Semantic Elements and Values for Geographic Concepts

	ORIGINAL CATEGORIES	IS-A	PART-OF	COVER	PURPOSE	NATURE	SIZE	SURROUND-NESS	LOCATION	FLOW
Ontology A	WATER	area		water						
	stream	body		fresh water		natural				flowing
	river	stream		water		natural	large volume			
	lake	body		water				land		
	canal	waterway			boats or irrigation	artificial				
Ontology B	WATER		the Earth's surface	water						
	watercourse	channel				natural or artificial				flows
	stream	watercourse				natural				flowing
	canal	waterway			transportation	man-made or improved natural				
	waterbody	body		water						
	lake	body		water				land		

SEMANTIC ELEMENTS

TABLE 16.9
Value Processing

	ORIGINAL CATEGORIES	IS-A	PART-OF	COVER	PURPOSE	NATURE	SIZE	SURROUND-NESS	LOCATION	FLOW
						SEMANTIC ELEMENTS				
Ontology A	WATER	area		water						
	stream	body		fresh water		natural				flowing
	river	stream		water		natural				
	lake	body		water			large	land		
	canal	way		water	transportation or irrigation	artificial				
Ontology B	WATER		the Earth's surface	water						
	watercourse	body		water		natural or artificial				flowing
	stream	body		water		natural				flowing
	canal	way		water	transportation	artificial or improved natural				
	waterbody	body		water					inland	
	lake	body		water				land		

V_{eC2} = "irrigation or drainage." Overlapping sets of values may also occur when $V_{e1C1} \supset V_{e1C2}$ and $V_{e2C1} \subset V_{e2C2}$, that is, the value of the semantic element e_1 of concept C_1 (V_{e1C1}) is more general than the corresponding value of concept C_2 (V_{e1C2}), whereas the value of another semantic element e_2 of concept C_1 (V_{e2C1}) is more specific than the corresponding value of concept C_2 (V_{e2C2}).

Semantic elements and values are not independent. Semantic elements provide the basis for comparing values. Obviously, values are compared only in terms of the same semantic elements. In case of different semantic elements (clear case of different concepts), values are not further examined.

Although many combinations between terms, semantic elements, and corresponding values may technically occur, in practice, comparison is meaningful mainly for semantically similar concepts, that is, concepts with the same or synonymous terms and concepts with common sets of semantic elements and values.

Table 16.10 includes indicative, meaningful combinations between T_{Ci}, E_{Ci}, V_{eiCi}, the comparison result, and the action required to resolve the case. Examples of some comparison cases between concepts of ontologies A and B are given in Table 16.11. This approach can also prove to be useful in cases where terms are neither equal nor synonymous but appear to present some similarity in certain semantic elements and their corresponding values. Some of these cases are straightforward and can be easily resolved. For example, concepts with $T_{C1} \neq T_{C2}$, $E_{C1} = E_{C2}$ result in:

- An overlap $C_1 \oplus C_2$ when $V_{eC1} \oplus V_{eC2}$
- A subsumption $C_1 \supset\subset C_2$ when $V_{eC1} \supset\subset V_{eC2}$

However, some other cases that involve different terms, overlapping sets of semantic elements, and overlapping values are complicated and require more detailed analysis and possibly an expert's involvement.

Some combinations have two possible comparison results. These also require further investigation and an expert's involvement in order to discover which is the right result for the specific concepts. For example, the combination $T_{C1} = T_{C2}$, $E_{C1} > E_{C2}$, $V_{eC1} = V_{eC2}$ can correspond to two cases. In the first case, concept C_1 is defined by more semantic elements and therefore, is more specific than concept C_2. In the second case, the additional semantic elements of C_1 do not necessarily determine its semantics but are rather context-specific, that is, they relate more to the context and scope of the ontology. Domain ontologies usually include specialized knowledge, which is not contained in general purpose ontologies. CORINE Land Cover is an example of this case since definitions describe the way concepts are identified from satellite images.

Table 16.10 gives an indication of the action required to resolve different comparison cases. These cases are resolved in the reconciliation process, whose purpose is to remove heterogeneities between concepts, in order to properly accommodate them in the integrated ontology. Each comparison case is dealt with differently. The first three (equivalence, difference, and subsumption) are easily resolved. In case of equivalence between two concepts, a direct correspondence (equality) between them

TABLE 16.10

Comparison Cases of T_{Ci}, E_{Ci}, V_{eiCi}

No.	Terms	Semantic Elements	Values	Comparison Results	Resolution Action
1	$T_{C1} = T_{C2}$	$E_{C1} = E_{C2}$	$V_{eC1} = V_{eC2}$	Equivalence	$C_1 = C_2$
2	$T_{C1} = T_{C2}$	$E_{C1} = E_{C2}$	$V_{eC1} \neq V_{eC2}$	Homonymy-difference	$C_1 \neq C_2$
3	$T_{C1} = T_{C2}$	$E_{C1} = E_{C2}$	$V_{eC1} \odot V_{eC2}$	Overlap	$C_1 \odot C_2$
4	$T_{C1} = T_{C2}$	$E_{C1} = E_{C2}$	$V_{eC1} > V_{eC2}$	Subsumption	$C_1 > C_2$
5	$T_{C1} = T_{C2}$	$E_{C1} = E_{C2}$	$V_{eC1} < V_{eC2}$	Subsumption	$C_1 < C_2$
6	$T_{C1} = T_{C2}$	$E_{C1} \neq E_{C2}$	-	Homonymy-difference	$C_1 \neq C_2$
7	$T_{C1} = T_{C2}$	$E_{C1} < E_{C2}$	$V_{eC1} = V_{eC2}$	More detailed definition or subsumption	$C_1 = C_2$ or $C_1 > C_2$
8	$T_{C1} = T_{C2}$	$E_{C1} < E_{C2}$	$V_{eC1} \neq V_{eC2}$	Homonymy-difference	$C_1 \neq C_2$
9	$T_{C1} = T_{C2}$	$E_{C1} < E_{C2}$	$V_{eC1} \odot V_{eC2}$	Overlap	$C_1 \odot C_2$
10	$T_{C1} = T_{C2}$	$E_{C1} < E_{C2}$	$V_{eC1} > V_{eC2}$	Subsumption	$C_1 > C_2$
11	$T_{C1} = T_{C2}$	$E_{C1} < E_{C2}$	$V_{eC1} < V_{eC2}$	Subsumption or overlap	$C_1 < C_2$ or $C_1 \odot C_2$
12	$T_{C1} = T_{C2}$	$E_{C1} > E_{C2}$	$V_{eC1} = V_{eC2}$	More detailed definition or subsumption	$C_1 = C_2$ or $C_1 < C_2$
13	$T_{C1} = T_{C2}$	$E_{C1} > E_{C2}$	$V_{eC1} \neq V_{eC2}$	Homonymy-difference	$C_1 \neq C_2$
14	$T_{C1} = T_{C2}$	$E_{C1} > E_{C2}$	$V_{eC1} \odot V_{eC2}$	Overlap	$C_1 \odot C_2$
15	$T_{C1} = T_{C2}$	$E_{C1} > E_{C2}$	$V_{eC1} > V_{eC2}$	Subsumption or overlap	$C_1 > C_2$ or $C_1 \odot C_2$
16	$T_{C1} = T_{C2}$	$E_{C1} > E_{C2}$	$V_{eC1} < V_{eC2}$	Subsumption	$C_1 < C_2$
17	$T_{C1} = T_{C2}$	$E_{C1} \odot E_{C2}$	$V_{eC1} = V_{eC2}$	Equivalence or overlap	$C_1 = C_2$ or $C_1 \odot C_2$
18	$T_{C1} = T_{C2}$	$E_{C1} \odot E_{C2}$	$V_{eC1} \neq V_{eC2}$	Homonymy-difference	$C_1 \neq C_2$
19	$T_{C1} = T_{C2}$	$E_{C1} \odot E_{C2}$	$V_{eC1} \odot V_{eC2}$	Overlap	$C_1 \odot C_2$
20	$T_{C1} = T_{C2}$	$E_{C1} \odot E_{C2}$	$V_{eC1} > V_{eC2}$	Subsumption or overlap	$C_1 > C_2$ or $C_1 \odot C_2$
21	$T_{C1} = T_{C2}$	$E_{C1} \odot E_{C2}$	$V_{eC1} < V_{eC2}$	Subsumption or overlap	$C_1 < C_2$ or $C_1 \odot C_2$

is specified, and they appear as one concept in the integrated ontology. In the opposite case, that is, when the concepts are different, no correspondence is specified, and the integrated ontology includes both concepts. In case a concept is more general than another, a subsumption (IS-A) relation is defined in the integrated ontology. The fourth case is the most difficult to resolve. In this case, it is necessary to split the common from the different parts of overlapping concepts.

Reconciliation is implemented by a conceptual analysis procedure known as Semantic Factoring. Semantic Factoring (Section 8.4) decomposes original concepts into a set of nonredundant, nonoverlapping conceptual building blocks (Sowa, 2000). These building blocks constitute concepts themselves and are called semantic factors. The procedure is based on the comparison results of the previous process.

Semantic Factoring proceeds bottom-up from specific to general concepts. At this point, it is necessary to rely on a general reference ontology, which will provide

TABLE 16.11
Examples of Some Comparison Cases

Ontology A	Ontology B

Equivalence ($C_1 = C_2$)

T_{C1}: canal	T_{C2}: canal
IS-A: stream	IS-A: stream
PURPOSE: irrigation and transportation	PURPOSE: irrigation and transportation
NATURE: artificial	NATURE: artificial

Difference ($C_1 \neq C_2$)

T_{C1}: canal	T_{C2}: canal
IS-A: stream	IS-A: path
PURPOSE: irrigation	PURPOSE: transportation
NATURE: artificial	NATURE: natural

Overlap ($C_1 \oplus C_2$)

T_{C1}: canal	T_{C2}: canal
IS-A: stream	IS-A: stream
PURPOSE: irrigation and transportation	PURPOSE: irrigation and drainage
NATURE: artificial	NATURE: artificial

Subsumption ($C_1 > C_2$)

T_{C1}: canal	T_{C2}: canal
IS-A: stream	IS-A: stream
PURPOSE: irrigation and transportation	PURPOSE: irrigation
NATURE: artificial	NATURE: artificial

More Detailed Definition ($C_1 = C_2$) or Subsumption ($C_1 > C_2$)

T_{C1}: canal	T_{C2}: canal
IS-A: stream	IS-A: stream
PURPOSE: irrigation and transportation	PURPOSE: irrigation and transportation
—	NATURE: artificial

Subsumption ($C_1 > C_2$)

T_{C1}: canal	T_{C2}: canal
IS-A: stream	IS-A: stream
PURPOSE: irrigation and transportation	PURPOSE: irrigation
—	NATURE: artificial

Subsumption ($C_1 < C_2$) or Overlap ($C_1 \oplus C_2$)

T_{C1}: canal	T_{C2}: canal
IS-A: stream	IS-A: stream
PURPOSE: irrigation	PURPOSE: irrigation and transportation
—	NATURE: artificial

TABLE 16.11 (continued)
Examples of Some Comparison Cases

Ontology A	Ontology B
	Overlap ($C_1 \oplus C_2$)
T_{C1}: canal	T_{C2}: canal
IS-A: stream	IS-A: stream
PURPOSE: irrigation and transportation	PURPOSE: irrigation and drainage
NATURE: artificial	AGENT: humans

the most specific concepts to initiate the comparison. For the purpose of the running example, WordNet is used as reference ontology.

Each concept with no equivalence (partial or exact) is assigned a semantic factor. "River" (Ontology A) is an example of a concept that has neither partially nor exactly equivalent concepts. Then, exactly equivalent concepts are also assigned a semantic factor. For example, according to Table 16.12, concept "lake" is equivalently defined between the two sources. Therefore, these two equivalent concepts correspond to one semantic factor, namely g_3 (Table 16.13).

Inexact equivalences are resolved with decomposition of overlapping concepts into three semantic factors: one corresponds to the common part and the other two to the different parts. For example, according to Table 16.14, concepts "canal" as defined by Ontology A and Ontology B are not exactly equivalent but overlap. More specifically, the overlap occurs because of the values of semantic properties PURPOSE and NATURE. The values of these semantic elements will determine the resultant semantic factors. Indeed, three semantic factors (g_4, g_5, and g_6) are defined: g_4 is an artificial transportation canal, g_5 is an artificial irrigation canal, and g_6 is an improved natural transportation canal. Therefore, two originally overlapping concepts are decomposed into three nonoverlapping semantic factors.

Once Semantic Factoring of one-factor concepts is completed, the process proceeds with more generic concepts, which consist of more than one semantic factor. At this phase, in order to properly assign semantic factors to more general concepts, it is necessary to know intra-ontology relations, that is, relations between concepts of the same ontology resultant from the original hierarchy. For example, concept "river" is a subclass of "stream" (Ontology A). Therefore, "stream" consists of semantic factors g_1 and g_2, which correspond to those subconcepts.

TABLE 16.12
Semantic Elements and Values
of Concept "Lake"

Equivalence	IS-A	Cover	Surrounded By
lake (Ontology B)	body	water	land
lake (Ontology B)	body	water	land

TABLE 16.13

Semantic Factoring

	ORIGINAL CATEGORIES	SEMANTIC FACTORS						
		g_1	g_2	g_3	g_4	g_5	g_6	g_7
Ontology A	**WATER**	x	x	x	x	x		
	stream	x	x					
	river	x						
	lake			x				
	canal				x	x		
Ontology B	**WATER**	x	x	x	x		x	x
	watercourse	x	x		x		x	
	stream	x	x					
	canal				x		x	
	waterbody			x				x
	lake			x				

TABLE 16.14

Semantic Elements and Values of Concepts "Canal"

Overlap	IS-A	Cover	Purpose	Nature
Canal (Ontology A)	way	water	transportation or irrigation	artificial
Canal (Ontology B)	way	water	transportation	artificial or improved natural

16.2.3.3   Integration

The third process consists in building the integrated ontology from the semantic factors. The proposed methodology is based on *Formal Concept Analysis* (Wille 1992; Ganter and Wille, 1999). Formal Concept Analysis (FCA), as explained in Chapter 8, is a theory for the formal representation of conceptual knowledge. It models a specific context, namely formal context, as consisting of a set of objects and a set of attributes and the binary relation between them. In our case, the formal context is given by the sets of semantic factors and their corresponding semantic properties and relations. Therefore, in order to apply FCA, it is necessary to perform a preliminary analysis of semantic factors and their properties-relations. This relies on an approach described in Schmitt and Saake (1997). First, Table 16.15 is derived from Table 16.9. Its rows correspond to the original concepts and its columns to the values of semantic properties-relations. A cross (x) in Table 16.15 indicates that an original concept has a specific value for a semantic property-relation. For example, "transportation" (m_8) is the value of property PURPOSE for concept "canal" from Ontology B.

TABLE 16.15

Transformation of the Many-Valued Context to a One-Valued Context

ORIGINAL CATEGORIES	IS-A body m_1	IS-A way m_2	IS-A stream m_3	IS-A area m_4	PART-OF the earth's surface m_5	COVER water m_6	COVER fresh water m_7	PURPOSE transportation m_8	PURPOSE transportation or irrigation m_9	NATURE natural m_{10}	NATURE artificial m_{11}	NATURE natural or artificial m_{12}	NATURE artificial or improved natural m_{13}	SIZE large m_{14}	SURROUND-NESS land m_{15}	LOCATION inland m_{16}	FLOW flowing m_{17}
WATER (Ontology A)	x			x		x											
stream						x	x			x							x
river			x			x				x				x			
lake	x					x									x		
canal		x				x			x		x						
WATER (Ontology B)	x				x	x											
watercourse	x					x											x
stream						x				x		x					x
canal		x				x		x					x				
waterbody	x					x										x	
lake	x					x									x		

TABLE 16.16
Cross-Table of the Integrated Context

	m_1	m_2	m_3	m_4	m_5	m_6	m_7	m_8	m_9	m_{10}	m_{11}	m_{12}	m_{13}	m_{14}	m_{15}	m_{16}	m_{17}
g_1	X		X	X	X	X	X			X		X		X			X
g_2	X			X	X	X	X			X		X					X
g_3	X			X	X	X									X	X	
g_4	X	X		X	X	X		X	X		X	X	X				
g_5		X		X		X			X		X						
g_6	X	X		X	X		X					X	X				
g_7	X			X	X											X	

This table is subsequently combined with Table 16.13 in order to form a single table (16.16) whose rows correspond to semantic factors and its columns to values of semantic properties-relations. A cross (x) in Table 16.16 indicates that the specific value describes the corresponding semantic factor. Semantic factors inherit the values of properties-relations of all the original concepts to which they belong. Table 16.16 constitutes the cross-table of the integrated context and is the necessary prerequisite to apply FCA. Semantic factors, semantic elements and values will be subsequently used to derive the formal concepts of the integrated context.

According to Chapter 8, in order to derive all concepts of a given small context, it is preferable to use

$$A'_t = \bigcap_{t \in T} \{g\}' \tag{8.5}$$

or

$$B'_t = \bigcap_{t \in T} \{m\}' \tag{8.5'}$$

and then to form (A'', A') or (B', B'') correspondingly (Wille, 1992). However, prior to forming (8.5) or (8.5'), the list of concept intents or concept extents should be drawn correspondingly. The object intents of the objects $g_1, \ldots, g_7$ of the context are formulated as follows:

$$g'_1 = \{m_1, m_3, m_4, m_5, m_6, m_7, m_{10}, m_{12}, m_{14}, m_{17}\}$$

$$g'_2 = \{m_1, m_4, m_5, m_6, m_7, m_{10}, m_{12}, m_{17}\}$$

$$g'_3 = \{m_1, m_4, m_5, m_6, m_{15}, m_{16}\}$$

$$g'_4 = \{m_1, m_2, m_4, m_5, m_6, m_8, m_9, m_{11}, m_{12}, m_{13}\}$$

$$g_5' = \{m_2, m_4, m_6, m_9, m_{11}\}$$

$$g_6' = \{m_1, m_2, m_5, m_6, m_8, m_{12}, m_{13}\}$$

$$g_7' = \{m_1, m_5, m_6, m_{16}\}$$

Then, for each object $g \in G$ we take the following substeps (Ganter and Wille, 1999). The objects are processed in an arbitrary order.

Substep 4.1. The intent M is entered into the list.

Substep 4.m. For each set A' entered into the list in an earlier step, we form the set:

$$A' \cap g'$$

and include it in the list, provided that it is not already contained within it. At the end, the list contains those sets that are the intersections of object intents

$$A_t' = \bigcap_{t \in T} \{g\}' \tag{8.5}$$

which constitute the concept intents (Table 16.17). Then, by using Table 16.16, the concept extent A'' of each concept intent A' is derived. Thus, we obtain the list of all concepts (A'', A') of the context.

TABLE 16.17
Derivation of Concept Intents A'

Object Intent g'	Concept Intents A'
1	$\{m_1, \ldots, m_{17}\}$
2 $\quad g_1' = \{m_1, m_3, m_4, m_5, m_6, m_7, m_{10}, m_{12}, m_{14}, m_{17}\}$	$\{m_1, m_3, m_4, m_5, m_6, m_7, m_{10}, m_{12}, m_{14}, m_{17}\}$
3 $\quad g_2' = \{m_1, m_4, m_5, m_6, m_7, m_{10}, m_{12}, m_{17}\}$	$\{m_1, m_4, m_5, m_6, m_7, m_{10}, m_{12}, m_{17}\}$
4 $\quad g_3' = \{m_1, m_4, m_5, m_6, m_{15}, m_{16}\}$	$\{m_1, m_4, m_5, m_6, m_{15}, m_{16}\}$
	$\{m_1, m_4, m_5, m_6\}$
5 $\quad g_4' = \{m_1, m_2, m_4, m_5, m_6, m_8, m_9, m_{11}, m_{12}, m_{13}\}$	$\{m_1, m_2, m_4, m_5, m_6, m_8, m_9, m_{11}, m_{12}, m_{13}\}$
	$\{m_1, m_4, m_5, m_6, m_{12}\}$
6 $\quad g_5' = \{m_2, m_4, m_6, m_9, m_{11}\}$	$\{m_2, m_4, m_6, m_9, m_{11}\}$
	$\{m_4, m_6\}$
	$\{m_1, m_2, m_5, m_6, m_8, m_{12}, m_{13}\}$
	$\{m_1, m_5, m_6, m_{12}\}$
7 $\quad g_6' = \{m_1, m_2, m_5, m_6, m_8, m_{12}, m_{13}\}$	$\{m_1, m_5, m_6\}$
	$\{m_2, m_6\}$
	$\{m_6\}$
8 $\quad g_7' = \{m_1, m_5, m_6, m_{16}\}$	$\{m_1, m_5, m_6, m_{16}\}$

TABLE 16.18

Concepts of the Running Example

Formal Concept	Term
$C_1 = ((,\{m_1, ..., m_{17}\})$	least concept
$C_2 = (\{g_1\}, \{m_1, m_3, m_4, m_5, m_6, m_7, m_{10}, m_{12}, m_{14}, m_{17}\})$	river
$C_3 = (\{g_1, g_2\}, \{m_1, m_4, m_5, m_6, m_7, m_{10}, m_{12}, m_{17}\})$	stream
$C_4 = (\{g_3\}, \{m_1, m_4, m_5, m_6, m_{15}, m_{16}\})$	lake
$C_5 = (\{g_1, g_2, g_3, g_4\}, \{m_1, m_4, m_5, m_6\})$	—
$C_6 = (\{g_4\}, \{m_1, m_2, m_4, m_5, m_6, m_8, m_9, m_{11}, m_{12}, m_{13}\})$	artificial transportation canal
$C_7 = (\{g_1, g_2, g_4\}, \{m_1, m_4, m_5, m_6, m_{12}\})$	—
$C_8 = (\{g_4, g_5\}, \{m_2, m_4, m_6, m_9, m_{11}\})$	canal (Ontology A)
$C_9 = (\{g_1, g_2, g_3, g_4, g_5\}, \{m_4, m_6\})$	water (Ontology A)
$C_{10} = (\{g_4, g_6\}, \{m_1, m_2, m_5, m_6, m_8, m_{12}, m_{13}\})$	canal (Ontology B)
$C_{11} = (\{g_1, g_2, g_4, g_6\}, \{m_1, m_5, m_6, m_{12}\})$	watercourse
$C_{12} = (\{g_1, g_2, g_3, g_4, g_6, g_7\}, \{m_1, m_5, m_6\})$	water (Ontology B)
$C_{13} = (\{g_4, g_5, g_6\}, \{m_2, m_6\})$	waterway
$C_{14} = (\{g_1, g_2, g_3, g_4, g_5, g_6, g_7\}, \{m_6\})$	largest concept (WATER)
$C_{15} = (\{g_3, g_7\}, \{m_1, m_5, m_6, m_{16}\})$	water body

The context of the running example results in the following concepts (Table 16.18). C_1, C_5, C_6, C_7, C_{13}, and C_{14} are new concepts generated by the algorithm either as the union or as the intersection of original concepts due to their combinations of objects and attributes. Some of them correspond to meaningful concepts and are included in the final concept lattice; others are neither meaningful nor useful and therefore are removed. For example, C_6 refers to the intersection of the overlapping concepts "canal" (Ontology A) and "canal" (Ontology B); it corresponds to the concept "artificial transportation canal" and is preserved in the final concept lattice. C_{13} is also preserved because it resulted from the union of the concepts "canal" (Ontology A) and "canal" (Ontology B) and refers to the meaningful concept "waterway." On the contrary, C_5 and C_7 do not correspond to any useful new concepts and therefore are removed to simplify the resulting concept lattice. The least concept C_1 does not correspond to a meaningful concept. It may only be included for the sake of completeness to form the bottom of the concept lattice. If it is removed, the lattice becomes a semilattice. The largest concept C_{14} is preserved to form the root of the resulting lattice. The generated concepts form the concept lattice shown in Figure 16.16. The figure shows whether the final concepts originate from Ontology A or B, are common to both, or are generated by the algorithm.

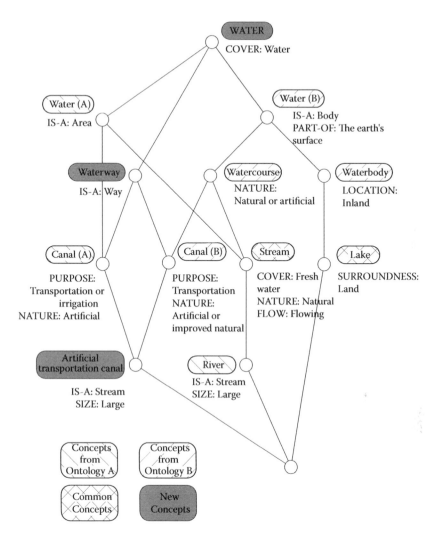

FIGURE 16.16 Concept lattice of the integrated ontology.

REFERENCES

Barriere, C. 1997. From a children's first dictionary to a lexical knowledge base of conceptual graphs. Ph.D. thesis, School of Computing Science, Simon Fraser University, British Columbia, Canada.

CL Research. 2001. DIMAP-4, Dictionary Maintenance Programs. http://www.clres.com (accessed 2 May 2007).

Dolan, W. B., L. Vanderwende, and S. D. Richardson. 1993. Automatically deriving structured knowledge base from on-line dictionaries. In *Proc. Pacific Association for Computational Linguistics*, Vancouver, British Columbia.

Ganter, B. and R. Wille. 1999. *Formal concept analysis, mathematical foundations.* Berlin: Springer-Verlag.

Gennari, J., M. A. Musen, R. W. Fergerson, W. E. Grosso, M. H. Crubezy, H. Eriksson, N. F. Noy, and S. W. Tu. 2003. The evolution of Protégé: An environment for knowledge-based systems development, *International Journal of Human-Computer Studies* 58 (1): 89–123.

Jensen, K. and J. L. Binot. 1987. Disambiguating prepositional phrase attachments by using on-line dictionary definitions. *Computational Linguistics* 13 (3-4): 251–260.

Kokla, M. 2005. Semantic interoperability in geographic information science. Ph.D. thesis, National Technical University of Athens, Athens, Greece (in Greek).

Kokla, M. and M. Kavouras. 2005. Semantic information in geo-ontologies: Extraction, comparison, and reconciliation. *Journal on Data Semantics*, Lecture Notes in Computer Science, Vol. 3534, 125–142. Berlin: Springer.

Mourafetis, G. 2005. Automated extraction and comparison of geographic information from definitions. M.Sc. thesis, Geoinformatics Postgraduate Course, National Technical University of Athens, Greece (in Greek).

Rodríguez, A. and M. Egenhofer. 2003. Determining semantic similarity among entity classes from different ontologies. *IEEE Transactions on Knowledge and Data Engineering* 15(2): 442–456.

Rodríguez, A. and M. Egenhofer. 2004. Comparing geospatial entity classes: An asymmetric and context-dependent similarity measure. *International Journal of Geographic Information Science* 18(3): 229–256.

Schmitt, I. and G. Saake. 1997. *Merging inheritance hierarchies for schema integration based on concept lattices.* Technical Report, Faculty of Information, University of Magdeburg.

Sowa, J. F. 2000. *Knowledge representation: Logical, philosophical and computational foundations.* Pacific Grove, CA: Brooks Cole Publishing.

Vanderwende, L. 1995. The analysis of noun sequences using semantic information extracted from on-line dictionaries. Ph.D. thesis, Faculty of the Graduate School of Arts and Sciences, Georgetown University, Washington, D.C.

Wille, R. 1992. Concept lattices and conceptual knowledge systems. *Computers and Mathematics with Applications* 23(6-9): 493–515.

Part 5

Post-Review

17 Epilogue

17.1 CLOSING ISSUES

Having gone through the book chapters, the reader has hopefully gained a good appreciation of the reasons that have led to semantic-aware interoperability approaches and ontology integration in the geospatial domain. This appreciation is subsequently supported by a balanced mixture of: (a) theoretical principles, (b) structures and tools, and (c) practical approaches to ontology integration, enriched with guidelines and examples. This chapter starts with a retrospective (and naturally more knowledgeable) view of the main issues addressed in the book and their effectiveness. It then proceeds with some areas requiring extensive research.

17.2 HINDSIGHT

The early days of geographic information systems were characterized by a general shortage of geographic data. The little existing was collected in the context of specific programs or applications and was used by a limited number of specialists who had (or thought they had) a common understanding of the represented phenomena and their representations. There were numerous technical problems to solve, which may sound simple today but troubled GI scientists in those days. Geometrical and topological algorithmic computations and data structure conversions are well-known examples. Another characteristic of the early GIS developments was the fact that representations and the associated data files pivoted around few geometric primitives (points, lines, polygons, etc.) (Figure 17.1). Semantic information could only be implicitly derived by considering the attached "feature codes" and attributes. It was much later that the modeling paradigm was reversed and centered around concepts and objects that possessed spatial and/or aspatial properties and relations. As a result, research on concepts lagged behind. However, this deficiency was not widely apparent, except to some scientists from narrow application domains, who experienced semantic integration problems and somehow managed them with ad hoc approaches, and to a few vanguard members of the scientific community who simply anticipated the problem and expected it to grow in the future.

Nevertheless, the picture changed gradually with the development of GIS systems and applications on the one hand and the great increase of data acquisition campaigns on the other. The former helped to solve tedious technical problems at the explication level, whereas the latter brought up issues of incompatibility. The common approach, that is, to deal with the issue with standards and feature catalogues, had only limited effectiveness. Semantic integration became, thus, a central issue to interoperability. The different, but not necessarily complementary, views on

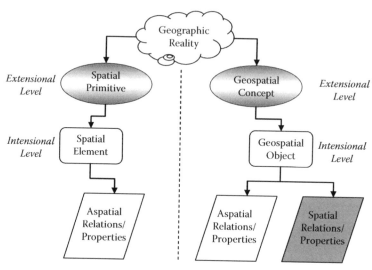

FIGURE 17.1 Two different paradigms in geospatial representation. Spatial primitive-centered (left) and concept-centered (right).

geographic reality created different representations. On the other hand, the number of users presented an unprecedented increase, especially in the Internet era. Data is not anymore used solely by those who collected it or other domain experts. The intended meaning is something difficult to be sure about and the number of data misuse cases increases, especially in the absence of sufficient and proper documentation. This need for a proper elicitation of knowledge has led to ontologies and ontological research.

In the geospatial domain, work on interoperability, ontologies, and semantic integration has shown very few practical results. From all reasons attributing to this fact, three appear to be quite serious (Figure 17.2).

The first reason is the actual complexity of the problem. People are not always sure what they mean when they ask something and the same is true in this case as well. The loose, unaware, or conflicting use of the notions of *integration, knowledge, concepts, semantics, similarity, context,* and *interoperability* blur the objective, and it becomes difficult to design an appropriate approach.

The second reason is the lack of an advanced, well-documented, and widely accepted corpus of geographic knowledge. Of course, there has been a lot of important independent research towards the understanding of geospatial issues and the development of appropriate theories and methodologies. There has not been, however, a necessary synthesis of the research, while many issues remain open to research. In order for GIScience to survive as a real science, establishing such a universal corpus is essential.

The third reason is that traditional GI structures, methodologies, and tools are not suited to accommodate the demanding complexity of geo-knowledge systems. Some of this work has been developed quite recently or is still under development; thus, it is not at all clear to the user which conceptual structure is suited for a particular task, how it can be practically applied, to what degree it is equivalent to other

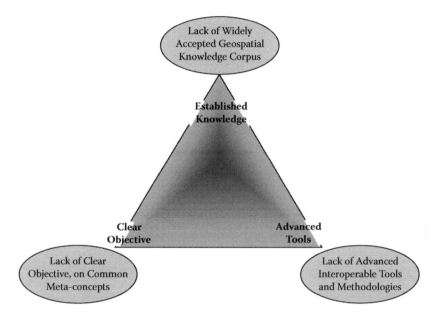

FIGURE 17.2   Major impediments to geospatial semantic integration.

structures, and whether there are ways of converting one into the other. Moreover, it is not widely known what methodology/technique can be employed to derive semantic information from different sources, such as structured or free text.

These reasons, among others, have been thoroughly discussed in the previous chapters. The primary objective of the book was to provide an effective collection of theories, tools, and semantically sensitive approaches to geo-ontology integration. Thus, regarding the well-researched problems, the reader has been supplied with specific solutions and guidelines. As for the more intricate problems, the reader has been armed with the available theoretical background, other related knowledge, and several hints to pursue a future solution.

More specifically, the semantic interoperability problem in the geospatial domain and the justification and expediency of establishing and developing such a research direction were introduced in Part 1 of the book (Chapters 1 to 3). In order, however, to establish theoretically sound integration approaches, it was essential to resort to the general theories that are available. These were presented in Part 2 (Chapters 4 to 6). From the practical point of view, the implementation of useful approaches necessitates the understanding of basic conceptual methodologies, tools, and structures. These were presented in Part 3 (Chapters 7 to 13). Because there is a lot of active research on semantics, ontology, and integration, it is essential to generalize/classify and understand the main aspects of the available approaches using a proper framework. Such an analysis together with examples and guidelines was employed in Part 4 (Chapters 14 to 16).

It is important to note here that the subjects presented in the book, especially in the Chapters of Parts 2 and 3, play a twofold role. First, they are essential to understanding existing integration approaches, which make use of some of them. Second, and perhaps more important, they help ambitious researchers to obtain the necessary background in order to engage in the multitude of remaining unsolved research problems (Section 17.3).

17.3 FORESIGHT

Some of the inherent difficulty of semantic integration in the geospatial domain was attributed to the three reasons presented in the previous section. From these, the first two reflect a weak level of current awareness and appear to be the most persevering. In fact, a lot depends on to what extent the domain knowledge is developed. Once these issues are settled, it is usually a matter of time to work out an appropriate solution to the problem. Given what has been so far presented, in terms of well-described and understood problems and available formal approaches (i.e., what can be done), it is now time to present a number of open research issues. These are grouped into three main categories, similarly to the way they were presented in the book. That is, (a) theoretical issues and knowledge elicitation, (b) formal instruments, and (c) integration approaches. From these, the theoretical issues are harder to solve; some are not even expected to be resolved in the foreseeable future. All these, of course, are considered with the geospatial context in mind. There are of course many theoretical issues related to computability/computational complexity that are suitable for research in other domains. These do not practically or directly contribute to the vital issues of GI interoperability and were not in the scope of the book. Thus, only a short note is made to algorithm design.

17.3.1 Theoretical Issues and Knowledge Elicitation

17.3.1.1 Ontological Research

Although ontological research has spread impressively and quickly across all domains, the difficult issues to tackle lie ahead. Most of what has been accomplished so far is the development of awareness of the necessity for better and more systematic documentation of "deeper" information about domain knowledge. "Deeper" (also called "semantic" by many) is contrasted to other, more primitive documentations of the past. There are not many full geographic ontologies, except some that resulted from ad hoc or standardization efforts. It is expected that in the near future there are going to be several geographic ontologies. Today, geospace presents a notion of wider interest, which surpasses the traditional domain of spatial sciences and engineering. This widespread interest introduces the necessity of using "lighter" ontologies based more on common sense and less on expertise; these light ontologies also develop dynamically. Developments in the area of semantic web and the proposed Geospatial Semantic Web (Egenhofer, 2002) present such requirements. In light of these developments, however, it is realized that light geographic ontologies based on cognition and common sense may prove to be not that "common," exhibiting, thus, greater variability and heterogeneity. Specialized scientific ontologies will always have their purpose and particular scientific audience, also serving as a more objective grounding/reference to reality when diverse views are to be integrated.

Each ontology expresses a way of carving the "reality" of a domain into its representation constituents. In our present context, such important constituents are the geographic concepts. Consequently, it would be of paramount importance if geographic ontologies could be engineered in a way that geographic concepts were essentially described. This requires the determination of an adequate set of principal semantic dimensions where every concept is projected. An appropriate universal set of such dimensions is proposed by the top/upper level ontology initiatives (Chapter 4).

In these initiatives, GIScience, with its extensive work on geospatial aspects, can significantly contribute to the enrichment of the relatively poor properties/relations of geographic concepts. After so many years of thorough research on geospatial information theory, it would be an extremely useful contribution to attempt a synthesis and acceptance of a corpus of geographic knowledge. For example, this would entail the development of ontologies such as those mentioned in Chapter 4, about: (a) spatiotemporal relations/properties and location, (b) place, (c) geographic boundaries, (d) semantic-thematic relations/properties, spatiotemporal entities, location, etc., (e) affordances, (d) natural entities, (e) domain specific concepts, (f) spatial operators, (g) thematic operators, and others.

Working towards this objective of establishing a number of standard geospatial ontologies, one does not only have to face heterogeneities across domains but also across different languages, societies, cultures, legal systems, and others. A simple but illustrative example would be the difficulty to reach a univocal definition of a *cadastral parcel* across the different members of the European Union. Notice that this is not the hardest case given that (a) there is an objective physical reality that provides similar grounding, (b) the cadastre has traditionally involved certain disciplines (which utilize some universal scientific practices), (c) there are commonalities in the legal systems due to historical reasons, and (d) most EU members have linguistically and culturally developed a similar spatial understanding. The difficulty in dealing with such issues in the process of integrating different ontologies should not be regarded as a weakness of ontologies. On the contrary, ontology integration brings to the surface latent problems not previously touched upon and may provide a systematic basis for their resolution.

There are also various theoretical issues touching upon questions that have puzzled philosophers for centuries. Although these are likely to find a positive and final answer in the near future, small practical steps are the only way to approach them. One of these, for example, is the problem of ontological vagueness (see Chapter 7).

17.3.1.2   Concepts

There are various conflicting theories, according to different perspectives, about the way concepts are formed. Geographic concepts in particular constitute a very important class of concepts that has not received the necessary attention. Recent advances in metaphysics, cognitive science, and epistemology can potentially contribute to this objective. Of course, it is unrealistic to expect the development of a universal theory of reality that encompasses all views as partial, consistent, and complementary. Based on the aforementioned theories, there can only be different postulations about geographic reality, which influence the way reality is perceived and understood and the way the associated concepts are represented. This influence, however, has not been sufficiently investigated, understood, and explained in practical terms.

17.3.1.3   Semantics

There are various views on semantics, as thoroughly discussed in the chapters of Parts 1 and 2. Despite the increasing amount of work on geospatial semantics, there has not been a wide agreement or even a prevailing view on what describes the semantics of geoconcepts, which are the principal semantic dimensions one can distinguish, and what is the role of relations, properties, and other signifying elements. As a result, there is no systematic and common way of comparing concepts from different ontologies, due to the different semantics employed. Establishing a general set of semantic relations/properties for geospatial concepts, like the ones discussed and proposed in Parts 2 and 4 of this book, would constitute a great step forward. In order to increase the chances of successfully attaining such a goal, one probably has to adopt one of the major philosophical theories dealing with issues of essential properties/relations and identity.

Similar approaches have been employed in the development of top-level ontologies. An advance in this direction would help develop better semantic reference systems, which in turn would present other important benefits. They could be used to (a) evaluate existing ontologies and possibly augment or detail the available semantics and (b) help to build better (and more standardized) new ontologies.

Despite opposing the "realistic" perspective of this book (see Chapters 1 and 6), an interesting direction is the one associated with *cognitive linguistics–cognitive semantics* (Lakoff, 1987; Smith, 1995; Talmy, 2000; Gärdenfors, 2000) and its application in GIScience (Kuhn, 2003).

Another theoretical semantic issue that has not been sufficiently researched and affects semantic interoperability is the fact that emphasis is primarily put on ontologies, concepts, and instances. Their semantic dimensions (e.g., relations, properties) are not examined in the same depth, as if they were semantically straightforward. If it is proven that this is not the case and associated properties exhibit heterogeneity and strange overlappings, the comparison of concepts can get quite messy.

17.3.2   FORMAL APPROACHES

A number of popular formal approaches and tools were presented in Part 3. In principle, formal approaches, conceptual structures, and tools—similarly to conventional data structures—are independent of the *referent*, and also, of the *thought of reference* (see the meaning triangle in Section 6.2). They may differ in expressivity, ease of use, clarity, availability of tools, or efficiency, but in principle again, they could also be considered equivalent. This is the reason why there are often ways of converting one structure or formalism to another. In practice, however, knowledge representation (KR) formalisms and structures differ significantly because knowledge may differ a lot and presents different needs. The application of such advanced formalisms in GIScience has not been sufficient. In Formal Concept Analysis (FCA), for instance, it seems that all attributes (relations/properties and their values) are treated equivalently, without the possibility of defining importance, a weighted influence, etc. Also, many-valued contexts have not been exhaustively solved, with the danger of neglecting semantic issues over technical ones. Moreover, regarding the adoption of formalisms such as Conceptual Graphs (CG), there has been a lack of easy procedures that would help build "automatically" a representation from a geo-knowledge source with an equivalent outcome.

An active area of research here would be to develop effective frameworks, which would unify different formalisms and structures. In this direction, the Information Flow Framework (IFF) for the standardization activity of Standard Upper Ontology (SUO) (Kent, 2004) attempts to unify Information Flow with Formal Concept Analysis. Also, Wille (1997) has proposed the association of Formal Concept Analysis to Conceptual Graphs.

Formal approaches also have to tackle the difficult issue of determining not fully known but indeterminate, vague, or inexact concepts on the basis of partial information about them. This requires the development of *concept approximation* methodologies such as those employing *rough set theory* (Hoa and Son, 2004) coupled with FCA (Kent, 1996) and similarity measures (Saquer and Deogun, 2001), proposing also ontology mapping (Zhao, Wang, and Halang, 2006).

Another very important research direction is towards the development of advanced processing methods for geospatial knowledge in natural language, as elaborated in the subsequent paragraphs.

Advances in Natural Language Processing: Natural language remains one of the main carriers of geospatial knowledge. Therefore, it is essential that there are effective means of exploiting this carrier; that is, understanding the meaning conveyed, extracting it in a systematic/automatic manner, and representing it in a formal language understood by computers. The importance and role of Natural Language Processing (NLP) techniques have been demonstrated in Chapter 12, particularly with regard to their application to structured text (geoconcept definitions). NLP is currently a very active research field, with many applications. A lot remains to be done, however. The focus here is on how geoknowledge is expressed in natural language so that appropriate techniques are developed to extract it or make it formal. Processing such language on the basis of geospatial aspects could form the basis of "Geo-NLP" techniques.

Here also, one could directly find analogies between natural language expressing geospace and one of the traditional geospatial "languages," that is, the cartographic language. From this point of view, some NLP issues also find their analogy. Text segmentation, for instance, would probably resemble feature extraction since both identify boundaries via parsing. Word sense disambiguation—another NLP technique—attempts to find the right meaning of a word from several possible meanings, based on context. This is also the case with geospatial semantics, which also exhibit polysemy. The map analogy here is the ability to understand what geofeatures represent in reality on the basis of contextual geospatial information. Context is also employed to resolve other situations of ambiguity in natural language such as syntactic ambiguity.

NLP Text Summarization: Another interesting NLP technique is *text summarization* (Mani and Maybury, 1999; Mani, 2001); namely, the automated reduction of a piece of text so that its semantically important parts are preserved. Effective and coherent text summarization is crucial in knowledge-discovery and data mining in order to deal with the problem of *information overflow*, also known as *information pollution* or *signal-to-noise ratio*. The power of summarization is evident everywhere. People have always relied greatly on summaries in order to acquire, comprehend, and convey all kinds of knowledge. It is even more so now with the rapid growth of available data. The classical quote "We see what we see because we miss all of the finer details" by Alfred Korzybski (1994, 376) expresses very lucidly what is also common sense. Abstraction is also the main tenet of cartography. It is very difficult to understand without summarization or abstraction. Freely speaking, a summary plays a similar role to that of categorization. In a broader sense, metadata and context also serve the same cause. In the era of information overflow, forecasters like Paul Saffo (1994) have advocated that context (e.g., point of view) will be the scarce resource, and as such, it will gain importance over content.

In summarization, a distinction is often made between an *extract* and an *abstract* (Mani, 2001). An *extract* is a summary produced entirely and directly from the source text. An abstract is at least partly based on additional outside knowledge. As a result, abstract summaries are considered semantically richer but are methodologically more demanding. On the other hand, it is evident that summarization is mostly context-sensitive, that is, it has to take into account a certain perspective/focus (e.g., that of a user, a task, or a domain). When this is not specified or it is very broad, summaries have a more generic (general purpose) character.

The map analogy to text summarization is map generalization. Summarization, similar to generalization, is a very ambitious, useful but also very complex task. It has become widely accepted (Hahn and Mani, 2000) that better summaries are yielded when the meaning of text is known. In order to be able to make notable progress, it is first essential to identify the semantic elements of natural text and build appropriate formal semantic structures. The approaches reviewed in this book contribute significantly towards richer knowledge approaches. Then, it is essential to develop mechanisms for "pruning" the semantic structure, reducing thus, the text at coarser levels of detail, while maintaining corresponding levels of meaning.

Summarization in the milieu of geospatial knowledge should take into account several points such as: (a) the notion of scale, (b) generalization hierarchies of existing geoconcepts, (c) ways of reducing detailed spatial relations to coarser ones, (d) generalization of thematic properties and relations, (e) associating context with appropriate generalization levels, and (f) spatialization methods which are often used to graphically summarize large amounts of multidimensional information. Notice that in all the above cases external knowledge is required in order to produce advanced geospatial summaries. Namely, these summaries cannot be produced solely from the input information.

NLP Controlled "Natural" Languages: Man-machine communication that matches human reasoning and language is an interesting but complex issue. Such an achievement would have obvious advantages. According to Rich (1983, 295), "Computers will not be able to perform many of the tasks people do every day until they, too, share the ability to use language." The truth is that complete "understanding" of natural language by a machine remains impossible even today due to the complexity of all its aspects: syntactic, semantic, and pragmatic. The term "understanding" stands for the transformation of natural language into another form of knowledge representation that may result in a computer's response, for example, the execution of a computer program.

In order to partially overcome this difficulty, several approaches have been proposed, such as (a) a semiautomatic language processing requiring human intervention or (b) the creation of controlled languages that constitute subsets of the natural ones with vocabulary and syntax constraints. Controlled languages can be directly transformed into any kind of structured logical expression like first-order logic, programming languages, etc. In this direction, Sowa (2004) has proposed the *Common Logic Controlled English* (CLCE). "Anyone who can read ordinary English can read sentences in CLCE with little or no training" (Sowa, 2004). CLCE can be subsequently translated into first-order logic expressions. Another example, the *Attempto Controlled English*, which has been developed as a subset of the English natural language, can be used in the requirements specification and can be translated to executable Prolog programs (Fuchs, Schwertel, and Schwitter, 1999).

In the geospatial domain, people often express and understand spatial relations through natural language instead of metric measurements (Aleman-Meza et al., 2004). When these relations are described through natural language (spoken or written), they seem to be closer to the model humans adopt for spatial reasoning and, therefore, the user can conceptualize, compare, and identify them more easily (Arpinar et al., 2007). On the contrary, when a user submits questions or requests regarding quantitative data (e.g., measures or coordinates) to systems that host geospatial data, spatial relations are not so easily conceptualized. Furthermore, information retrieval from geographical databases commonly entails (and presupposes the knowledge of) a specialized language for communication and query formation (e.g., SQL). Normally, this requires the user to be an experienced programmer. On the contrary, geographical information retrieval from the World Wide Web does not require any familiarization with a specialized language because the search is executed through the usage of keywords. However, this procedure is not very effective because, as a result of such a query, a very large percentage of the Web pages retrieved are not relevant to the information searched. In addition, this retrieval method does not offer any support to deep structures "hidden" in geospatial data, so that the user often misses critical information (Egenhofer, 2002).

To accomplish the retrieval of geospatial information in both a human friendly and an effective way, Kontaxaki and Kavouras (2005) have proposed an approach based on CG and a specific Controlled English Query Language enriched with geospatial concepts and relations in a way that supports geospatial reasoning.

NLP Qualitative Linguistic Terms (QLTs): Definitions constitute an invaluable source of semantics for concepts. NLP approaches were introduced in Chapters 12, 15, and 16 demonstrating how semantic relations/properties can be extracted. In order, however, to make the best use of definitions, it is necessary to have ways of handling supplementary and non-straightforward linguistic items of qualitative nature. Qualitative linguistic terms (QLTs) work as degree modifiers in the semantics of geographic concepts. They denote the degree of possessing a property; therefore, their modeling is necessary in concept disambiguation. Tomai and Kavouras (2005) report an extensive use of QLTs in geospatial concept definitions. Although a lot of NLP research has been directed towards nouns (Vanderwende, 1995) and verbs (Fellbaum, 1990), their modifiers have not been sufficiently treated. According to Tomai and Kavouras (2005), QLTs are often encountered as *qualifiers* such as (a) nouns functioning as adjectives, for example, "school building" or (b) noun-modifying adjectives, for example, "brackish water" or (c) adverbs that modify verbs, adjectives, or other adverbs, for example, "frequently includes," "very large," and "quite often." Less often, but still importantly, QLTs are encountered in definitions as *quantifiers*, for example, "all," "each," "every," "some," "most," "many," and "several."

Methods for modeling quantifiers include the *generalized quantifiers theory* by van der Does (1996), van der Does and van Eijck (1996), and van der Does and de Hoop (2001), as well as work conducted in the field of *computational linguistics* by Clark (2004a, 2004b) for the quantifiers automation and in the field of *semantic automata* by van Benthem (1986). Tomai and Kavouras (2005) have applied the *generalized quantifiers theory* to geospatial definitions adopting a regular tree structure known as the *tree of numbers*, which subsequently translates to strings of binary values. In this way, it becomes possible to determine similarities among quantifiers and assign appropriate numeric values.

Semantic Similarity: As discussed in Chapter 13, a pivotal research issue on semantic integration is the ability to measure and reason about the similarity of geographic concepts/ontologies. Similarity measures are often used as an indication of relevance in information retrieval and web services. Here, one can isolate at least two sets of open issues.

The first set has to do with the actual methodology of measuring similarity. Some critical research questions are:

- What do similarity measures really measure?
- Which semantic elements are supposed to participate in the similarity measure, and how should they be combined? What is the role of spatial properties in the measure?
- How does context affect similarity measures?
- How do we measure similarity in intensional and extensional cases?
- How do we compare two concepts that do not have the same dimensions (e.g., the same semantic properties)? On the intersection or on the union set of properties?
- What is the cost of concluding that two concepts are similar, if they are actually not, and what is the cost of concluding that they are not similar, when in fact they are?
- How does conceptual vagueness affect similarity?
- When does similarity change through time?
- How can we measure "meta similarity," that is, similarity between different similarity measures?

The second set of questions refers to the integration issue. Here are some critical issues:

- What is the practical use of a similarity value in an integration task?
- If a threshold-like decision about concept equivalence is proposed as a solution, how does one reach that decision, and how is a sorites-like case avoided?
- How can ontological similarity be effectively conveyed via visualization? How does one deal with the reduction of many dimensions of a concept to the limited dimensions in the visualization space?

17.3.3 INTEGRATION APPROACHES

The integration framework of Chapter 14 provides a basis for understanding, evaluating, and comparing any proposed integration approach. Based on this, a number of characteristic integration approaches were presented and evaluated (Chapter 15). This evaluation has presented a number of open issues for research. Overall, many of the so-called "semantic" integration approaches are far from what they promise. A major realization is that at the practical level most approaches attempt a shallow identification of various inconsistencies, not necessarily essential, between concepts. Then, the user is asked to make a rather coarse decision of some merging that does not allow backward interoperability. There is, therefore, a pressing need to develop more true-like integration approaches, which maintain the initial ontologies. In critical cases we may need to verify whether two concepts are exactly identical by getting into great detail of examination; while other times we may be contented with looser identicalness. This resembles the lumping and splitting problem (Section 16.1).

Also, the approaches that are currently being used to pursue extensional-based integration in the geospatial domain have not been sufficiently investigated and are often based on questionable assumptions. The same actually holds for most types of integration approaches such as those relying on top-level or reference ontologies.

17.3.4 ALGORITHM DESIGN

The realization of every integration approach, either as one step or as a number of substeps, is usually accomplished by a straightforward procedure; that is, a finite set of clear instructions for completing some task, commonly known as an *algorithm*. Such algorithms were introduced and demonstrated in Chapters 15 and 16. Explanatory details about implementing appropriate algorithms were not in the scope of the present book. The design of algorithms, nonetheless, is of interest. Many of the approaches reported in the literature are only vaguely described. The process of designing an appropriate algorithm forces us to examine, account for, and test many special cases, which otherwise might have slipped undetected. Algorithms, thus, help to verify (at least to some degree) the solvability of a problem and the effectiveness of an integration approach.

An important concern in designing any algorithm is its *computational complexity*. All three subprocesses of the integration process (see Chapter 14) and the associated formal tools present situations of some complexity. If, for instance, a knowledge base is expressed as a graph (conceptual graph, lattice, etc.), and a specific query is looking for a part of a certain "spatial" arrangement in the graph, then this requires a "pattern matching" algorithm. Another intricate problem case is when there are several ontologies, many concepts, properties, and contexts to be compared and integrated. It is important not only to get an answer/solution but also to make sure that this is efficient. The success of such algorithms requires a thorough consideration of complexity in the design phase. Modern algorithmic design theory (Goodrich and Tamassia, 2002; Dasgupta, Papadimitriou, and Vazirani, 2006) has established the appropriate theoretical background and developed various tools, methods, and techniques in order to classify problems and algorithms in complexity classes.

17.3.5 EVALUATION OF ONTOLOGY INTEGRATION

A fundamental engineering principle practiced by various professionals is that a solution to a problem is not very enlightening or useful unless it is accompanied by a quality/confidence/reliability measure. Correspondingly, the result of an ontology integration process also requires some evaluation. Notice that this is not an evaluation (or comparison) of available integration approaches—already presented in Chapter 15—but an evaluation of the result of the integration. However, while ontology integration approaches grow, evaluation approaches are still sparse (Ehrig and Euzenat, 2005; Euzenat et al., 2006). Most work has focused on the related but more general issue of ontology evaluation and the development of evaluation metrics (Brank, Grobelnik, and Mladenić, 2005; Alani and Brewster, 2006; Kalfoglou and Hu, 2006). It is thus anticipated that evaluation research efforts will gain importance so as to satisfy this need. Any evaluation process requires an evaluation context, which is primarily summarized in the purpose/objective of the integration (see the general evaluation framework in Chapter 14 and the Comparative Presentation in Section 15.14). That is, the result of an integration may be acceptable in one context but totally unacceptable in another. These "how to evaluate" techniques, among other issues, may attempt to determine the following:

- Do the resulting concepts cover the entire domain without overlappings (at the same ontological level)?
- Does the integration approach provide mappings for all concepts involved?
- Have all concepts found their place in the final schema?
- What is the degree of alteration of the initial (source) ontologies?
- Is backward interoperability possible?
- To what extent was the process neutral (without intervention)?
- Do experts agree with (a) the suggestions of the approach when their intervention is required or (b) with the result of the unsupervised integration?
- Can the choices made during the process be stored as preferences to be used for similar integration tasks in the future?

17.4 AFTERWORD

In a nutshell, it is obvious that the issue of semantic interoperability and integration in the geospatial domain is a major ontological issue. It brings to the surface all weaknesses in dealing with geographic information and knowledge. These include imperfect representations of reality, nonstandard, subjective, and ad hoc practices, and a serious lack of a systematic way to deal with conflicting and vague views of reality. The progress made so far is a significant one, yet the road is long. Some problems will find their solution soon, while some others will take much longer. Problems associated with major philosophical questions are not likely to be fully resolved.

The objectives of this book were (a) to reveal the extent of the problem with a number of critical questions, (b) to provide a rather succinct collection of related theories and formal tools, and (c) to review and compare representative approaches dealing with the issue, providing also some practical guidelines. More than that, however, it was an honest attempt to understand what we are doing with this difficult issue, and possibly help other researchers, especially young ones, do the same and make better progress.

REFERENCES

Alani, H. and C. Brewster. 2006. Metrics for Ranking Ontologies. In *Proc. 4th International EON Workshop (EON2006): Evaluation of Ontologies for the Web*. http://ftp. informatik.rwth-aachen.de/Publications/CEUR-WS/Vol-179/eon2006alanietal.pdf (accessed 9 September 2007).

Aleman-Meza, B., C. Halaschek, A. Sheth, I. B. Arpinar, and G. Sannapareddy. 2004. SWETO: Large-Scale Semantic Web Test-bed. Paper presented at International Workshop on Ontology in Action, Banff, Alberta, Canada, June 20-24.

Arpinar, B., A. Sheth, C. Ramakrishnan, L. Usery, M. Azami, and M. Kwan. 2007. Geospatial ontology development and semantic analytics. In *Handbook of Geographic Information Science*, ed. J. P. Wilson and A. S. Fotheringham. Blackwell Publishing (in print).

Brank, J., M. Grobelnik, and D. Mladenić. 2005. A survey of ontology evaluation techniques. In *Proc. Conference on Data Mining and Data Warehouses (SiKDD 2005)*, Ljubljana, Slovenia. http://kt.ijs.si/dunja/sikdd2005/Papers/BrankEvaluationSiKDD2005.pdf (accessed 30 April 2007).

Clark, R. 2004a. *Generalized quantifiers and semantic automata: A tutorial*. ftp://www.ling. upenn.edu/facpapers/robin_clark/tutorial.pdf (accessed 1 May 2007).

Clark, R. 2004b. *Programming semantic automata*. ftp://www.ling.upenn.edu/facpapers/ robin_clark/sam.pdf (accessed 1 May 2007).

Dasgupta, S., C. Papadimitriou, and U. Vazirani. 2006. *Algorithms*. New York: McGraw-Hill.

Egenhofer, J. M. 2002. Toward the Semantic Geospatial Web. In *Proc. Tenth ACM International Symposium on Advances in Geographic Information Systems*, McLean, Virginia.

Ehrig, M. and J. Euzenat. 2005. Relaxed precision and recall for ontology matching. In *Proc. K-Cap 2005 Workshop on Integrating Ontology*, Banff, Alberta, Canada, ed. B. Ashpole, J. Euzenat, M. Ehrig, and H. Stuckenschmidt, 25–32.

Euzenat, J., M. Mochol, P. Shvaiko, H. Stuckenschmidt, O. Šváb, V. Svátek, W. R. van Hage, and M. Yatskevich. 2006. Results of the Ontology Alignment Evaluation Initiative 2006. http://www.dit.unitn.it/~p2p/OM-2006/7-oaei2006.pdf (accessed 30 April 2007).

Fellbaum, C. 1990. English verbs as a semantic net. *International Journal of Lexicography* 3: 278–301.

Fuchs, N. E., U. Schwertel, and R. Schwitter. 1999. Attempto Controlled English: Not just another logic specification language. *Lecture Notes in Computer Science 1559*. Berlin: Springer.

Gärdenfors, P. 2000. *Conceptual spaces*. Cambridge, MA: MIT Press.

Goodrich, M. T. and R. Tamassia. 2002. *Algorithm design: Foundations, analysis, and internet examples*. New York: John Wiley & Sons.

Hahn, U. and I. Mani. 2000. The challenges of automatic summarization. *Computer,* IEEE Computer Society Press, 33(11): 29–36.

Hoa, N. S. and N. H. Son. 2004. Rough set approach to approximation of concepts from taxonomy. Knowledge Discovery and Ontologies (KDO-2004), Pisa, Italy, September 24. http://olp.dfki.de/pkdd04/cfp.htm (accessed 01 May 2007).

Kalfoglou, Y. and B. Hu. 2006. Issues with evaluating and using publicly available ontologies. In *Proc. 4th International EON Workshop (EON2006): Evaluation of Ontologies for the Web.* http://ftp.informatik.rwth-aachen.de/Publications/CEUR-WS/Vol-179/eon2006kalfoglouetal.pdf (accessed 9 September 2007).

Kent, R. 1996. Rough concept analysis: A synthesis of rough sets and formal concept analysis. *Fundamenta Informaticae,* 27(2): 169–181.

Kent, R. E. 2004, The IFF Foundation for ontological knowledge organization. In *Knowledge organization and classification in international information retrieval,* ed. N. J. Williamson and C. Beghtol. Cataloging and Classification Quarterly Binghamton, NY: Haworth Press.

Kontaxaki, S. and M. Kavouras. 2005. Geo-Q: Spatial knowledge extraction based on a Controlled English Query Language and conceptual graphs. In *Proc. GIS Planet 2005, International Conference and Exhibition on Geographic Information,* May 29–June2, Estoril, Portugal.

Korzybski, A. 1994. Science and sanity: An introduction to non-Aristotelian systems and general semantics, with a preface by R. P. Pula, 5th ed. Institute of General Semantics.

Kuhn, W. 2003. Why information science needs cognitive semantics—and what it has to offer in return. Draft paper, MUSIL Lab, Institute for Geoinformatics, University of Münster, Germany. http://musil.uni-muenster.de/documents/WhyCogLingv1.pdf (accessed 30 April 2007).

Lakoff, G. 1987. *Women, fire and dangerous things: What categories reveal about the mind.* Chicago: University of Chicago Press.

Mani, I. 2001. *Automatic summarization, (Natural language processing),* Vol. 3. Amsterdam: John Benjamins Publishing.

Mani, I. and M. T. Maybury. 1999. *Advances in automatic text summarization.* Cambridge, MA: MIT Press.

Rich, E. 1983. *Artificial intelligence.* New York: McGraw-Hill.

Saffo, P. 1994. It's the context, stupid. *Wired Magazine,* Issue 2.03. http://www.wired.com/wired/archive/2.03/context.html (accessed 1 May 2007).

Saquer, J. and J. S. Deogun. 2001. Concept approximations based on rough sets and similarity measures. *International Journal of Applied Mathematics and Computer Science* 11(3): 655–674.

Schwitter, R., N. E. Fuchs, and U. Schwertel. 1998. Attempto: Controlled English (ACE) for software specifications. Paper presented at Second International Workshop on Controlled Language Applications, Language Technologies Institute, Carnegie Mellon University, Pittsburgh, May.

Smith, B. 1995. Formal ontology, common sense and cognitive science. *International Journal of Human-Computer Studies* 43: 641–667.

Sowa, J. F. 2004. *Common logic controlled English specifications.* http://www.jfsowa.com/clce/specs.htm, (accessed 30 April 2007).

Talmy, L. 2000. *Toward a cognitive semantics.* Cambridge, MA: MIT Press.

Tomai, E. and M. Kavouras. 2005. Qualitative linguistic terms and geographic concepts-quantifiers in definitions. *Transactions in GIS* 9(3): 277–290.

van Benthem, J. 1986. *Essays in logical semantics.* Dordrecht: Reidel.

van der Does, J. 1996. Lectures on quantifiers. Sixth European Summer School in Logic, Language, and Information. http://www.science.uva.nl/pub/theory/pionier/ESSLLI_96/Qlectures.ps (Accessed 9 September 2007).

van der Does, J. and H. de Hoop. 2001. Generalized quantifiers. *Glot International* 5: 3–12.

van der Does, J. and J. van Eijck. 1996. Basic quantifier theory. In *Quantifiers, Logic and Language*, ed. J. van der Does and J. van Eijck, 1–45. CSLI Lecture Notes 54. Stanford, CA: CSLI Publications.

Vanderwende, L. H. 1995. The analysis of noun sequences using semantic information extracted from on-line dictionaries. Ph.D. thesis, Georgetown University.

Wille, R. 1997. Conceptual graphs and formal concept analysis. In *Proc. International Conference on Conceptual Structures ICCS'97*, ed. D. Lukose, H. S. Delugach, M. Keeler, L. Searle, and J. F. Sowa, 290–303. Lecture Notes in Computer Science, 1257. Berlin: Springer.

Zhao, Y., X. Wang, and W. Halang. 2006. Ontology mapping based on rough formal concept analysis. Telecommunications, International Conference on Internet and Web Applications and Services/Advanced International Conference on Internet and Web Applications and Services (AICT-ICIW '06), 180.

Index